ASCE Manuals and Reports on Engineering Practice No. 79

Steel Penstocks

Second Edition

Prepared by
the Task Committee on Steel Penstock Design of
the Pipeline Planning and Design Committee of
the Pipeline Division of
the American Society of Civil Engineers

Edited by John H. Bambei Jr., P.E.

Published by the American Society of Civil Engineers

Cataloging-in-Publication Data on file with the Library of Congress.

Published by American Society of Civil Engineers
1801 Alexander Bell Drive
Reston, Virginia 20191
www.pubs.asce.org

MANUALS AND REPORTS ON ENGINEERING PRACTICE

(As developed by the ASCE Technical Procedures Committee, July 1930, and revised March 1935, February 1962, and April 1982)

A manual or report in this series consists of an orderly presentation of facts on a particular subject, supplemented by an analysis of limitations and applications of these facts. It contains information useful to the average engineer in his everyday work, rather than the findings that may be useful only occasionally or rarely. It is not in any sense a "standard," however; nor is it so elementary or so conclusive as to provide a "rule of thumb" for nonengineers.

Furthermore, material in this series, in distinction from a paper (which expressed only one person's observations or opinions), is the work of a committee or group selected to assemble and express information on a specific topic. As often as practicable the committee is under the direction of one or more of the Technical Divisions and Councils, and the product evolved has been subjected to review by the Executive Committee of the Division or Council. As a step in the process of this review, proposed manuscripts are often brought before the members of the Technical Divisions and Councils for comment, which may serve as the basis for improvement. When published, each work shows the names of the committees by which it was compiled and indicates clearly the several processes through which it has passed in review, in order that its merit may be definitely understood.

In February 1962 (and revised in April 1982) the Board of Direction voted to establish:

A series titled "Manuals and Reports on Engineering Practice," to include the Manuals published and authorized to date, future Manuals of Professional Practice, and Reports on Engineering Practice. All such Manual or Report material of the Society would have been refereed in a manner approved by the Board Committee on Publications and would be bound, with applicable discussion, in books similar to past Manuals. Numbering would be consecutive and would be a continuation of present Manual numbers. In some cases of reports of joint committees, bypassing of Journal publications may be authorized.

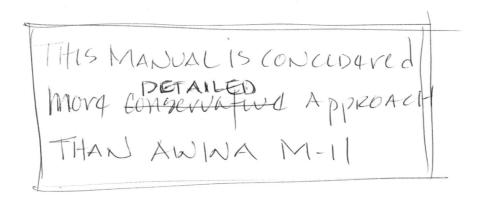

MANUALS AND REPORTS ON ENGINEERING PRACTICE CURRENTLY AVAILABLE

No.	Title
28	Hydrology Handbook, Second Edition
45	How to Select and Work Effectively with Consulting Engineers: Getting the Best Project, 2012 Edition
50	Planning and Design Guidelines for Small Craft Harbors, Revised Edition
54	Sedimentation Engineering, Classic Edition
60	Gravity Sanitary Sewer Design and Construction, Second Edition
62	Existing Sewer Evaluation and Rehabilitation, Third Edition
66	Structural Plastics Selection Manual
67	Wind Tunnel Studies of Buildings and Structures
71	Agricultural Salinity Assessment and Management, Second Edition
73	Quality in the Constructed Project: A Guide for Owners, Designers, and Constructors, Third Edition
74	Guidelines for Electrical Transmission Line Structural Loading, Third Edition
77	Design and Construction of Urban Stormwater Management Systems
79	Steel Penstocks, Second Edition
81	Guidelines for Cloud Seeding to Augment Precipitation, Second Edition
85	Quality of Ground Water: Guidelines for Selection and Application of Frequently Used Models
91	Design of Guyed Electrical Transmission Structures
92	Manhole Inspection and Rehabilitation, Second Edition
94	Inland Navigation: Locks, Dams, and Channels
96	Guide to Improved Earthquake Performance of Electric Power Systems
97	Hydraulic Modeling: Concepts and Practice
98	Conveyance of Residuals from Water and Wastewater Treatment
99	Environmental Site Characterization and Remediation Design Guidance
100	Groundwater Contamination by Organic Pollutants: Analysis and Remediation

No.	Title
101	Underwater Investigation: Standard Practice Manual
102	Design Guide for FRP Composite Connections
103	Guide to Hiring and Retaining Great Civil Engineers
104	Recommended Practice for Fiber-Reinforced Polymer Products for Overhead Utility Line Structures
105	Animal Waste Containment in Lagoons
106	Horizontal Auger Boring Projects
107	Ship Channel Design and Operation, Revised Edition
108	Pipeline Design for Installation by Horizontal Directional Drilling
109	Biological Nutrient Removal (BNR) Operation in Wastewater Treatment Plants
110	Sedimentation Engineering: Processes, Measurements, Modeling, and Practice
111	Reliability-Based Design of Utility Pole Structures
112	Pipe Bursting Projects
113	Substation Structure Design Guide
114	Performance-Based Design of Structural Steel for Fire Conditions: A Calculation Methodology
115	Pipe Ramming Projects
116	Navigation Engineering Practice and Ethical Standards
117	Inspecting Pipeline Installation
118	Belowground Pipeline Networks for Utility Cables
119	Buried Flexible Steel Pipe: Design and Structural Analysis
120	Trenchless Renewal of Culverts and Storm Sewers
121	Safe Operation and Maintenance of Dry Dock Facilities
122	Sediment Dynamics upon Dam Removal
123	Prestressed Concrete Transmission Pole Structures: Recommended Practice for Design and Installation

TASK COMMITTEE ON STEEL PENSTOCK DESIGN

John H. Bambei Jr., P.E., Chair, Denver Water
Kyle Couture, P.E., Secretary, American SpiralWeld Pipe, Birmingham, AL
Charles S. Ahlgren, P.E., Pacific Gas and Electric Company, San Francisco, CA
Richard E. Barrie, P.E., MWH Americas, Inc., Chicago, IL
Bob Card, P.E., Lockwood, Andrews & Newnam, Inc., Houston, TX
Thomas Charles, P.E., CDM Inc., Denver, CO
Dennis Dechant, P.E., Dechant Infrastructure Services, Aurora, CO
Thomas Kahl, P.E., Kleinschmidt Associates, Pittsfield, ME
Brent Keil, P.E., Northwest Pipe Company, Inc., Pleasant Grove, UT
John Luka, P.E., American SpiralWeld Pipe, Columbia, SC
Nathan Nakamoto, P.E., U.S. Bureau of Reclamation, Denver, CO
George Ruchti, Lockwood, Andrews & Newnam, Inc., Houston, TX
Brett Simpson, P.E., American SpiralWeld Pipe, Birmingham, AL
Dick Stutsman, P.E., Consultant, San Anselmo, CA
Chris Sundberg, S.E., CH2M HILL, Bellevue, WA
Bruce VanderPloeg, P.E., Northwest Pipe Company, Portland, OR
Jerry Westermann, P.Eng., MEng, Hatch Ltd, Niagara Falls, ON, Canada
B. Nash Williams, National Welding Corporation, Midvale, UT
Jim Witnik, P.E., MWH Americas, Inc., San Diego, CA
Roger Wood, P.E., CDM Inc. (retired), Ft. Collins, CO

Administrative Support   Sandra Seiger, Denver Water
CAD Figures   Bruce G. Schulte, Denver Water

BLUE RIBBON REVIEW PANEL

CONTENTS

PREFACE

This *Manual of Practice for Steel Penstocks* covers the design, manufacture, installation, testing, start-up, and maintenance of steel penstocks, including branches, wyes, associated appurtenances, and tunnel liners. Standards presented in this manual are applicable only to steel penstocks. As used in the manual, a *penstock* is defined as a closed water conduit located between the first free water surface and a hydroelectric power station.

The *Manual of Practice for Steel Penstocks* was first published in 1993. This 2012 edition is an update of that publication. This update was prepared by a special ASCE task committee under the supervision of the ASCE Pipeline Division with assistance from the ASCE Energy Division. The task committee was charged with updating the manual to match current practice, codes, and standards.

In 1985, the Hydropower Committee began work on the *Civil Engineering Guidelines for Planning and Designing Hydroelectric Developments*, commonly referred to as the *Guides*. Upon the work's completion, the committee decided to expand the section on penstocks into a nationally accepted standard or code.

The penstock design documents most commonly used by the penstock industry are the ASME Boiler and Pressure Vessel Code (ASME Code) and the AISI Steel Plate Engineering Data—Volume 4, Buried Steel Penstocks.

The ASME Code has been in existence since 1914 and originally was based on an allowable stress of one-fifth the tensile strength of the steel. An ASME Code section specifically for pressure vessels came out in the mid-1920s. In the 1950s, the stress basis was changed to one-fourth of the tensile strength of the steel. The high-stress alternative rules, based on one-third of the tensile strength of the steel, came out in the 1960s, although the theory and stress basis had been in use for some time before that. Fabrication based on these rules has resulted in high-quality structures. The rules for both divisions of Section VIII of the ASME Code are accepted worldwide and have been applied to many types of structures other than pressure vessels, including penstocks. The alternative rules of Section VIII, Divisions 1 and 2, have been used extensively in preparing this manual.

The development of Volume 4 was based primarily on the 40-plus years of laboratory research and practical hydroelectric experience of the U.S. Bureau of Reclamation. The manual provides more specific penstock design information than the ASME Code but refers to the latter for fabrication and welding.

Although the procedures outlined in this manual are intended to be clear enough for the inexperienced engineer, many of the decisions on methodology of design and safety factors should be discussed and decided with the aid of an experienced design engineer. Avoiding risk requires a knowledge of diameter, pressure, potential consequence, and many other factors. Proper design of penstocks requires a certain amount of experience.

The design procedures presented in this document are recommended by the task committee. However, the committee recognizes that there are other methods and procedures that may be equally valid. This document does not preclude their use, provided the engineer is satisfied with their validity and applicability.

ACKNOWLEDGMENTS

I wish to thank the task committee members who dedicated many hours of their personal time to this undertaking. Also, the other committee members and I wish to thank the organizations, both public and private, that supported the participation of the committee members, allowing them to attend meetings, and that provided the office support required for updating this manual.

Special thanks go to Kyle Couture, who as the secretary for the task committee, administered and managed the documents and numerous meetings.

Finally, special gratitude goes to Sandi Seiger at Denver Water for her efforts in handling the word processing and document assembly, including many revisions, and to Bruce Schulte at Denver Water for his work in CAD producing the figures for the manual.

John H. Bambei Jr., P.E.
Chair, Task Committee for Revision of
ASCE Manual for Steel Penstocks

ABBREVIATIONS AND ACRONYMS

°F	degrees Fahrenheit
AC	alternating current
ACI	American Concrete Institute
AISC	American Institute of Steel Construction
AISI	American Iron and Steel Institute
ANSI	American National Standards Institute
API	American Pipeline Institute
ASCE	American Society of Civil Engineers
ASFM	acoustic scintillation flowmeter
ASME	American Society of Mechanical Engineers
ASME BPVC	American Society of Mechanical Engineers Boiler Pressure Vessel Code
ASNT	American Society for Nondestructive Testing
ASTM	formerly known as American Society for Testing and Materials
ATC	Applied Technology Council
AWS	American Welding Society
AWWA	American Water Works Association
BSSC	Building Seismic Safety Council
BSTC	bolted sleeve type coupling
CE	carbon equivalent
CFMP	Civil Features Monitoring Program
CJP	complete joint penetration
CP	cathodic protection
CSE	copper/copper sulfate reference electrode
CVN	Charpy V-notch
CWI	certified welding inspector
DC	direct current
DFT	dry film thickness
ECS	emergency closure systems
EPRI	Electric Power Research Institute
ESW	electroslag welding
FCAW	flux cored arc welding
FEA	finite element analysis
FEMA	Federal Emergency Management Agency
FERC	Federal Energy Regulatory Commission
ft	feet
GMAW	gas metal arc welding
HAZ	heat-affected zone
HDPE	high-density polyethylene

HSLA	high-strength, low alloy
IARC	International Agency for Research on Cancer
IBC	International Building Code
in.	inch
ISG	intake shutoff gate
ISO	International Organization for Standardization
kip	one thousand pounds
kJ	kilojoules
ksi	kips per square inch
kW	kilowatt
lb	pound
LRFD	load resistance factor design
LST	lowest service temperature
MCE_R	maximum considered earthquake risk
min	minimum
min.	minute
mph	miles per hour
MSDS	material safety data sheet
MSE	mechanically stabilized earth
MSS	Manufacturers Standardization Society
MT	magnetic particle examination
NACE	National Association of Corrosion Engineers
NDE	nondestructive examination
NEHRP	National Earthquake Hazards Reduction Program
NIST	National Institute of Standards and Technology
NPS	nominal pipe size
OD	outside diameter
OSHA	Occupational Safety and Health Administration
P.E.	professional engineer
PG&E	Pacific Gas and Electric Company
PQR	procedure qualification record
PRL	power revenue loss
PRV	pressure reducing valve
psi	per square inch
PSV	penstock shutoff valve
PT	liquid penetrant examination
PVQ	pressure vessel quality
PWF	present worth factor
QA/QC	quality assurance/quality control
ROV	remote operated vehicles
R.P.E.	registered professional engineer
RPM	revolutions per minute
RT	radiographic examination

s	second
SAW	submerged arc welding
Se3	equivalent stress
SI	stress intensity
SMAW	shielded metal arc welding
SNL	speed no load
SPFA	Steel Plate Fabricators Association
SSPC	Society for Protective Coatings
TBM	tunnel boring machine
TSV	turbine shutoff valve
UHMW	ultra-high-molecular weight
USBR	U.S. Bureau of Reclamation
UT	ultrasonic examination
UV	ultraviolet
VT	visual examination
WPQ	welder performance qualification
WPS	welding procedure specification

CHAPTER 1

General

Source: Photograph courtesy of National Welding Corp.; reproduced with permission.

This section presents an overview of general penstock design and maintenance issues. More detailed descriptions can be found in subsequent sections. The following pages address general design considerations; internal and external pressure design; economic diameter; shutoff systems; prevention of vibration; wind, snow, and ice loading; earthquake loading; and geologic considerations.

1.1 DESIGN CONSIDERATIONS

The design of a safe and cost-effective penstock system requires the consideration of technical, environmental, economic, and constructibility factors. Figure 1-1 shows the flow of considerations that affect the design of a penstock system. Detailed considerations, guidelines, criteria, and design methodology are presented in subsequent sections.

1.1.1 Required Information

It is important for the designer to gather as much information as possible about the installation. Essential information includes (1) owner requirements and (2) site-specific requirements.

1.1.1.1 Owner Requirements The owner's installation preference and requirements must be clearly delineated because these preferences affect the overall approach to the penstock system, including methodology, material selection, and design. Consideration must be given to the following:

1. Preferred material and design type;
2. Plant operation base load, voltage control peaking, or a combination of both; also important is the likely number of unit operating cycles (i.e., daily start–stop and load change requirement);

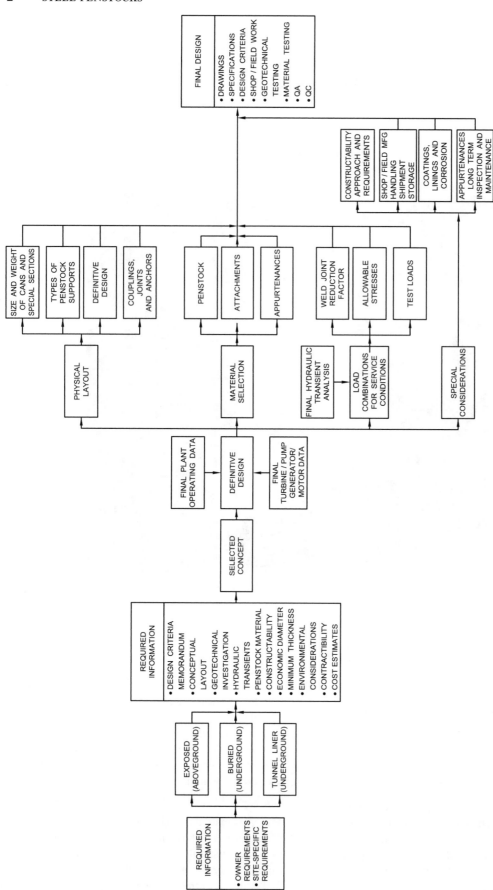

Fig. 1-1.   General considerations for penstock design

3. Parameters and criteria for determining the annual cost of capital investment and the annual cost of power revenue loss;
4. Inspection and maintenance philosophy;
5. Applicable internal and governmental guidelines, criteria, and design requirements; and
6. Legal and political issues, including environmental, permit, and licensing issues.

1.1.1.2 Site-Specific Requirements Equally important are site-specific requirements, which affect design by imposing environmental restraints, limitations on size and weight of penstock sections, geologic restraints, hydrologic considerations, and limitations (alignment and support) to the penstock physical layout.

Consideration must be given to

1. Land ownership, right-of-way limitations, mineral rights, and limitations relating to excavation and/or quarrying operations;
2. Environmental restraints, including aesthetics; fish, game, and wildlife preservation; archaeological excavations; disposal of material; clearing; and erosion;
3. Terrain configuration;
4. Site geology, hydrology (groundwater conditions), and soils;
5. Applicable codes and mandatory requirements; and
6. Other site-specific considerations.

1.1.2   Type of Installation

The type of installation selected must reflect the previously stated considerations. A given penstock installation could include all of the following types:

1. Exposed penstock (aboveground),
2. Buried pipeline (underground), and
3. Steel tunnel liner (underground).

Each penstock type has associated design, material, and construction costs.

1.1.3   Preliminary Study

The preliminary study phase is an important phase of the general design effort and requires an experienced engineer. The final penstock configuration, alignment, design, and other key requirements and parameters must be determined during this study phase.

The engineer must investigate the site conditions and make several layouts of various alignments. Terrain, geologic characteristics, and foundation conditions play important roles during this study phase. Since the ultimate goal of this study phase is to determine the most economic and constructible alignment, it is not necessary to approach the study phase with great precision.

1.1.4   Selected Concept

The selected penstock configuration should incorporate material and designs that minimize life-cycle costs, with prudent consideration given to technical, environmental, constructibility, safety, and maintainability issues.

1.1.5   Definitive Design

The definitive design phase consists of compiling final design-related data, finalizing the conduit alignment and layout, and confirming the final plant operating characteristics. Turbine, generator, governor, and closure valve characteristics that influence water-hammer analysis also must be determined.

Definitive guidelines and design criteria must be prepared for

1. Physical layout, conduit alignment, supports, and anchor blocks;
2. Material selection for the penstock, tunnel liners, and appurtenances (see Chapter 2);
3. Design and service loads (see Chapter 3); and
4. Special considerations for manufacturing and field installation.

Within each of these definitive design groups are the following subgroups that form the basis for the design.

1.1.5.1 Physical Layout The effort expended during the preliminary study phase for the selected conduit arrangement, layout, and alignment now must be refined to achieve a working concept. These refinements primarily apply to the following topics.

1.1.5.1.1 Pipe Sections The size and weight of "can" sections must be selected to be compatible with the needs and difficulty of the field installation. Traffic, shipping, and access limitations and restraints should be considered. Customarily, can sections up to 12 ft in diameter, 60 ft long, and weighing from 25 to 40 tons can be shipped and handled reasonably in the field for installation.

1.1.5.1.2 Specialty Sections Similar consideration must be given to specialty sections, such as mitered bends, wyes, and tee connections. Special attention must be given to the detailing of these sections, particularly wyes with external support (girder beam and stiffener rings). These sections quickly become oversize and heavy, making shipment and handling difficult. The designer must be willing to accept some heavy field welding for these structures and should take this into account when preparing specifications and for QA/QC requirements.

1.1.5.1.3 Supports Supports must be refined to reflect environmental considerations, can size, geology, and local foundation conditions. Reinforced concrete saddles and steel ring girders are the two most commonly used support types for exposed aboveground installations. Anchor and thrust blocks are special applications of localized support systems.

Cable harnesses for suspended pipe that bridges long spans and steel or concrete bent supports are some specialty types of supports. The spacing of ring girders is governed by practical considerations. Shell diameter, thickness, and material type have a major influence on the spacing, which may range from 40 to 200 ft.

Saddles are used in conjunction with stiffened or unstiffened steel pipe. Factors to consider in the spacing of saddles include the profile slope, the maximum acceptable deflection of the pipe when acting as a beam, the shell stress conditions at the saddle support horns, and the foundation conditions for the saddle supports. Also, the designer must be aware that above each saddle support, circumferential bending moments develop in the penstock shell and cause the upper portion of an unstiffened pipe to deform. This situation results in an upper shell portion that is ineffective as a beam, reducing the effective cross section of the member, i.e., moment of inertia.

For pipe buried in soil, key considerations are the flexibility of the pipe, the allowable vertical pipe deflection due to settlement, the ability of the surrounding soil to resist lateral loads, and the type of lining used. Accepted practice and methods should be applied with regard to trenching and backfilling for steel pipe based on soil mechanics, consistent with field experience, for this type of pipe support.

Support requirements and considerations for a steel liner relate primarily to the interaction between the steel shell, concrete lining, and surrounding rock. Special supports, which may be required during field installation, are customarily of a temporary nature. However, the designer is cautioned about the potential for "hard spot" formation, resulting in potentially high, localized shell stresses. This problem is particularly applicable to penstock sections subjected to hydrostatic pressure testing and concrete encasement.

1.1.5.1.4 Geology and Foundation Characteristics Geology and foundation characteristics are important for all types of installations. For ring girder and saddle support designs, the foundation is required to resist primarily downward loads, some small loads parallel to the pipe due mainly to frictional loads, and some loads that result from temperature effects. In earthquake-prone areas, support foundations and sloping terrain also must safely withstand seismic loadings.

Penstock installations on sloping terrain are subject to potential slope stability problems. Slope instability must be addressed and investigated to determine geologic and soil properties of the terrain and the potential for slope movement and potential differential settlement. Long-term monitoring of slopes or the incorporation of structural improvements should be considered in the foundation analysis. Ice and snow, as well as avalanche conditions, also must be considered.

The cumulative loads acting in line with the pipe should be resisted by appropriate anchor blocks. It is recommended that the foundation conditions at each anchor block location be determined individually for use in the design of anchor blocks. If geologic and soil conditions permit, supports can be divided into groups with like foundations, and appropriate conservative foundation designs can be prepared for each grouping, resulting in an economy of design. See Section 1.6 for relevant geologic considerations.

1.1.5.1.5 Mechanical Joints Mechanical joints are used in both exposed and buried penstocks to accommodate longitudinal movement caused by temperature changes and/or to facilitate construction. The joints also allow for movement where differential settlement or deflections are anticipated. Should the engineer decide not to use mechanical joints, the penstock and anchor blocks must be adequately designed to handle loads due to temperature changes, settlement, or deflections.

Slide joints fall into two major categories: (a) expansion joints and (b) sleeve-type couplings.

Expansion joints permit only longitudinal movement. The joints are used primarily with aboveground installations and are located between girder or saddle supports at the point where the penstock deflections are of equal magnitude and direction. If used in a buried installation, the joints must be located in a vault to permit inspection and maintenance. Expansion joints also may be used where warranted to permit expansion and contraction.

Sleeve-type couplings are used in both aboveground and underground installations. For underground installations, the penstock must be pressurized and the couplings must be checked for leakage before the penstock is buried. Sleeve-type couplings allow for longitudinal, angular, and some differential settlement movements.

For sleeve-type couplings used on very steep slopes, consideration must be given to preventing the coupling from sliding downslope as the penstock sections move with temperature changes. This step is accomplished by using a pipe stop, internal to the coupling, or by anchoring the coupling at some point to the outside of the penstock shell. In either case, consideration must be given to the need to slide the coupling beyond the joint for future maintenance.

Other types of joints and couplings are available, particularly for underground installations. These types include O-ring joints and mechanical restrained couplings.

Slide (expansion or contraction) joints that keep resulting forces from reaching operating equipment (valves, pumps, etc.) should also be considered.

1.1.5.1.6 Anchor Blocks Anchor blocks fix the pipe in place during installation and operation. They are designed to resist the loads resulting from gravitational, hydrodynamic, hydrostatic, and earthquake forces, as well as from temperature changes and changes in alignment.

For aboveground designs, anchor blocks are normally located at all points of significant change in slope or horizontal alignment and sometimes at intermediate points in long tangent runs. For installations using expansion joints, an anchor spacing of up to 500 ft may be used between anchors and expansion joints, depending on the design temperature range.

Welded, buried penstocks generally do not require anchors at points of minor vertical and/or horizontal alignment changes. This situation must be verified by analysis. Buried penstocks with expansion joints require anchors similar to those used for aboveground installations.

1.1.5.2 *Special Considerations*

The engineer, in conjunction with personnel experienced in field installations, constantly must ask how the installation will be fabricated and built in the field. Constructibility is an important aspect that is often overlooked. A design may appear good on paper but may be extremely difficult to construct. The engineer should think out the entire installation, item by item. Can it be handled in the shop and field? What are the size and weight limitations for shipment? How will each piece be stored? Are there special storage and/or cribbing requirements? What about internal supports and stiffeners? Will special welding and heat treatment procedures be needed in the shop or field? What are the specifications for coating and painting? Are provisions incorporated in the design to permit or accommodate long-term inspection and maintenance requirements?

Corrosion is also an important factor. Water analysis is necessary to determine the chemical content and hardness of the water. Corrosion from possible other sources (i.e., dissimilar metals, stray electrical currents, corrosive soils, high flow velocity, etc.) must be evaluated. Also, special attention must be given to corrosion and erosion protection at bends, wyes, and other special components. Cathodic protection may be required for some penstock types. Special consideration must be given to accessories, such as pressure-relief valves, standpipes, air valves, penstock access ways, and penstock flow shutoff systems.

1.2   ECONOMIC DIAMETER

The referenced equations for selection of an economic diameter for a hydroelectric penstock serve as good starting points for initial project layouts and planning level cost estimates, but caution should be applied in relying on them too greatly. The power industry today reflects the dramatic changes in world economies and major commodity markets experienced daily, which in turn translate into rapidly changing values for energy, fuel, steel, copper, concrete, and other common construction materials. This dynamic forces the engineer to be more diligent in terms of economic penstock sizing, to apply material cost and value of energy parameters in the best possible manner.

Energy markets now reflect seasonal variations in energy values, peak vs. off-peak values, time-of-day fluctuations, and so on, reflecting the complications of today's technology-driven society and actual energy consumption fluctuations. The point of mentioning this is to emphasize the importance of capturing the variables of most critical importance for a given penstock application. A simplified average or "blended composite" cost of energy may not accurately reflect either the near-term or future condition for which the penstock size will be determined. Design engineers should strive to understand all of the economic parameters unique to their circumstances and to properly reflect and incorporate regional, local, and client-specific factors in setting up economic diameter and pipe selection models. In portions of the United States, time-of-day and time-of-year energy price variations are substantial and can skew the optimum economic pipe size to an unexpected extreme. It should also be recognized that monthly flow averages again may not reflect the swings and variations that daily flow values often do, and neglecting these variations can result in significant miscalculation and dramatic effects on energy production for a given hydropower facility.

Engineers are encouraged to develop adaptable pipe sizing models that can be updated and modified as project designs evolve through planning, permitting, design, and bidding phases. Because of the factors mentioned above, the economic diameter selected initially may prove unwise at a later date. Capturing the picture for today and neglecting future conditions or constraints may also require a change in thinking and approach that should be properly reflected in this type of analysis. This caution almost forces the engineer to get out a crystal ball to look into the future to predict the types of variations described. This prediction is obviously not possible, but the point is that engineers must look ahead to consider how material and energy price variations may differ and they should develop sensitivity functions in their pipe selection models to account for these future conditions. Ultimately, the most economic penstock diameter is one that maximizes the value of energy production for a given facility.

Other large-cost items, such as valves and trash racks, must be evaluated by separate value engineering.

There are several published economic diameter equations available to the engineer. Examples may be found in ASCE's *Civil Engineering Guidelines for Panning and Designing Hydroelectric Developments* (1989).

This section shows equations that are practical and convenient to use. These equations apply to penstocks conveying water to conventional power plants. By moving the optimal efficiency term (E) to the denominator, the equations are equally applicable to pumping plant discharge lines.

These equations are not recommended for sizing pump storage penstocks because major modifications are necessary to account for the changes in parameter values, such as the composite power values and hours per year of operation, during the pumping and power generation modes. Additional information for sizing penstocks for pump storage plants may be found in Sullivan (1971).

Economic diameter equations are developed for the following two cases:

- Case 1: Minimum Penstock Shell Thickness for Shipping and Handling
- Case 2: Internal Pressure Governs

1.2.1 Installation Cost

The installation cost for a Case 1 penstock is calculated by the following equation:

$$U1 = \pi WCtD \qquad (1\text{-}1)$$

where $U1$ = capital cost of the penstock in Case 1 in dollars ($);

  W = specific weight of steel (lb/ft^3);
  C = installation cost of penstock ($/lb);
  t = shell thickness (ft); and
  D = nominal pipe diameter (ft).

Eq. (1-1) gives the cost of the penstock per unit length. The penstock length cancels out in developing the economic diameter equations. The Pacific Gas and Electric Company formula for determining the required minimum shell thickness in terms of diameter is

$$t = \frac{D}{288} \qquad (1\text{-}2)$$

If $D/288$ is substituted for t in Eq. (1-1) and the constant terms are combined, the installation cost equation may be rewritten as

$$U1 = 0.0109 WCtD \qquad (1\text{-}3)$$

Additional information about minimum shell thickness may found in Chapter 4.

For a Case 2 penstock, the shell thickness (t) is calculated by the hoop stress equation:

$$t = \frac{FHD}{2S} \qquad (1\text{-}4)$$

where F = the conversion factor for ft of water to lb/in.^{2}, 0.4333 lb/ft/in.^{2};

  H = design head (weighted average) (ft);
  D = inside diameter (ft); and
  S = allowable stress (lb/in.^{2}).

By substituting this expression for t into Eq. (1-1) and simplifying, the installation cost ($U2$) for a Case 2 penstock may be written as

$$U2 = \frac{0.6806 WCHD^2}{S} \qquad (1\text{-}5)$$

1.2.2 Power Revenue Loss

The power revenue loss (PRL) of a penstock is calculated by using the following equation from USBR (1976) for the power-plant capacity (kW) and inserting the head loss term (H_L) for head.

$$\text{kW} = 0.0846 Q\, E\, (H_L) \qquad (1\text{-}6)$$

where Q = design flow (ft^3/sec), and

  E = overall turbine/generator efficiency.

The Darcy–Weisbach equation is used to calculate the head loss (H_L) caused by turbulent flow in the penstock. The equation is given as

$$H_L = \frac{fLV^2}{D2g} \qquad (1\text{-}7)$$

where f = friction factor;

  L = length (ft);
  V = average velocity (ft/sec);
  D = inside diameter (ft); and
  g = acceleration due to gravity (32.2 ft/sec^2).

The equation is modified by replacing the velocity term (V) with

$$V = \frac{4Q}{\pi D^2} \qquad (1\text{-}8)$$

A unit length is used that results in an expression that computes head loss per unit length of penstock. This modified version of the Darcy-Weisbach equation is given as

$$H_L = \frac{16 fQ^2}{2g\pi^2 D^5} = \frac{0.02517 fQ^2}{D^5} \qquad (1\text{-}9)$$

Substituting this head loss expression into Eq. (1-6), the power-plant capacity equation may be rewritten as

$$\text{kW} = \frac{0.002129 fEQ^3}{D^5} \qquad (1\text{-}10)$$

To calculate the annual PRL, Eq. (1-10) is multiplied by composite value of power (M) and hours of operation per year (h).

The annual PRL then may be written as

$$PRL = \frac{0.002129 fEQ^3 hM}{D^5} \qquad (1\text{-}11)$$

It is important to note that the value of power is a single composite value obtained by adding the capacity and energy components. The composite value of power is expressed in units of $/kWh. Often the capacity is expressed in $/kW/yr, in which case it is necessary to divide it by the number of hours per year of power generation.

The present worth of the annual power revenue loss (X) is obtained by multiplying Eq. (1-11) by the present worth factor (pwf) for a given interest rate and repayment period, resulting in

$$X = \frac{0.002129 fEQ^3 hM (pwf)}{D^5} \qquad (1\text{-}12)$$

1.2.2 Economic Diameter Equations

The economic diameter equations are developed by adding the Case 1 or Case 2 installation cost equation to Eq. (1-12) to obtain an equation for the total cost. The total cost equation is differentiated with respect to diameter (D), set equal to zero, and then solved for D. The resulting economic diameter equations for Case 1 and Case 2 penstocks are as follows.

For Case 1, when the shell thickness (t) is determined by $D/288$, the economic diameter (D) is given as

$$D = 0.9025 \left[\frac{fhMEQ^3(pwf)}{WC} \right]^{0.1429} \qquad (1\text{-}13)$$

When a specific value is used for t, the economic diameter is

$$D = 0.3867 \left[\frac{fhMEQ^3(pwf)}{WCt} \right]^{0.1667} \qquad (1\text{-}14)$$

In Case 2, the economic diameter (D) for a penstock is given as

$$D = 0.5 \left[\frac{SMhfMEQ^3(pwf)}{WCH} \right]^{0.1429} \qquad (1\text{-}15)$$

It is important to know how the power plant will be operated when determining the penstock diameter. Some of the parameter values, such as the composite value of power and the number of hours of power generation, will vary greatly for a base-load power plant compared to a peaking-load power plant. Selecting the design flow (Q) is most important because this term is cubed in the economic diameter equations. Variations in the design flow value have a significant impact on the resulting diameter calculations. For example, in Eq. (1-15), if the value of Q is doubled and the other parameter values remain unchanged, the diameter increases by approximately 35%. By doubling any one of the other parameter values and leaving the flow term Q unchanged, the diameter increases by approximately 10%.

After the economic diameter has been determined, it can be shown that for a Case 1 penstock the installation cost is approximately five times the present worth of the annual power revenue loss. Similarly, for a Case 2 penstock, the installation cost is approximately 2.5 times the present worth of the power revenue loss.

1.2.3   Economic Diameter Examples

The following examples illustrate the use of the equations for calculating the present worth factor and economic diameter for Case 1 and 2 penstocks.

1.2.3.1   Case 1 Penstock

Here are the given data:

f = 0.0l, friction factor;
h = 6,500, hours per year of operation;
M = 0.05 \$/kWh, composite value of power;
E = 0.85, turbine/generator efficiency in decimal form;
i = 8.75%, interest rate;

n = 50 years, repayment period;
Q = 3,000 ft³/sec, design flow;
t = D/288, the penstock shell thickness;
W = 490 lb/ft³, specific weight of steel; and
C = 2.00 \$/lb, capital cost of penstock installed.

The present worth factor (PWF) for both Case 1 and Case 2 penstocks is calculated by the standard equation:

$$pwf = \frac{(i+1)^n + 1}{i(i+1)^n} \qquad (1\text{-}16)$$

where i = interest rate (%) and

n = repayment period (years).
Substituting in the values gives

$$pwf = \frac{(1.0875)^{50} - 1}{0.0875(1.0875)^{50}} = 11.26$$

The economic diameter is calculated by using Eq. (1-13):

$$D = 0.9025 \left[\frac{fhmEQ^3(pwf)}{WC} \right]^{0.1429}$$

$$D = 0.9025 \left[\frac{(0.01)(6,500)(0.05)(0.85)(3,000^3)(11.26)}{(490)(2)} \right]^{0.1429}$$

$$= 17.06 \text{ ft} \qquad \text{Substituting in the values gives}$$

1.2.3.2   Case 2 Penstock

Here are the given data:

f = 0.01 (dimensionless) friction factor;
h = 7,075 h, annual hours of operation;
M = 0.05 \$/kWh, composite value of power;
E = 0.85, turbine/generator efficiency, in decimal form;
i = 8.75%, interest rate;
n = 50 years, repayment period;
Q = 2,900 ft³/sec, design flow;
W = 490 lb/ft³, specific weight of steel;
C = 3.50, capital cost of penstock installed;
H = 442.4 ft, design head (weighted average); and
S = 20,000 lb/in.², allowable stress.

The present worth factor (PWF) equals 11.26, the same as in Case 1. The economic diameter is calculated using Eq. (1-15):

$$D = 0.5 \left[\frac{SMhfEQ^3(pwf)}{WCH} \right]^{0.1429}$$

Substituting in the values gives

$$D = 0.5 \left[\frac{(20,000)(0.05)(7,075)(0.01)(0.85)(2,900^3)(11.26)}{(490)(3.50)(442.4)} \right]^{0.1429}$$

$$= 15.00 \text{ ft}$$

1.3 SHUTOFF SYSTEMS

Flow shutoff control systems are installed in hydraulic conduit systems because of the potential for tunnel blowouts, penstock ruptures or major leaks, or blocked wicket gates. Shutoff systems can be installed at the upper end, an intermediate location, or the lower end of the hydraulic system. In the event of an excess flow signal, the shutoff system automatically closes under the transient flow conditions.

For a sudden penstock failure or a major penstock leak, the automatic shutoff system

1. Prevents rapid dewatering of the tunnel,
2. Minimizes the loss of water from the upstream reservoir,
3. Minimizes the resulting damage or loss of life, and
4. Provides a means of regaining control of the system.

1.3.1 Intake Shutoff Gates

For hydraulic conveyance systems, such as steel pipelines, penstocks, or tunnels, intake shutoff gates (ISGs) must be installed at the intake structure to provide protection in the event of a system failure. When the intake is deeply submerged in the reservoir, gates are usually installed in a shaft or chamber downstream of the intake. Gate types include slide gates, fixed-wheel gates, roller gates, radial gates, tainter gates, coaster gates, and tractor gates. For safety in case of power failure or other emergency, gates must close under their own weight rather than being forced down by the operator. An adequate air venting system must be designed and installed downstream of intake gates to prevent the development of negative pressures that could result in tunnel collapse.

1.3.2 Penstock Shutoff Valves

For hydraulic systems with tunnels and an aboveground penstock, penstock shutoff valves (PSVs) must be located downstream of the tunnel portal but just upstream of the penstock. There must be no unprotected steel liner between the tunnel and the PSV, unless an intake shutoff gate is installed upstream. This type of system has the added benefit that the penstock can be dewatered without dewatering of the tunnel.

As with the intake gate system, an adequate air venting system must be designed and installed downstream of both the valve(s) and gates to prevent development of downstream negative pressures that could result in penstock or tunnel collapse.

1.3.3 Automatic Shutoff Control System

Emergency shutoff valves and gates usually are unattended and are often located some distance from the powerhouse operator. Therefore, a local shutoff control system, which can sense an increase in flow and automatically close a gate or valve, must be installed. This automatic control system also must be controlled remotely from the powerhouse.

Examples of sensing systems used for automatic control systems are differential float-well systems, Rolex devices, and sonic flow measuring systems. These sensing systems usually activate the gate or valve operating system when the tunnel or penstock velocities exceed a certain percentage over the maximum normal transient velocity. Normal transient velocities are caused by rapid loading or unloading of the unit. The valve or gate operating system usually is AC electrically powered, or hydraulically operated with either a stored energy (nitrogen) backup, or mechanically counterweighted.

For exposed penstocks in seismic zones, seismically triggered shutoff systems must be considered.

1.3.3.1 Turbine Shutoff Valve   A turbine shutoff valve (TSV) must be located at the lower end of long penstocks (just upstream of the turbine). The valve must be designed to close during excessive flow conditions, such as when turbine wicket gates become blocked and cannot be closed. Generally, the operating system for a TSV incorporates water-operated cylinders, mechanical counterweighted systems with either an electric or hydraulic operator.

A TSV also is used for maintenance of the turbine equipment (upstream clearance point). Use of the penstock shutoff valve at the upper end of the penstock for this purpose would require draining of the penstock each time maintenance is performed.

Numerous large projects with short, large-diameter penstocks, such as Grand Coulee, Robert Moses, Chief Joseph, Guri (Venezuela), and Itaipu (Brazil and Paraguay), operate successfully without turbine shutoff valves. For short and especially large-diameter penstocks, use of TSVs may be uneconomical. Spherical valve diameters are limited to approximately 15 ft, and butterfly valves usually are limited to 25 ft. Since these types of projects have short penstocks, uncontrolled turbine flow can be shut off by closing the intake shutoff gates.

1.4 PREVENTION OF VIBRATION

Pressure pulsations generated within the hydraulic conduit system may cause vibration of a freestanding penstock if the frequency of the pressure pulsations within the system is close to one of the natural frequencies of the penstock. If the forcing frequency of the pressure pulsation is exactly equal to a natural frequency in one or more of the penstock's modes of vibration, a state of resonance may develop, and severe vibration may occur. When the amplitude of the vibrations is excessive, fatigue and sudden failure of the penstock shell is a possibility since an excessive number of oscillations may occur in a relatively short period. Failure is most likely near supports where high localized restraint is introduced.

1.4.1 Sources of Pressure Pulsations

Known sources of pressure pulsations during operation of the turbines include

1. Draft tube pulsations associated with vorticity and vortex shedding in the draft tube,
2. Turbine blade passing frequency,
3. Rotational speed-dynamic imbalance, and
4. Water-hammer pressure waves within the penstock.

1.4.1.1 Draft Tube Pulsations The forcing frequency (F_1) of draft tube pulsations associated with vorticity and vortex shedding (vortex rotating core phenomenon) has been reported to vary from about

$$\frac{\text{RPM}}{60} \times \frac{1}{3.6} \text{ to } \frac{\text{RPM}}{60} \times \frac{1}{2.8}$$

in the part load range, and from about

$$\frac{\text{RPM}}{60} \times \frac{1}{1.3} \text{ to } \frac{\text{RPM}}{60} \times 1.0$$

in the full load to overload operation range, where RPM = unit operating speed in revolutions per minute and F_2 = frequency in Hz.

1.4.1.2 Turbine Blade Passing Frequency The turbine blade passing frequency (F_2) can be determined by

$$F = \frac{\text{RPM}}{60} \times B_1 \qquad (1\text{-}17)$$

or

$$F = \frac{\text{RPM}}{60} \times B_1 \times B_2 \qquad (1\text{-}18)$$

where B_1 = number of turbine runner blades on Francis turbine and B_2 = number of wicket gates.

1.4.1.3 Rotational Speed-Dynamic Imbalance The frequency is determined from

$$F_3 = \frac{\text{RPM}}{60} \qquad (1\text{-}19)$$

1.4.1.4 Water-Hammer Pressure Waves within Penstock Water-hammer pressure waves are generated within a penstock by changes in flow caused by turbine wicket gate movements or valve movements. The intensity of the pressure wave produced is proportional to the speed of propagation of the pressure wave and the change in water velocity. These pressure waves normally damp out rapidly when the turbine wicket gate movements terminate. However, in the presence of governor hunting, for example, resonance could develop. The theoretical frequency of these pressure surges generated within the penstock is determined by

$$F_p = \sum_{i=1}^{i=n} \frac{a_i}{4L_i} = \frac{a_1}{4L_1} + \frac{a_2}{4L_2} + \dots + \frac{a_n}{4L_n} \qquad (1\text{-}20)$$

where L_i = length of pipe sections of different diameter (ft) and a_i = velocity of water-hammer wave in penstock section of length L (ft/s).

When a penstock has varying diameters, the apparent frequency is slightly less than the theoretical frequency. The higher frequency harmonics of the penstock differ slightly from exact integer multiples of either the theoretical or the apparent frequency.

1.4.2 Vibration Frequency of Penstocks

Figure 1-2 shows the vibration mode shapes of simply supported, unstiffened cylindrical shells. In addition to the obvious factors of penstock diameter, plate thickness, and support span, other factors that affect the natural frequencies of penstocks include the internal water pressure, axial tension, and specific support geometry. It has been noted that penstock spans along steeply inclined sections can have different vibration frequencies due to differences in internal pressure.

Circumferential mode flexing and vibration can be particularly troublesome for penstocks with low head and large discharge with their relatively large diameters and thin shell thicknesses (large D/t ratio).

The most appropriate method of determining the vibrating frequencies of penstocks is to perform a three-dimensional finite element static and dynamic analysis of the penstock that carefully models the support conditions and geometry. A three-dimensional finite element program that provides static and dynamic stresses and displacements can be used.

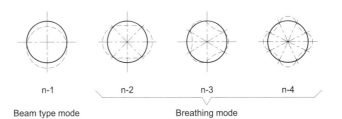

n-1   n-2   n-3   n-4

Beam type mode   Breathing mode

(A) Circumferential Mode

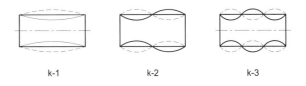

k-1   k-2   k-3

(B) Longitudinal Mode

Fig. 1-2. Vibration Mode Shapes of Simply Supported, Unstiffened Cylindrical Shells

The selected final penstock design must ensure that the natural frequencies of the penstock are different by at least 20% from the expected forcing frequencies of pressure pulsations generated within the hydraulic system.

Circumferential ring stiffeners spaced at intervals along the span of a penstock are useful in changing the frequency of the penstock in the circumferential, breathing modes of vibration.

1.5 WIND, SNOW, AND ICE LOADING

Loading on the penstock due to a combination of wind and snow should be considered during the design of the penstock. As a conservative practice, it is recommended that the combination of both loads be taken into account and provided for in the penstock design. Both loads are considered to be equivalent static loads in the following sections.

1.5.1 Wind Loading

Wind loading and accompanying design criteria must comply with local codes and regulations.

As a minimum, the design wind force should be determined following the procedure outlined in Chapter 29, Table 29.1-1 and Formula 29.5-1 of ASCE Standard 7-10 (ASCE 2010), as follows:

$$\text{Design wind force } F = q_z G C_f A_f$$

where q_z = velocity pressure,

G = gust-effect factor,

C_f = force coefficient (since steel penstocks are round, C_f is typically <1.0), and

A_f = projected area.

Wind-loading design must include the diameter of the penstock plus the ice layer. Local codes and site-specific weather conditions must be reviewed and incorporated as required.

Although there is little chance that wind vortex shedding will force an empty penstock into resonance, it is recommended that the penstock be checked for this condition, using the available methods for calculating resonant wind velocity.

1.5.2 Snow and Ice Loading

Snow-loading design must comply with local codes and, as a minimum, with requirements given in Chapter 7 of ASCE Standard 7-10 (ASCE 2010). Values listed for flat or rise less than 4 in. per foot must be used, as a minimum. A typical assumption is to apply snow loads to the top 120 degrees of the penstock. Lighter loads are not recommended even in areas without snow because of the live loads that must be resisted during construction and also in service. Ice loads in localized areas may be significant. Regions adjacent to portals may be subjected to drifting loads. Also portal areas and expansion joints or mechanical couplings could be subjected to significant ice buildup. In these areas, the weight of the ice must be included in the dead-load assessment.

1.6 GEOLOGIC CONSIDERATIONS

This section briefly describes the geologic aspects of a penstock installation and is not intended to serve as a complete reference on the subject. Suggestions for further reading include ASCE (1989), Leggett and Karrow (1983), and Hunt (1984). The terms *rock* and *soil* are used here in an engineering context. Namely, *rock* is a hardened, coherent, solid mass of earth material, and *soil* is a loose and consistent particulate earth material.

1.6.1 Investigations and Related Designs

One convenient method for establishing the scope of geotechnical investigations and designs is to identify all the geologic considerations pertinent to a given penstock installation. These considerations involve relationships between engineering and the geologic domains that must be considered when planning, designing, constructing, and maintaining a project. Each geologic feature becomes an individual design consideration objective within the overall geotechnical program. Section 1.6.2 identifies relevant geologic considerations for a penstock installation. Identification of geologic considerations should be assessed by a competent geotechnical or geological engineer.

A principal challenge for each geologic consideration is the prediction of ground behavior, which does not necessarily require complex mathematical analysis. Ground behavior may be analyzed by inference using existing knowledge of past activity at the site and previous experience in similar geologic conditions. Selection of the proper methods of analysis depends on various factors, such as the existence of suitable numerical modeling, available geologic and geotechnical data, knowledge of the project site, regulatory requirements, acceptable risks, and cost. Timely evaluation and selection of the appropriate means of behavioral analysis are crucial because the type of analysis determines the nature and quantity of data required to perform the evaluation. Data will be collected during different phases of the geotechnical exploration program.

The geology of the site dictates the ground behavior. Specific concerns mark geologic conditions that are particularly favorable or adverse to the resolution of the geotechnical or geologic design. For each identified geologic design issue, background and site-specific information are collected to characterize the geologic physical properties and the way subsurface strata will behave related to the constructed facility. Also, the geologic information collected must support the predicted ground behavior at an acceptable level of confidence. Geologic assessments may incorporate surface and subsurface investigations, which can include geologic mapping, test pits, and soil and rock coring; in situ testing, which may include standard penetration testing, packer permeability testing, dilatometer testing, and pressure meter testing; geophysics investigations, including shallow, deep, and cross-hole reflection and refraction surveys, electrical resistivity surveys, and gravity surveys; and laboratory testing, all to determine the strength, compressibility, and permeability of the geologic strata present at the site.

If the site geology does not provide suitable conditions for the construction of a penstock, temporary and permanent ground modifications can be done to enhance the strength, deformation, and permeability characteristics of the natural ground and to control or cut off groundwater inflows to the site. The methods and results of these explorations, tests, analyses, and designs must be fully documented. If the consequences of a faulty prediction of ground behavior are severe or unacceptable, an inspection program during construction will be necessary to verify that the geotechnical and geologic design assumptions are met in relation to the exposed geologic conditions encountered during construction and in the final operation of the facility. If changes in the anticipated geologic conditions are encountered, the analyses and design must be reevaluated for structural competency.

Preliminary geologic investigations should be performed early in the project to assist the owner in the preliminary layout of the project and to help identify geologic issues that will affect the final design, construction, and operation of the facility. Also during this preliminary phase, future geotechnical and geologic phases of the project should be identified and planned to provide information necessary for the final design and construction of the project.

1.6.2 Geologic Design and Construction Issues for Consideration

Certain general geologic design and construction issues must be considered in the development of steel penstocks and tunnel liner projects. Geologic considerations for tunneling are not included in this summary because they are much more extensive in nature than the material covered in this text. The applicability of these considerations depends on the penstock's profile and geologic conditions present at the facility.

1. Foundation support for penstock and anchor blocks
   a. Allowable bearing capacity
   b. Liquefaction potential
   c. Settlement, swelling, and collapsibility potential
   d. Sliding resistance
   e. Active, passive, and at rest design parameters
   f. Modulus of soil reaction
   g. Rock and soil anchor design parameters
   h. Structurally supported foundations (piles or drilled piers)
      (1) Depth to sound bearing strata
      (2) End bearing capacity
      (3) Skin friction capacity
      (4) Uplift capacity
      (5) Down drag
2. Permanent excavations
   a. Cut-slope stability (includes falling rocks that may hit penstock)
   b. Bottom heave
   c. Rockfall prevention
      (1) Netting
      (2) Berms
      (3) Jersey barriers
      (4) Mechanically stabilized earth (MSE) walls
      (5) Retaining walls
3. External pressures and loads for buried steel penstocks
   a. Groundwater (from any source) pressures
   b. Earth pressures
4. Steel tunnel liner, including concrete backfill
   a. External pressures and loads, such as groundwater (from any source) pressures, swelling soils, and other unstable ground loads
   b. Load-carrying characteristics of the ground, e.g., modulus of deformation used in composite design method for tunnels and susceptibility to hydraulic fracturing
5. Availability and quality of earth construction materials
6. Effects of geology on construction operations
   a. Rippability or excavatability of subsurface strata
   b. Pile driving
   c. Grout injection
   d. Dewatering
   e. Abrasiveness of rock structure
   f. Processing of earth materials for aggregates, etc.
   g. Stability and maintenance of temporary cut-slopes, trench walls, staging, and stockpile areas
7. The alteration of shallow geologic materials (e.g., weathering) and the deterioration of engineering materials (e.g., subsurface corrosion and erosion, alkali-aggregate reaction) subjected to geologic agents and ground currents
8. Natural geologic hazards
   a. Earthquakes and fault movements
   b. Volcanic lava, ash flows, and lahars (debris flows)
   c. Landslides and hillside instability (including rock falls that may hit the penstock)
   d. Natural (and human-caused) ground subsidence and collapse
   e. Erosion and sedimentation

The geologic considerations for any needed ground modification or improvement work must be determined in the early phases of the project because extensive modification programs can add significant costs to a project.

1.6.3 Adverse Geologic Conditions

The following are common geologic features that may adversely affect the behavior of the ground:

1. Presence of weak, low-strength, and loose soils;
2. Low-strength or slickensided seams or weak zones in rock and soil formations;
3. Soluble and partially soluble stratigraphy;
4. Volcanic sequences;
5. Permafrost;
6. Highly fractured and jointed rock zones;
7. Inactive faults and shear zones, including bedding or foliation shears;
8. Active faults; and
9. Irregular weathering patterns in bedrock stratigraphy caused by decomposition, disintegration, and solutioning that result in highly variable geologic profiles and material properties.

1.6.4 Engineering Solutions for Ground Modification or Improvement

The following are several methods of ground improvement that can be used to improve the geologic conditions at sites:

1. Removal and replacement of any unsuitable subsurface materials by excavation and replacement, washing, jetting, open annulus around structural support for isolation from unstable ground;
2. Filling of larger sink holes and voids in the ground with backfill concrete, sand–cement–flyash grouts or select earth material;
3. Replacement of unsuitable ground with suitable materials: jet grouting stone columns; deep soil mixing; bentonite–soil, cement–bentonite, and cement-slurry walls; sheet piling; secant pile walls;
4. Protection of the ground surface from excessive water erosion or concentrated mechanical, thermal, or chemical loads and agents: geotextiles; bearing pads; surface coatings; vegetation covers; and surface drainage;
5. Solidification or cementation of the ground: cement and chemical grouting, jet grouting, and deep soil mixed walls;
6. Reinforcement of the ground for compression, tension, and shear resistance: soil nails, rock bolts, driven piles, drilled piers, stone columns, posttensioned soil and rock anchors, or shear trench;
7. Groundwater control: dewatering wells, well points, interceptor trenches and ditches, drainage tunnels, surface diversions, and permanent and temporary ground freezing;
8. Densification of the ground: vibrocompaction, heavy tamping, compaction grouting, preconsolidation using preloading in combination with wick drains or sand drains, soil replacement;
9. Chemical treatment of clays: liming stabilization; and
10. External stabilization or support: retaining walls, MSE walls, tieback walls, soil nail walls, earth berms, reinforced shot-concrete tunnel lining, rib and lagging tunnel lining, and steel set supports.

1.7 EARTHQUAKE LOADING

1.7.1 General Loading Criteria

ASCE Standard 7-10 (ASCE 2010) states, "Earthquakes 'load' structures indirectly. As the ground displaces, a building (structure) will follow and vibrate. The vibration produces deformations with associated strains and stresses in the structure. Computation of dynamic response to earthquake ground shaking is complex." The seismic design criteria for a penstock (including supports, anchor blocks, and the related structural system) are intended to (1) protect the economic investment in the facility, (2) provide acceptably high functional reliability within a short time following an earthquake, and (3) ensure safety to life.

It is recommended that the engineer follow the seismic loading methodology outlined in ASCE Standard 7-10 (ASCE 2010) for the minimum penstock seismic loading. This document is based on earthquake engineering information from the Federal Emergency Management Agency's (FEMA) sponsored Building Seismic Safety Council's (BSSC) National Earthquake Hazards Reduction Program (NEHRP) and the Applied Technology Council (ATC). ASCE Standard 7-10 (ASCE 2010) is the seismic load source document for the International Building Code (IBC), which is presently the most recognized building code in the United States.

ASCE Standard 7-10 (ASCE 2010) determines that the maximum considered earthquake risk (MCE_R) target for seismic loads is based on the simplified concept of a response spectrum. This standard states, "A response spectrum for a specific earthquake ground motion does not reflect the total time history of response, but only approximates the maximum value of response for simple structures to that ground motion. The design response spectrum is a smoothed and normalized approximation for many different ground motions, adjusted at the extremes for characteristics of larger structures." Further information on the development of the design response spectrums is presented in this standard and NEHRP documents.

ASCE Standard 7-10 (ASCE 2010) also states, "The seismic requirements of ASCE 7 are stated in terms of forces and loads. However, the user should always bear in mind that there are no external forces applied to the aboveground portions of a structure during an earthquake. The design forces are intended only as approximations to produce the same deformations, when multiplied by the Deflection Amplification factor C_d, as would occur in the same structure should an earthquake ground motion at the design level occur."

ASCE Standard 7-10 (ASCE 2010) requires a more detailed site-specific design response spectrum or ground motion analysis for areas of high seismic activity. Site-specific analysis may also be warranted for penstocks having unusually high economic value, stringent reliability requirements, or where seismically induced damage or failure could result in the uncontrolled release of stored water, with the possibility of flooding with loss of life or significant economic loss. If a penstock failure is a potential dam safety concern, then the level of penstock seismic analysis should appropriately match the seismic analysis for the associated site-specific dam or water-retaining structure.

1.7.2   General Penstock Seismic Considerations

Seismic evaluation considerations and complexity vary depending on whether the penstock is buried, continuously supported on the ground surface, or elevated above ground and surge tanks.

1.7.2.1 Buried Penstocks   Penstocks buried in stable compacted trenches accelerate with the surrounding ground and do not experience an increase in shell stresses. The primary earthquake-induced failure mechanisms for buried penstocks are shell rupture due to differential shell or appurtenance displacements from seismic-induced slope instability, ground faulting, or lateral spreading. Appurtenances, such as attached pipes, are a concern if differential displacement relative to the penstock shell can induce localized shell rupture at attachment locations. Buried penstocks with shells spanning active faults require specialized evaluation and considerations to provide the shell flexibility needed to accommodate differential seismic ground movements.

1.7.2.2 Continuous   Surface   Support   Although uncommon, some steel penstocks are continually supported by the ground surface. Because of the low profile and continuous shell bottom support, these penstocks are similar to buried penstocks, with ground movements transferred directly to the penstock shell, and the primary failure mechanism is shell rupture caused by differential shell or appurtenance displacement. Although ground-supported penstocks are more susceptible than buried penstocks to loss of support and differential earth pressures from seismic-induced slope instability, some penstocks spanning active seismic faults have been designed with sufficient flexibility in sections of continuously ground-supported penstock to accommodate seismic ground surface fault movement.

1.7.2.3 Elevated   Penstocks   Elevated penstocks experience the highest seismic reactions and therefore typically warrant a quantitative evaluation. The shells of elevated penstock experience additional longitudinal bending moments caused by lateral and vertical seismic accelerations. The shell longitudinal bending moments caused by horizontal lateral loads are perpendicular to gravity loads, and maximum orthogonal moments are not directly superimposed.

The supports of elevated penstocks experience concentrated reactions from contributory portions of the supported penstocks. The primary concerns at the supports are that these additional seismic support reactions do not overload the penstocks' shell, saddles, support structure, or foundation.

Elevated penstocks can also be susceptible to collateral earthquake damage caused by earthquake impacts, such as dislodged rock and earth slides, ground slope failures, and differential ground movement.

Similar to buried and surface-supported penstocks, appurtenances, such as attached pipes and valves, are a seismic concern if differential displacement relative to the penstock shell can induce localized shell rupture at attachment locations.

1.7.2.4 Surge Tanks   Because surge tanks are typically tall, slender structures, they can be susceptible to earthquake lateral loads, large anchor bolt reactions, and foundation overturning. Surge tanks typically have free water surfaces, so they are also susceptible to sloshing loads.

1.7.3   Elevated Penstock Seismic Analysis Method

This section outlines the general penstock seismic analysis methodology as presented in ASCE Standard 7-10 (ASCE 2010).

1.7.3.1 Determine   Site-Specific   Conditions   and Requirements   The first step in a seismic analysis is to determine the site-specific seismic design criteria. This step means quantifying the design spectral response accelerations, S_{DS} and S_{D1}, to be used in the analysis and determining the seismic design category. The seismic design category determines seismic requirements based on the structure's importance, type, and overall seismic severity. The procedure for determining the S_{DS}, S_{D1}, and seismic design category are presented in Chapter 11 with example calculations presented in Chapter 18. Some specific seismic considerations related to penstocks are the following:

- The size and purpose of a specific hydroelectric power station and of its penstock needs to be considered when determining the structure's risk category in Table 1.5-1. A hydroelectric power station would typically have a risk category of either III (structures that represent a substantial risk or economic loss) or IV (structures designated as essential facilities). This determination means that if a power station is required to operate in an emergency, such as providing black start capability or to provide "island" operation for a locality in the event of a grid disconnect, it should be risk category IV. Smaller stations that do not need to provide these functions and whose operation should only have a marginal effect on the operation of the local utility's grid operation may be classified as risk category III.

- Section 11.8 requires that power stations located in higher seismic geographic areas, particularly if they are risk category IV, be classified as seismic design categories C through F and that they require site-specific geotechnical investigation reports to evaluate items, such as slope stability, liquefaction, differential settlement, and surface displacements. The relevance of this requirement to a station's penstock depends on the type and configuration of a penstock's support system. For example, elevated penstocks, particularly traversing steep ground slopes are much more vulnerable and require a more extensive evaluation than buried penstocks traversing a flat topography.

1.7.3.2 Determine the Seismic Forces   ASCE Standard 7-10 (ASCE 2010) allows the three seismic analysis methods of (1) equivalent lateral force, (2) modal spectrum response, and (3) seismic response history procedures. For typical penstocks, this manual recommends following the requirements of Section 15.7.14 for the seismic evaluation of elevated penstocks.

For penstocks in locations that do not require site-specific response or ground motion seismic analysis, the minimum seismic lateral forces should not be less than the requirements of Section 12.8, which is the equivalent lateral force procedure.

The equivalent lateral force procedure states that the seismic base shear shall be determined by Formula 12.8-1 of ASCE Standard 7-10 (ASCE 2010) as follows:

$$\text{Seismic base shear } V = C_s W$$

where C_s = seismic response coefficient and W = effective seismic weight.

The seismic response coefficient, C_s, is determined in accordance with the requirements of Sections 12.8 and 15.5.4 and Eq. (15.4.1).

The effective seismic weight, W, is determined assuming the penstock's water is a rigid mass acting at the volumetric center of gravity, as required by Section 15.7.14.2.

Vertical seismic reactions need to be determined and appropriately combined as required by Section 12.4.2.

Chapter 18 includes example calculations for the seismic reactions of an elevated ring girder supported penstock in a moderate or lower seismic area using the equivalent lateral force procedure from ASCE Standard 7-10 (ASCE 2010). Chapter 18 was prepared using ASCE Standard 7-10 (ASCE 2010), which was the current version at the time this chapter was prepared; chapter references and/or methodology may be altered in subsequent versions.

1.7.4   Component and Support Design

The additional seismic reactions are then combined with the gravity and other component reactions to complete the penstock and support system design. For items that are designed by the allowable stress method, the seismic and other stresses are combined and compared with the member allowable stresses. Similarly, for members designed by the load resistance factor design (LRFD) or ultimate load procedure, the seismic reactions are combined with the other member reactions and compared with the member capacities. The design and analysis need to follow the penstock support system load path from the penstock shell through the foundation bearing.

1.7.5   Surge Tank Seismic Analysis Method

Chapter 15.7.7.1 of ASCE Standard 7-10 (ASCE 2010) allows the seismic analysis of flat bottom ground supported steel surge tanks to be performed in accordance with the requirements of AWWA D100 (AWWA 2011). This pseudo-dynamic procedure accounts for the fundamental period of the convective mode of vibration, which is the sloshing of surface liquid against the tank shell. Further background is also available from Chapter 14.4.7 of FEMA P-750 (FEMA 2009).

REFERENCES

ASCE. (1989). *Civil engineering guidelines for planning and designing hydroelectric developments*, American Society of Civil Engineers, New York.

ASCE. (2010). "Minimum design loads for buildings and other structures," *ASCE 7-10*, Reston, VA.

AWWA. (2011). *AWWA standard for welded carbon steel tanks for water storage, ANSI/AWWA D100*, Denver.

FEMA. (2009). *NEHRP recommended provisions and commentary for seismic regulations for new buildings and other structures, FEMA P-750*, Washington, DC.

Hunt, Roy E. (1984). *Geotechnical engineering investigation manual*, McGraw-Hill, New York.

International Building Code (IBC). International Code Council, Washington, DC.

Leggett, Robert F., and Karrow, Paul F. (1983). *Handbook of geology in civil engineering*, McGraw-Hill, New York.

Sullivan, R. K., Jr. (1971). "Sizing of pumped storage conduits." *J. Power Div.*, 97(3), 667–673.

USBR. (1976). *Selecting hydraulic reaction turbines*, Engineering monograph no. 20, U.S. Bureau of Reclamation, Denver.

CHAPTER 2

Materials

Source: Photograph courtesy of American Cast Iron Pipe Co.; reproduced with permission from American SpiralWeld Pipe LLC.

Material selection is an important aspect of a penstock design. The specific materials chosen for various portions of a penstock impact the manufacturing and installation costs and are critical to successful service throughout the life of the penstock.

2.1 GENERAL

Steel materials used in the fabrication of penstocks, including pressure-carrying components and nonpressure-carrying attachments, such as flanges, ring girders, stiffener rings, thrust rings, lugs, support systems, and other appurtenances, must be manufactured and tested in strict accordance with

appropriate ASTM standards, or their equivalent engineer-accepted international standards, and as set forth in this manual.

The properties of steels are governed by their chemical composition, by the processes used to form the base metal, and by their heat treatment. The effects of these parameters on the properties of steels affect material selection and are discussed in the following sections.

2.1.1 Material Selection

The various available materials provide the engineer a wide range of attributes necessary to fulfill the service requirements of any steel penstock. Having an understanding of the basics of steel materials allows the engineer to choose the appropriate material for the specific application in addition to appreciate the economic impact of such choices.

2.1.1.1 Steel Selection The selection of the steel for the design and construction of penstocks, tunnel liners, water conduits, and appurtenances depends on thickness, availability, service temperature, particular use, economics, and the ease of shop and field fabrication. Material thickness is important because it affects heat treatment requirements, mechanical testing, nondestructive testing examination needs, material and fabrication cost, and method of fabrication. Pressure-vessel-quality steels or the equivalent are normally used for higher pressure and larger diameter penstocks. In some cases, structural quality steels can be used; such cases may include penstocks for low head applications or where design stress is not the critical factor in determining the resulting penstock thickness. Regardless, the engineer must evaluate all aspects of material selection relative to the service conditions of the penstock. In addition to yield and tensile strength characteristics, the following important steel properties require careful consideration by the engineer: elongation, ductility, fracture toughness, weldability, and cost. For steel penstock requirements, many possible materials may be suitable for projects, depending on application. See Table 2-1 for a listing of some acceptable materials.

The selection of steel material, specified nondestructive examination (NDE) requirements, and hydrostatic pressure testing becomes more critical for special sections, such as wyes, bends, thrust rings, and expansion and reducer sections. Special steels and more stringent NDE and QA/QC requirements may be justified for these sections.

Attachments, such as ring girders, stiffener rings, and support systems, may be designed for and fabricated from plate or structural shapes made from structural quality steel. Caution should be exercised, however, because the selection of some high-strength alloy steel materials for the penstock may require special considerations if attachments are made from different materials.

2.1.1.2 Steel Characteristics As stated previously, the steel must be manufactured and tested in accordance with ASTM steel standards. ASTM supplemental requirements warranted for a particular installation also must be specified. If "foreign" steel is used for the penstock, review the applicable foreign standards (in English) and compare them with the applicable ASTM standards. Subtle differences may exist among standards and must be identified; "special" requirements must be called out in the project specifications. Similarly, foreign testing methods may differ from U.S. practices. In some instances, foreign standards may require substantial, additional testing. The weldability of foreign steels also must be considered.

Proper steel selection for penstock application is related to the manufacturing process and the way the penstock is loaded during installation and service. From a design per-

spective, because penstocks may be stressed in all directions, it is desirable, to the extent feasible, to use steel that has consistent mechanical properties in all directions.

Steels for penstocks can be furnished as structural grade with testing in accordance with ASTM A6 (ASTM 2010a) or pressure vessel grade with testing in accordance with ASTM A20 (ASTM 2010b). Steels tested in accordance with ASTM A6 are tested once on a heat lot basis, whereas steels tested in accordance with ASTM A20 are subjected to more extensive testing. To meet the requirements of ASTM A20 (ASTM 2010b), each plate is tested once, and each coil is tested three times; one test each at the beginning, middle, and end of the coil. The engineer must ensure that the level of testing required is commensurate with the service conditions of the finished product.

2.1.2 Chemical Composition

Steels are a mixture of iron and carbon with varying amounts of other elements, primarily manganese, phosphorus, sulfur, and silicon. These and other elements are either unavoidably present or intentionally added in various combinations to achieve specific characteristics and properties of the finished steel products. Table 2-2 identifies common chemical elements and their effects on the properties of hot-rolled and heat-treated carbon and alloy steels. The effects of carbon, manganese, sulfur, silicon, and aluminum are of primary interest. As part of the cooling and solidification process all steels exhibit some degree of nonuniformity of chemical composition, known as segregation, but the degree of this effect varies based on the casting process used to manufacture the steel.

2.1.3 Casting

Today's steelmaking practice typically uses the continuous casting process, which includes the direct casting of steel from the ladle into coils or slabs that are subsequently converted to plate or coils. This steelmaking development has significantly improved the quality of the final product by reducing the amount of segregation and providing greater uniformity in composition relative to steel made from ingots. The outer laps of coils cool more rapidly than the central laps. Although the minimum required mechanical properties are maintained throughout, the difference in cooling rates can result in measurable differences in the mechanical properties along the length of the coil. Section 2.2.1 defines two specific levels of testing that can be used to monitor the material properties relative to the engineer's requirements. Historically, the mechanical differences have been small and have not been an issue, but the engineer can contact the manufacturer regarding the degree of difference to understand if it is critical to the design.

For very thick steel plates that cannot be produced by the continuous casting process, the traditional steelmaking pro-

Table 2-1 Steels Used in the Manufacture of Penstocks and Associated Structural Supports

Specification no.	Grade	Minimum yield strength (ksi)	Tensile strength (ksi) (min or min to max)	Elongation[a] in 2-in. (percentage)	Remarks
			Coil		
ASTM A1011-SS, ASTM (2010k)	30	30	49	25	
	33	33	52	23	
	36 Type 1	36	53	22	
	36 Type 2	36	58–80	21	
	40	40	55	21	
ASTM A1011-HSLAS, ASTM (2010k)	45 Class 1	45	60	25	
	45 Class 2	45	55	25	
	50 Class 1	50	65	22	
	50 Class 2	50	60	22	
	55 Class 1	55	70	20	
	55 Class 2	55	65	20	
	60 Class 1	60	75	18	
	60 Class 2	60	70	18	
	65 Class 1	65	80	16	
	65 Class 2	65	75	16	
ASTM A1011-HSLAS-F, ASTM (2010k)	50	50	60	24	
ASTM A1018-SS, ASTM (2010l)	30	30	49	22	
	33	33	52	22	
	36 Type 1	36	53	21	
	36 Type 2	36	58–80	21	
	40	40	55	19	
ASTM A1018-HSLAS, ASTM (2010l)	45 Class 1	45	60	22	
	45 Class 2	45	55	22	
	50 Class 1	50	65	20	
	50 Class 2	50	60	20	
	55 Class 1	55	70	18	
	55 Class 2	55	65	18	
	60 Class 1	60	75	16	
	60 Class 2	60	70	16	
	65 Class 1	65	80	14	
	65 Class 2	65	75	14	
ASTM A1018-HSLAS-F, ASTM (2010l)	50	50	60	22	
			Plate		
ASTM A36, ASTM (2008a)	—	36	58–80	23	Structural

(Continued)

Table 2-1   Steels Used in the Manufacture of Penstocks and Associated Structural Supports *(Continued)*

Specification no.	Grade	Minimum yield strength (ksi)	Tensile strength (ksi) (min or min to max)	Elongation[a] in 2-in. (percentage)	Remarks
ASTM A283, ASTM (2007a)	A	24	45–60	30	Structural
	B	27	50–65	28	
	C	30	55–75	25	
	D	33	60–80	23	
ASTM A285, ASTM (2007b)	A	24	45–65	30	Pressure vessel quality (PVQ)
	B	27	50–70	28	
	C	30	55–75	27	
ASTM A516, ASTM (2010i)	55	30	55–75	27	PVQ over 1-1/2 in. is normalized
	60	32	60–80	25	
	65	35	65–85	23	
	70	38	70–90	21	
ASTM A517 Quenched and tempered, ASTM (2010j)	—	100	115–135	16	2-1/2 in. and under
	—	90	105–135	14	Over 2-1/2 in. to 6 in.
ASTM A537 CL-1 is normalized CL-2 is quenched and tempered, ASTM (2008b)	CL-1	50	70–90	22	2-1/2 in. and under
	CL-1	45	65–85	22	Over 2-1/2 in. to 4 in.
	CL-2	60	80–100	22	2-1/2 in. and under
	CL-2	55	75–95	22	Over 2-1/2 in. to 4 in.
	CL-2	46	70–90	20	Over 4 in. to 6 in.
ASTM A572 S90 All grades killed fine grain, ASTM (2007c)	42	42	60	24	Structural
	50	50	65	21	Over 3/4 in.
	50 (S81)	50	70	21	3/4 in. and under
	60	60	75	18	1-1/4 in. maximum
ASTM A738, ASTM (2007d)	A	45	75–95	20	
	B	60	85–102	20	
Pipe					
ASTM A53, ASTM (2010c)	B	35	60	See ASTM A53	Type E & S pipe specification
ASTM A106, ASTM (2010e)	B	35	60	See ASTM A106	
ASTM A139, ASTM (2010f)	B	35	60	30	Pipe specification: specify 0.25% maximum carbon Con Cast
	C	42	60	25	
	D	46	60	23	
	E	52	66	22	
API-5L, API (2009)	B	35	60		Pipe specification includes 100% NDE
	X42	42	60		
	X46	46	63		
	X52	52	66		
	X56	56	71		
	X60	60	75		

[a]Values presented in Table 2-1 are base minimum values identified in ASTM standards. Elongation requirements vary based on thickness, width, and shape of material. Refer to the current referenced specification to verify specific elongation values for a given material.

Table 2-2 Alloying Elements and Their Effects on Steel Properties

Element	Effect
Carbon (C)	Principal hardening element in steel
	Increases strength and hardness
	Decreases ductility, toughness, and weldability
	Moderate tendency to segregate
Manganese (Mn)	Increases strength
	Controls harmful effects of sulfur
Phosphorus (P)	Increases strength and hardness
	Decreases ductility and toughness
	Considered an inpurity, but sometimes added for atmospheric corrosion resistance
	Strong tendency to segregate
Sulfur (S)	Considered undesirable except for machinability
	Decreases ductility, toughness, and weldability
	Adversely affects surface quality
	Strong tendency to segregate
Silicon (Si)	Used to deoxidize or kill molten steel
Aluminum (Al)	Used to deoxidize or kill molten steel
	Refines grain size, thus increasing strength and toughness
Vanadium (V) and Niobium (Nb)[a]	Small additions increase strength
Nickel (Ni)	Increases strength and toughness
Chromium (Cr)	Increases strength
	Increases atmospheric corrosion resistance
Copper (Cu)	Primary contributor to atmospheric corrosion resistance
Nitrogen (N)	Increases strength and hardness
	Decreases ductility and toughness
Boron (B)	Small amounts (0.0005%) increase hardenability in quenched and tempered steels
	Used only in aluminum-killed steels
	Most effective at low carbon levels

[a]Niobium is also commonly referred to, documented and reported as, Columbium (Cb)

cess is still used, in which molten steel is poured (teemed) into a series of molds to form castings known as ingots. The ingots are removed from the molds, reheated, and then rolled into products with square or rectangular cross sections. This hot-rolling operation elongates the ingot and produces semifinished products known as blooms, slabs, or billets.

Certain elements tend to segregate more readily than others. Sulfur segregates to the greatest extent. The following elements also segregate, but to a lesser degree, and in descending order: phosphorus, carbon, silicon, and manganese. The degree of segregation is influenced by the composition of the liquid steel, the liquid temperature, and ingot size. The most severely segregated areas of the ingot are removed by cropping, which is the cutting and discarding of unacceptable material during rolling.

2.1.4 Killed and Semikilled Steels

The primary reaction involved in most steelmaking processes is the combination of carbon and oxygen to form carbon monoxide gas. The solubility of this and other gases dissolved in the steel decreases as the molten metal cools to the solidification temperature range. Thus, excess gases are expelled from the metal and, unless controlled, continue to evolve during solidification. The oxygen available for the reaction can be eliminated and the gaseous evolution can be inhibited by deoxidizing the molten steel using additions of silicon or aluminum, or both. Steels that are strongly deoxidized do not evolve any gases and are called killed steels because they lie quietly in the mold. Increasing amounts of gas evolution results in semikilled, capped, or rimmed steels.

In general, killed steel ingots are less segregated and contain negligible porosity when compared to semikilled steel ingots. Consequently, killed steel products usually exhibit a higher degree of uniformity in composition and properties than semikilled steel products.

2.1.5 Heat Treatments for Steels

Steels respond to a variety of heat treatments, which can be used to obtain certain desirable characteristics. These heat treatments can be divided into slow cooling treatments and rapid cooling treatments. The slow cooling treatments, such as annealing, normalizing, and stress relieving, decrease hardness and promote uniformity of structure; rapid cooling treatments, such as quenching and tempering, increase strength, hardness, and toughness. Heat treatments of base metals are generally mill options or ASTM requirements. Following are some of the more common heat treatments for steels.

2.1.5.1 Annealing Annealing consists of heating the steel to a given temperature followed by slow cooling. The temperature, the rate of heating and cooling, and the time the metal is held at temperature depends on the composition, shape, and size of the steel product being treated and the desired properties. Usually steels are annealed to remove stresses, induce softness, increase ductility and toughness, produce a given microstructure, increase uniformity of microstructure, improve machinability, and facilitate cold forming.

2.1.5.2 *Normalizing* Normalizing consists of heating the steel to between 1,650°F and 1,700°F followed by slow cooling in air. This heat treatment is commonly used to refine the grain size, improve uniformity of microstructure, and improve ductility and fracture toughness.

2.1.5.3 *Stress Relieving* Stress relieving of carbon steels consists of heating and holding the steel to between 1,000°F and 1,200°F to equalize the temperature throughout the piece, followed by slow cooling. The stress-relieving temperature for quenched and tempered steels must be maintained below the tempering temperature for the product. The process relieves internal stresses induced by welding, normalizing, cold working, cutting, quenching, and machining. It is not intended to alter significantly the microstructure or the mechanical properties.

2.1.5.4 *Quenching and Tempering* Quenching and tempering consists of heating and holding the steel at the proper austenitizing temperature (about 1,650°F) for a significant time to produce a desired change in microstructure, then quenching by immersion in a suitable medium (such as water for bridge steels). After quenching, the steel is tempered by reheating to a temperature between approximately 800°F and 1,200°F, holding for a specified time at temperature, and then cooling under suitable conditions to obtain the desired properties. Quenching and tempering increase the strength and improve the toughness of the steel but may decrease ductility.

2.1.5.5 *Controlled Rolling* Controlled rolling is a thermomechanical treatment at the rolling mill that tailors the time–temperature–deformation process by controlling the rolling parameters. The parameters of primary importance are (1) the temperature at the start of controlled rolling in the finishing stand, (2) the percentage reduction from the start of controlled rolling to the final plate thickness, and (3) the plate finishing temperature.

Hot-rolled plates are deformed as quickly as possible at temperatures above 1,800°F to take advantage of the hot workability of the steel at high temperatures. In contrast, controlled rolling incorporates a hold or delay time to allow the partially rolled slab to reach a desired temperature before the start of final rolling. Controlled rolling involves deformations at temperatures between 1,500°F and 1,800°F. Because rolling deformation at these low temperatures increases the mill loads significantly, controlled rolling usually is restricted to plates that are less than 2 in. thick. Controlled rolling increases strength, refines the grain size, improves toughness, and may eliminate the need for normalizing.

2.1.5.6 *Controlled Finishing–Temperature Rolling* Controlled finishing–temperature rolling is a less severe practice than controlled rolling and is aimed primarily at improving notch toughness of plates up to 2 1/2 in. in thickness. The finishing temperatures (approximately 1,600°F) in this practice are higher than those required for controlled rolling. However, because heavier plates are involved than with controlled rolling, mill delays still are required to reach the desired finishing temperatures. Fine grain size and improved notch toughness can be obtained by controlling the finishing temperature.

2.1.6 Mechanical Properties

The mechanical properties of a material are those properties that characterize its elastic and inelastic (plastic) behavior under stress or strain. These properties include parameters related to the material's strength, ductility, and hardness. The strength and ductility parameters under tensile loading can be defined and explained best by analyzing the tensile stress–strain curve for the material.

2.1.6.1 *Stress–Strain Curves* Fig. 2-1 shows an idealized tensile stress–strain curve for constructional steels. The curve is obtained by the tensile loading to failure of specimens that have rectangular or circular cross sections. A detailed discussion of definitions and details for tension test specimens and test methods is available in ASTM A370 (ASTM 2010h).

The curve shown in Fig. 2-1 is an engineering tensile stress–strain curve, as opposed to a true tensile stress–strain curve, because the plotted stresses are calculated by dividing the instantaneous load on the specimen by its original, rather than reduced, cross-sectional area. Also, the strains are calculated by dividing the instantaneous elongation of a gauge length of the specimen by the original gauge length.

The initial straight-line segment of the stress–strain curve, where stress is linearly related to strain, represents the elastic behavior of the specimen. In this region, the strain is fully recoverable and the specimen returns to its original length when the load is removed. The slope of the line, which is the ratio of stress to strain in the elastic region, is the modulus of elasticity, or Young's modulus, and is approximately equal to 29×10^6 psi for steels. As the load increases, the stress and strain relationship becomes nonlinear and the specimen experiences permanent plastic deformation. The stress cor-

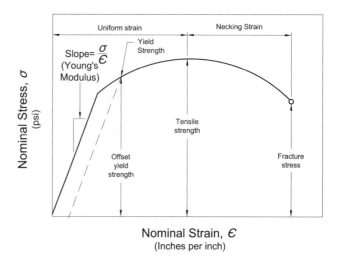

Fig. 2-1.   Idealized tensile stress–strain curve

responding to the initial deviation from linearity represents the yield strength of the material and the beginning of the plastic region. Usually, the stress required to produce additional plastic strain increases with increasing strain, and thus the steel strain hardens. The rate at which stress increases with plastic strain is the strain-hardening modulus.

Fig. 2-2 shows tensile stress–strain curves for constructional steels that exhibit two types of behavior in the plastic region. The curve in Fig. 2-2(a) exhibits a smooth deviation from linearity, and the stress continuously increases to a maximum value and then decreases until the specimen fractures. On the other hand, the stress for the curve in Fig. 2-2(b) reaches a peak immediately after the stress–strain curve deviates from linearity, dips slightly, and then remains at a constant value for a considerable amount of additional strain. Thereafter, the steel strain hardens and the stress increases with strain to a maximum and then decreases until the specimen fractures. The stress corresponding to the peak value represents the yield strength and is the stress at which

the material exhibits a specific limiting deviation from linearity of stress and strain. The deviation may be expressed as a 0.2% offset or a 0.005 in./in. total extension under load (Fig. 2.2(a)).

The maximum stress exhibited by the engineering stress–strain curve corresponds to the tensile strength of the steel.

2.1.6.2 Ductility and Toughness   The tensile stress–strain curve can be divided into a uniform strain region and a nonuniform strain region, which combine to give the total strain to fracture (see Fig. 2-1). In the uniform strain region, the cross-sectional area along the entire gauge length of the specimen decreases uniformly as the specimen elongates under load. Initially, the strain hardening compensates for the decrease in cross-sectional area and the engineering stress continues to increase with increasing strain until the specimen reaches its tensile strength. Beyond this point, the plastic strain becomes localized in a small region of the gauge length and the specimen begins to neck locally, with a corresponding decrease in total stress, until the specimen fractures.

Ductility is a material's ability to undergo plastic deformation. The total percent elongation and the total percent reduction of area at fracture are two measures of ductility that are obtained from the tension test. The percent elongation is calculated from the difference between the initial gauge length and the gauge length after fracture. Similarly, the percent reduction of area is calculated from the difference between the initial and the final cross-sectional area after fracture. Both elongation and reduction of area are influenced by gauge length and specimen geometry. The conversion between elongation of an 8-in. gauge length strap specimen and a 2-in. gauge length round specimen can be found in ASTM A370 (ASTM 2010h).

Ductility is an important material property because it allows the redistribution of high local stresses. Such stresses occur in welded connections and at regions of stress concentration, such as holes and changes in geometry.

Toughness is the ability of a material to absorb energy before fracturing and is related to the area under the stress–strain curve. The larger the area under the curve, the tougher the material.

2.1.6.3 Properties   Physical properties of a material are obtained by conducting a mechanical test. The most common mechanical test is a tension test. The procedures and definitions for the tension test methods are presented in ASTM A370 (ASTM 2010h). The general specifications and tolerances for acceptability of structural steel plates are presented in ASTM A6 (ASTM 2010a). Those for pressure vessel steel plates are presented in ASTM A20 (ASTM 2010b).

Steel producers are aware of the chemical segregation and the variability in the properties of steel products. Consequently, based on their experience, they establish target chemical compositions that ensure that the requirements of the material specifications are met. Because of the recognized variations, the target chemistries and processing

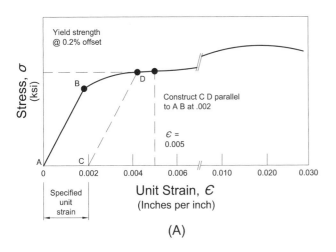

(A)

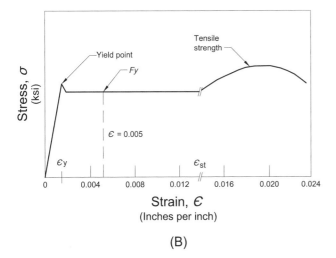

(B)

Fig. 2-2.   Schematic stress–strain curves for steels

practices are selected to achieve properties that exceed the minimum properties required by the material specifications.

2.1.7   Fracture Toughness

Steels may fracture either in a ductile or a brittle manner. The mode of fracture is governed by the temperature at fracture, the rate at which the loads are applied, and the magnitude of the constraints that would prevent plastic deformation. The effects of these parameters on the mode of fracture are reflected in the fracture-toughness behavior of the material. In general, the fracture toughness increases with increasing temperature, decreasing load rate, and decreasing constraint. Furthermore, there is no single unique fracture-toughness value for a given steel, even at a fixed temperature and loading rate.

Traditionally, the fracture toughness for low- and intermediate-strength steels has been characterized primarily by testing Charpy V-notch (CVN) specimens at different temperatures. The procedures and definitions for the Charpy V-notch test methods are presented in ASTM A370 (ASTM 2010h). However, the fracture toughness for materials can be established best by using fracture-mechanics test methods.

The Charpy V-notch impact specimen has been the most widely used specimen for characterizing the fracture-toughness behavior of steels. These specimens may be tested at different temperatures, and the impact fracture toughness at each test temperature may be determined from the energy absorbed during fracture, the percent shear (fibrous) fracture on the fracture surface, or the change in the width of the specimen (lateral expansion).

At low temperatures, constructional steels exhibit a low value of absorbed energy (about 5 ft-lb) and zero fibrous fracture and lateral expansion. The values of these fracture-toughness parameters increase as the test temperature increases until the specimens exhibit 100% fibrous fracture and reach a constant value of absorbed energy and lateral expansion. This transition from brittle-to-ductile fracture behavior usually occurs at different temperatures for different steels and even for a given steel composition. Consequently, like other fracture-toughness tests, there is no single unique CVN value for a given steel, even at a fixed temperature and loading rate. Therefore, when fracture toughness is an important parameter, the engineer must establish and specify the necessary level of fracture toughness for the material to be used in the particular structure at a given temperature or in a critical component within the structure.

2.2   TYPES OF MATERIALS

2.2.1   Coils

Steel coil has a proven record of performance for penstocks and provides an economically beneficial alternative to the use of plate steel. Coils produced on modern rolling mills from continuously cast slabs have proven to have the consistent chemical and mechanical properties desired throughout the coil. Depending on market conditions and material requirements, the maximum coil thickness available is limited to between 3/4 in. and 1 in. Heat treatment of coil steel (normalized or quenched and tempered) is not available.

When spiral weld pipe is produced directly from coils, the testing provisions required by the engineer for chemical and physical properties should be as follows. The mill producing the coil must furnish a certified chemical analysis of each heat. Required physical tests may be taken by the pipe manufacturer from the coil or from the completed pipe. The spiral weld pipe manufacturer must furnish certified reports of the physical tests. The engineer must define the material testing requirements to be either structural grade or pressure vessel grade. Structural grade conformance requires two sets of the required physical tests be taken for each heat or each 50 tons of each heat. Pressure vessel grade conformance requires that one set of the required physical tests be taken from each coil.

A set of physical tests must consist of three tests on a coil—one from the outside wrap of the coil, one from the middle third of the coil, and one from the inner wrap adjacent to the portion of the coil that is used. The set from the middle portion of the coil may be taken from the finished pipe. For test orientation, the longitudinal axis of the test specimens must be transverse to the final rolling direction of the coil.

2.2.2   Steel Plate

Steel plate is manufactured to two standards—ASTM A6 (ASTM 2010a) for structural use and ASTM A20 (ASTM 2010b) for pressure vessels. The physical and chemical properties of steel produced to these two standards can be similar, particularly when supplemental requirements are specified. The standards differ, however, in the amount of testing required to ensure uniform quality.

The individual ASTM standards for plates give the engineer a wide range of properties from which to select the economical material for a particular application. Structural steel (ASTM 2010a) is suitable for ring girders, stiffener rings, thrust rings, lugs, support systems, and for the pipe shell in many of the less critical penstocks, where design stress does not exceed 21,000 psi.

Pressure vessel quality plates (ASTM 2010b) normally are used in the fabrication of the pipe shell for penstocks. They may also be used for crotch plates, sickle beams, ring girders, or other structural parts if desired.

2.3   TEMPERATURE CONSIDERATIONS

The notch toughness of steel at low temperatures must be considered in penstock design. Notch toughness usually is measured by Charpy impact tests.

The lowest service temperature (LST) of a buried penstock is approximately 30°F. Also, the lowest service temperature of an aboveground penstock is close to 30°F for the pipe shell. Structural support systems, ring girders, and empty pipe may be subjected to lower temperatures.

Fine-grain killed steel, with a 55 ksi minimum specified yield strength or less and a thickness of 5/8 in. or less, has adequate toughness at a service temperature as low as 20°F. Normalized and quenched and tempered steels have improved toughness at even lower temperatures. See Section 2.6.3 for testing requirements for other steels and thicknesses.

2.4 ATTACHMENT MATERIALS AND CRITERIA

2.4.1 Nozzles

Nozzle openings in a penstock must be fabricated from suitable materials defined as follows:

1. Forged steel weld couplets, Weldolets, Sweepolets, or other fittings conforming to ASME B16.11 (ASME 2009) and made of material conforming to ASTM A105 (ASTM 2010d).
2. Steel pipe manufactured from material listed in Section 2.5 or fabricated in accordance with ASME Boiler and Pressure Vessel Code, Section II (ASME 2010).

2.4.2 Bends, Tees, Reducers, and Caps

Forged steel weld fittings must be in accordance with ASME B16.9 (ASME 2007), conforming to ASTM A234 (ASTM 2010g) for the schedule and grade required.

2.4.3 Dished Heads

Heads may be of ellipsoidal, torispherical, hemispherical, conical, or toriconical configuration, manufactured from materials listed in Section 2.5.

2.4.4 Attachments

Nonpressure-carrying attachments, such as ring girders, stiffener rings, thrust rings, lugs, support systems, and other appurtenances, may be fabricated from any of the materials listed in Section 2.5.

2.5 MATERIAL SPECIFICATIONS

Table 2-1 lists some of the steels used in the manufacture of steel penstocks and associated structural supports. These materials have been specified with and without supplemental requirements. Other steel types or grades may be acceptable for use, and it is recommended that the engineer consult with a metallurgist or other material expert as necessary to verify acceptability of the desired material.

Steels with physical characteristics within the same ranges as those in Table 2-1 but manufactured to different specifications are available for the manufacture of penstocks. The engineer should consult a metallurgist or fabricator when selecting steels.

2.6 MATERIAL TESTING REQUIREMENTS

The following sections contain testing requirements for materials used for pressure-retaining parts of exposed steel penstocks, tunnel linings, and attachments welded directly to these pressure-retaining parts.

Table 2-3 Required Charpy V-Notch Energy Values for Pressure-Retaining Material

Nominal wall thickness (in.)	Specified minimum yield strength					
	55 ksi or less		56 ksi to 75 ksi		76 ksi to 105 ksi	
	Energy for base materials[a]		Energy for base materials[a]		Energy for base materials[a]	
	Average of 3	Lowest 1 of 3	Average of 3	Lowest 1 of 3	Average of 3	Lowest 1 of 3
5/8 or less[b]	—	—	—	—	—	—
Over 5/8 to 1	20	15	25	20	30	25
Over 1 to 1 1/2	25	20	30	25	35	30
Over 1 1/2 to 2 1/2	35	30	40	35	45	40
Over 2 1/2	45	40	50	45	55	50

[a]Where two base materials with different required energy values are joined, the weld metal impact energy requirements of the procedure qualification tests must conform to those of the base material with the higher value.

[b]No test required.

2.6.1 Chemical Analysis

Chemical analysis must be performed to ensure conformance of material to the requirements of ASME (2010) or ASTM A specifications.

2.6.2 Tensile and Bend Tests

Tensile and bend tests must be performed to ensure conformance of material to the requirements of ASME (2010) or ASTM A specifications. Weld filler material must meet the same minimum strength requirements as the base metal.

2.6.3 Impact Tests

Materials other than those excluded in Section 2.3 must be impact tested and must meet the criteria specified in Table 2-3. Impact tests must be conducted at 5°F below the LST.

Weld filler material must meet the same minimum Charpy requirements as the base metal.

2.6.4 Special Carbon Equivalent Criteria Requirements

For materials with specified minimum yield strength over the indicated values, the following maximum carbon equivalent (CE) criteria apply.

1. For specified minimum yield strength over 55 ksi, CE = 0.45 (S20.2, ASTM 2010b).
2. For specified minimum yield strength over 75 ksi, CE = 0.53 (S20.2, ASTM 2010b).

REFERENCES

API. (2008). "Specification for line pipe," 44th Ed. *ANSI/API Spec 5L*, Washington, DC.

ASME. (2007). "Factory-made wrought buttwelding fittings." *ASME B16.9*, New York.

ASME. (2009). "Forged fittings, socket-welding and threaded." *ASME B16.11*, New York.

ASME. (2010). "Boiler and pressure vessel code, section II: Materials." New York.

ASTM. (2007a). "Standard specification for low and intermediate tensile strength carbon steel plates." *ASTM A283/A283M*, West Conshohocken, PA.

ASTM. (2007b). "Standard specification for pressure vessel plates, carbon steel, low- and intermediate-tensile strength." *ASTM A285/A285M*, West Conshohocken, PA.

ASTM. (2007c). "Standard specification for high-strength low-alloy columbium–vanadium structural steel." *ASTM A572/A572M*, West Conshohocken, PA.

ASTM. (2007d). "Standard specification for pressure vessel plates, heat-treated, carbon–manganese–silicon steel, for moderate and lower temperature service." *ASTM A738/A738M*, West Conshohocken, PA.

ASTM. (2008a). "Standard specification for carbon structural steel." *ASTM A36/A36M*, West Conshohocken, PA.

ASTM. (2008b). "Standard specification for pressure vessel plates, heat-treated, carbon–manganese–silicon steel." *ASTM A537/A537M*, West Conshohocken, PA.

ASTM. (2010a). "Standard specification for general requirements for rolled structural steel bars, plates, shapes, and steel piling." *ASTM A6/A6M*, West Conshohocken, PA.

ASTM. (2010b). "Standard specification for general requirements for steel plates for pressure vessels." *ASTM A20/A20M*, West Conshohocken, PA.

ASTM. (2010c). "Standard specification for pipe, steel, black and hot-dipped, zinc-coated, welded and seamless." *ASTM A53/A53M*, West Conshohocken, PA.

ASTM. (2010d). "Standard specification for carbon steel forgings for piping applications." *ASTM A105/A105M*, West Conshohocken, PA.

ASTM. (2010e). "Standard specification for seamless carbon steel pipe for high-temperature service." *ASTM A106/A106M*, West Conshohocken, PA.

ASTM. (2010f). "Standard specification for electric-fusion (arc)-welded steel pipe (NPS 4 and over)." *ASTM A139/A139M*, West Conshohocken, PA.

ASTM. (2010g). "Standard specification for piping fittings of wrought carbon steel and alloy steel for moderate and high temperature service." *ASTM A234/A234M*, West Conshohocken, PA.

ASTM. (2010h). "Standard test methods and definitions for mechanical testing of steel products." *ASTM A370*, West Conshohocken, PA.

ASTM. (2010i). "Standard specification for pressure vessel plates, carbon steel, for moderate- and lower-temperature service." *ASTM A516/A516M*, West Conshohocken, PA.

ASTM. (2010j). "Standard specification for pressure vessel plates, alloy steel, high-strength, quenched and tempered." *ASTM A517/A517M*, West Conshohocken, PA.

ASTM. (2010k). "Standard specification for steel, sheet and strip, hot-rolled, carbon, structural, high-strength low-alloy, high-strength low-alloy with improved formability, and ultra-high strength." *ASTM A1011/A1011M*, West Conshohocken, PA.

ASTM. (2010l). "Standard specification for steel, sheet and strip, heavy-thickness coils, hot-rolled, carbon, commercial, drawing, structural, high-strength low-alloy, high-strength low-alloy with improved formability, and ultra-high strength." *ASTM A1018/A1018M*, West Conshohocken, PA.

CHAPTER 3

Design Criteria and Allowable Stresses

Source: Photograph courtesy of Denver Water; reproduced with permission.

3.1 GENERAL CONSIDERATIONS

3.1.1 Certified Design Criteria

Penstocks must be designed in accordance with the requirements of the certified design criteria, which must be prepared by a qualified, experienced, and registered professional engineer who is knowledgeable in penstock design. The certified design criteria give the engineer the following input information:

1. Internal and external pressure profiles;
2. Dead loads;
3. Live loads;
4. Externally applied mechanical loads;
5. Wind, snow, and seismic loads;
6. Thermal loading conditions; and
7. Cyclic conditions, including dynamic loads.

These loads are combined as given in Section 3.3 to describe the various load combinations for the following conditions:

1. Normal operating conditions,
2. Intermittent conditions,
3. Emergency conditions,
4. Exceptional conditions,
5. Construction conditions, and
6. Hydrostatic test conditions.

3.1.2 Design Report

The design report is based on the requirements of the certified design criteria. The design report consists of comprehensive, detailed calculations that clearly describe the stresses in the penstock, support system, and appurtenances for all loads,

25

which must be combined as required. The resultant stresses must not exceed the allowable stress levels given in Section 3.5.

The design report must be certified by a registered professional engineer (RPE) experienced in the design of penstocks.

3.2 DESIGN LOADS

The design of pressure-carrying components of penstocks uses an allowable stress design method in which all loads (both live and dead) apply load factors equal to 1.0. This section discusses loads to be considered in the design of exposed and buried penstocks and tunnel liners. Loads include construction loads, live loads, dead loads, intermittent loads, and service loads, defined as follows.

3.2.1 Construction Loads

P_{C1} Loads during shipping, handling, storing, and erecting;

P_{C2} Uplift and external pressure from wet concrete; and

P_{C3} External grouting pressure.

3.2.2 Live Loads

$LL1$ Wind;

$LL2$ Snow and/or ice;

$LL3$ Vehicle loads;

$EQ1$ Design basis earthquake criteria—Follow requirements of ASCE Standard 7 (ASCE 2010); and

$EQ2$ In cases where a penstock failure is a potential dam safety concern, the earthquake analysis should match the appropriate state or Federal Energy Regulatory Commission (FERC) analysis for site-specific dam or water-retaining structure.

3.2.3 Dead Loads

$DL1$ Weight of structure and permanent accessories;

$DL2$ Weight of water when full or partially full;

$DL3$ Weight of soil or rock fill or soil surcharge;

$DL4$ External hydrostatic pressure, including flotation; and

$DL5$ Rock loads in tunnels.

3.2.4 Intermittent Loads

$DL6$ Weight of water when filling and draining penstock.

3.2.5 Service Loads—Internal and External Pressures

In addition to the above-defined loads, the penstock and tunnel liner must be designed for the following internal and external pressure service loads when applicable.

3.2.5.1 Internal Pressure Loads

P_{N1} The maximum static head without surge or water hammer based on water at the highest reservoir level.

P_{N2} The maximum static head minus the head loss plus the water hammer and surge for a plant operation load rejection (turbine shutoff valve (TSV) closure, wicket gate, pressure reducing valve (PRV), needle valve, etc.) when all units are operating with normal governor closure time. This maximum includes transient pressures that would be expected during normal operating conditions.

P_{N3} The minimum static head minus the water-hammer and down-surge pressures that would occur when all units operate from speed no load (SNL) to full load acceptance for all units.

P_{I1} The penstock full of water but at zero surcharge pressure.

P_{I2} The penstock half full of water.

P_{EM1} The water hammer and surge, calculated for the final part of gate closure to zero gate position at the maximum governor rate in $2L/a$ seconds, where L = length of the penstock and a = wave speed in the penstock. This equation simulates the condition of the governor cushioning stroke being inoperative.

P_{EX1} The internal pressure value, which includes full gate closure with malfunctioning control equipment in the most adverse manner.

P_{T1} The hydrostatic test condition internal pressure.

3.2.5.2 External Pressure Loads

P_{I3} The penstock dewatered and subjected to maximum groundwater seepage pressure, or for exposed penstocks, the maximum vacuum that can be generated by dewatering.

3.2.5.3 Construction External Pressure Loads

P_{C2} The equivalent external pressure caused by wet concrete, consisting of the depth of the pour times the equivalent density of the concrete.

P_{C3} The equivalent external pressure caused by the grouting of the space between the concrete and the steel tunnel liner.

3.2.5.4 Temperature Loads

$TL1$ The loads caused by the expansion and contraction from daily and seasonal temperature variation, including construction and hydrotest cases.

$TL2$ The temperature gradient across the penstock diameter.

3.2.5.5 Sliding and Friction Loads

EJL The longitudinal loads caused by the friction at the sliding expansion joints and mechanical couplings.

SFL The longitudinal loads caused by friction or sliding at the supports.

3.2.5.6 *Differential Settlement of Adjacent Supports*

DS1 The loads generated by the settlement of a single support.

3.2.6 Analysis of Loadings

The previously described loadings are to be applied to the penstock steel shell and support system by treating it as a continuous beam supported by ring girders, saddles, or soil. Analysis should incorporate consideration of the discontinuity of the system at the expansion joints and mechanical coupling. The effects of these loadings must be reflected in the beam analysis as follows.

3.2.6.1 *Beam and Support Ring Loadings* The commonly used beam equations must be used for calculating longitudinal stresses caused by loads for penstock span and support configuration.

3.2.6.2 *Bending Stress Imposed by Ring Girders and Concrete Saddle Supports* When ring girders and concrete saddle supports are used, they impart a local bending stress into the penstock shell, which must be added to the longitudinal and/or circumferential loading.

3.2.6.3 *Support Reactions* These loads vary according to the type of support normally encountered in conventional practice. For penstocks with welded field joints, the continuous beam method may be used in the analysis. Consideration must be given to the adjusted support reactions for the case in which a single support experiences a differential settlement.

3.3 SERVICE CONDITIONS AND LOAD COMBINATIONS

Penstocks must be designed for the load combinations included under each of the conditions in Table 3-1.

3.3.1 Normal Condition

The normal condition load combinations consist of the three normal pressures, dead loads, live loads, thermal loads, and friction loads listed for the three cases shown across the top of Table 3-1. The allowable stresses are given in Section 3.5.3.

3.3.2 Intermittent Condition

The intermittent condition loads and load combinations consist of the three normal pressures, three intermittent pressures, dead loads, live loads, earthquake loads, thermal loads, and friction loads listed for the eleven cases shown across the top of Table 3-1. The allowable stresses for this condition are given in Section 3.5.3.

3.3.3 Emergency Condition

The emergency condition loads and load combinations consist of the emergency pressure, dead loads, live loads, differential settlement loads, thermal loads, and friction loads listed for the three cases shown across the top of Table 3-1. The allowable stresses are given in Section 3.5.3.

3.3.4 Exceptional Condition

The loads and load combinations for the exceptional condition consists of one normal pressure, one exceptional pressure, dead loads, live loads, earthquake loads, and friction loads as listed for the two cases shown across the top of Table 3-1. The allowable stresses are given in Section 3.5.3.

3.3.5 Construction Conditions

Construction conditions include the dead loads, live loads, and handling loads described in Section 3.2. This list includes penstock fabrication, storage, shipping, handling, and installation. The major load to consider for this condition is the weight of the penstock itself. Precautions must be taken to prevent overstressing of the shell caused by the penstock supporting its own weight.

In addition, steps must be taken to prevent excessive deflection of the penstock, particularly after the lining and coating have been applied. Internal bracing may be used to prevent shell overstressing and excessive deflection. Padding supports and tie-down straps may be necessary to prevent damage to the lining and coating.

Internal bracing (pipe stulling) may be necessary for buried penstocks under backfill conditions.

Penstocks or tunnel liners to be encased in concrete must be designed with consideration given to the buoyant and concreting loads caused by concrete placement.

Loads and load combinations for construction conditions consist of the one case shown in Table 3-1. The allowable stresses are given in Section 3.5.3.

3.3.6 Hydrostatic Test Condition

The loads and load combinations for the hydrostatic test condition consist of the one case shown in Table 3-1. The allowable stresses for this condition are given in Section 3.5.3.

3.4 STRESS CATEGORIES

3.4.1 Theories of Failure

The stress at any point in a structure may be completely defined by giving the magnitudes and directions of the three principle stresses. When two or three of these stresses differ from zero, the proximity to yielding may be determined by means of a strength theory. The theories most commonly

Table 3-1 Load Combinations for Service Conditions

Load	Normal			Intermittent											Emergency			Exceptional		Construction	Hydro-test
	1	2	3	1	2	3	4	5	6	7	8	9	10	11	1	2	3	1	2	1	1
P_{N1}	•			•															•		
P_{N2}		•			•	•															
P_{N3}			•				•	•													
P_{I1}									•	•											
P_{I2}											•	•									
P_{I3}													•	•							
P_{EM1}															•	•	•				
P_{EX1}																		•			
P_{T1}																					•
DL1	•	•	•	•	•	•	•	•	•	•	•	•	•	•	•	•	•	•	•	•	•
DL2	•	•	•	•	•	•	•	•	•	•	•	•	•	•	•	•	•	•	•	•	•
DL3	•	•	•	•	•	•	•	•	•	•	•	•	•	•	•	•	•	•	•	•	•
DL4													•	•						•	
DL5	•	•	•	•	•	•	•	•	•	•	•	•	•	•	•	•	•	•	•	•	
DL6	•	•	•	•	•	•	•	•	•	•	•	•	•	•	•	•	•	•	•		•
LL1					•		•		•		•					•				•	
LL2	•	•	•	•	•	•	•	•	•	•	•	•	•	•	•	•	•	•	•	•	
LL3	•	•	•	•	•	•	•	•	•	•	•	•	•	•	•	•	•	•	•	•	
EQ1				•			•		•		•			•							
EQ2																			•		
DS1															•						
TL1	•	•	•	•	•	•	•	•	•	•	•	•	•	•	•	•	•	•	•		•
TL2													•	•							
EJL	•	•	•	•	•	•	•	•	•	•	•	•	•	•	•	•	•	•	•		•
SFL	•	•	•	•	•	•	•	•	•	•	•	•	•	•	•	•	•	•	•		•
P_{C1}																				•	
P_{C2}																				•	
P_{C3}																				•	

used are the maximum stress theory, the maximum shear stress theory (also known as the Tresca criterion) and the distortion energy theory (also known as the octahedral shear stress theory or the Huber–Hencky–von Mises criterion). Both the maximum shear stress theory and the distortion energy theory are much better than the maximum stress theory in predicting both yielding and fatigue failure in ductile materials in a biaxial or triaxial stress state. Failure is assumed to occur when the state of stress in the structure equals the same state of stress in the tension test at yield.

This manual recommends use of the maximum shear stress theory because it is easier to apply and offers advantages in some applications to the fatigue analysis. This manual also recognizes as acceptable the Huber–Hencky–von Mises theory based on the two-dimensional state of stress equation. The various strength theories are compared in

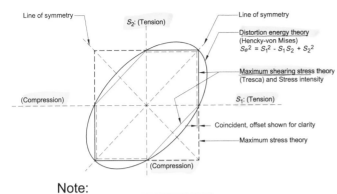

Note:

S_1 and S_2 Principal stresses: each theory depicts a constant equivalent stress.

Fig. 3-1. Comparison of theories of failure for biaxial state of stress

Fig. 3-1 for a two-dimensional (biaxial) stress state for a constant stress intensity (SI) or equivalent stress (S_e). S_1 and S_2 represent the principal stresses. In the figure, all theories give identical results for certain stresses. Also see Sections 3.4.13, 3.4.14, and USBR (1986).

The maximum shear stress at a point is equal to one-half of the algebraic difference between the largest and the smallest of the three principal orthogonal stresses. Thus, if the principal stresses are S_1, S_2, and S_3 and $S_1 > S_2 > S_3$ (algebraically), the maximum shear stress is $1/2 (S_1 - S_3)$. The maximum shear stress theory of failure states that yielding in a component occurs when the maximum shear stress reaches a value equal to the maximum shear stress at the yield point in a tensile test. In the tensile test, at yield, $S_1 = S_y$, $S_2 = 0$, and $S_3 = 0$; therefore, the maximum shear stress is $S_y/2$. The theory states that yielding in the component occurs when

$$\tfrac{1}{2}(S_1 - S_3) = \tfrac{1}{2}S_y \qquad (3\text{-}1)$$

To avoid the unfamiliar and unnecessary operation of dividing both the calculated and the allowable stresses by two before comparing them, a term called equivalent intensity of combined stress or, more briefly, stress intensity (SI) has been formulated. Stress intensity is defined as twice the maximum shear stress and is equal to the largest algebraic difference between any two of the three principal orthogonal stresses.

For a comparison of the Huber–Hencky–von Mises theory with the maximum shear stress theory in the two-dimensional state of stress, consider the following cases:

Case 1: Principal stresses of $S_1 = 10$ ksi, $S_2 = 6$ ksi, $S_3 = 0$
Equivalent stress: $S_e = 8.72$ ksi
Stress intensity: $SI = 10$ ksi
Case 2: Principal stresses of $S_1 = 5$ ksi, $S_2 = -5$ ksi, $S_3 = 0$
Equivalent stress: $S_e = 8.66$ ksi
Stress intensity: $SI = 10$ ksi

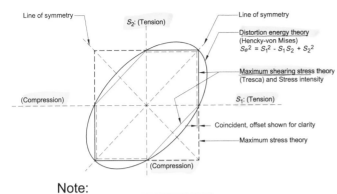

Case 3: Principal stresses of $S_1 = 10$ ksi, $S_2 = 10$ ksi, $S_3 = 0$
Equivalent stress: $S_e = 10$ ksi
Stress intensity: $SI = 10$ ksi

In Cases 1 through 3, the maximum shear stress theory gives a result equal to or more conservative than the Huber–Hencky–von Mises theory. However, the maximum stress theory is grossly unconservative for Case 2.

3.4.2 Stress Intensity Limits

The allowable stress limits are expressed as limits on stress intensity. Stress intensity limits are applicable to the shell of the penstock and to the biaxially stressed shell or flat plate elements. They are also applicable to the linear elements stressed principally in one direction, along the axis of the element. Examples of linear elements are columns, tension members, beams and rings stressed by loads, and moments lying in the plane of the ring. A linear element stressed along only its axis has the characteristic that the uniaxial stress is equal to the stress intensity.

Stress intensities as defined in this manual follow closely with those of ASME Boiler and Pressure Vessel design (ASME 2010a and 2010b). Other methodologies work well in many applications but may differ from those listed in this manual. In deciding on which design methodology to use for the design of a given water conveyance system, many factors should be considered. Size of the conduit; rated working pressure; consequences of potential failure; and location of the pipeline relative to other structures, highways, population, or critical infrastructure are all important issues. This manual is intended to design higher pressure, larger diameter, and/or more critical conveyance systems used for generation of power. When comparing different methodologies, it is important not to pick stress intensities from different manuals unless specifically advised that doing so is appropriate. Some methodologies are designed on tensile strength, whereas other designs are based on yield strength for the steels chosen. If design is not consistent, then improper material selection and design stresses can occur.

3.4.3 Membrane and Bending Stress Categories

Stresses or stress intensities are categorized as primary, secondary, or peak. Peak stresses are of concern only in relation to fatigue and are not given any further consideration in this section. However, if a fatigue analysis is considered necessary, refer to ASME BPVC, Section VIII, Division 2, Appendix 5 (ASME 2010b). For vibration fatigue, see Sections 1.4 and 17.4.3 of this book.

The primary category of stresses or stress intensities are general membrane (P_m), local membrane (P_l), and bending (P_b).

The secondary category (Q) includes stresses or stress intensities caused by restraint of the penstock itself and which are generally characterized as self-limiting. The total range of stress intensities, secondary stresses combined with primary stresses and peak stresses, is considered for fatigue. Secondary stress intensities combined with primary stress intensities are limited to twice the yield strength to ensure shakedown to elastic action as discussed in Section 3.4.11.

3.4.4 Shear Stress Category

Shear stress (expressed as an average) in the web of a beam, plate girder, web of a stiffener ring, and shear keys is limited to $0.6S$, where S is the allowable stress intensity. Peak shear stress, such as from torsion acting on a round bar, is limited to $0.8S$.

3.4.5 Weld Joint Reduction Factor (E)

The weld joint reduction factor (E) is imposed on the calculated tension stress. The value of E is 1.0 for joints in compression. Its value depends on the degree of examination the joint receives and the type of joint. See Section 3.5.1 for permissible E values and the example in Section 3.4.14.

3.4.6 Temperature Stress Intensities

Temperature stress intensities in penstocks are caused by several sources, including the following:

1. Solar sources, where the structure is restrained with respect to axial growth and/or lateral displacement and temperature gradients caused by solar effects develop;
2. Friction forces that develop from temperature changes at sliding and rolling supports, at sleeve-type expansion joints, and from the backfill or encasement of buried penstocks or liners;
3. Forces that are proportional to axial growth or angular rotation, such as the action in bellows-type expansion joints; and
4. Changes in ambient air or water temperatures.

Temperature stress intensities should be considered in the design of penstocks, supports, liners, and appurtenances when they are significant (stresses greater than 10% of the basic allowable SI).

Temperature stress intensities are generally considered secondary and placed in the Q category.

3.4.7 Plasticity Shape Factors (X)

The plasticity shape factor (X) is accounted for in the classification of SI. For solid rectangular sections, such as bending across the thickness of a flat plate or shell, a factor of 1.5 is permitted. For sections with a shape factor less than 1.5 (such as thin-walled pipe), a shape factor of 1.0 should be used. For sections with a shape factor greater than 1.5, the shape factor used should not exceed 1.5.

3.4.8 Stress Increase Factor (K)

Allowable stress increase factors (K) are given in Section 3.5.2. The K factor represents the proportional increase in SI allowed for loading cases calculated over the basic allowable SI.

3.4.9 Compression Stability

Shells, flat plates, and linear structural elements loaded to produce uniform compression stress across the thickness or section must be checked for buckling or instability. Appropriate changes must be made to the design to ensure an adequate factor of safety against buckling. Buckling and instability requirements in the ASME BPVC, Section VIII, Division 1 (ASME 2010a) for shells and ANSI/AISC "Specification for Structural Steel Buildings" (AISC 2010) for plate and linear structural elements must be met. For stiffener rings on penstocks, refer to Section 4.4. Refer to Chapter 6 for stiffened or unstiffened liners.

3.4.10 Basic Allowable Stress Intensity (S)

The basic allowable stress intensity (S) should be the lesser of the minimum specified tensile strength divided by 2.4 or the minimum specified yield strength divided by 1.5.

3.4.11 Stress Categories—Definitions and Limits

3.4.11.1 General Primary Membrane Stress (P_m) General primary membrane stress (P_m) is the stress (averaged across the thickness) necessary to satisfy the simple laws of equilibrium. The value of P_m is limited to KS, where K is the stress increase factor and S is the allowable stress intensity. In penstock shells, an example is hoop stress from internal pressure.

3.4.11.2 Local Primary Membrane Stress (P_L) Local primary membrane stress (P_L) is discussed in ASME BPVC, Section VIII, Division 2 (ASME 2010b) (the ASME Code uses S_m for KS):

Cases arise in which a membrane stress produced by a pressure or other mechanical loading and associated with a primary and/or a discontinuity effect would, if not limited, produce excessive distortion in the transfer of load to other portions of the structure. Conservatism requires that such a stress be classified as a local primary membrane stress even though it has some characteristics of a secondary stress. A stress region may be considered as local if the distance over which the stress intensity exceeds $1.1\,S_m$ does <u>not</u> extend in the meridional direction more than $1.0\sqrt{Rt}$, where R is the midsurface radius of curvature measured normal to the surface from the axis of rotation and t is the minimum thickness in the region considered. Regions of local primary membrane stress which exceed $1.1\,S_m$ shall not be closer in the meridional direction than

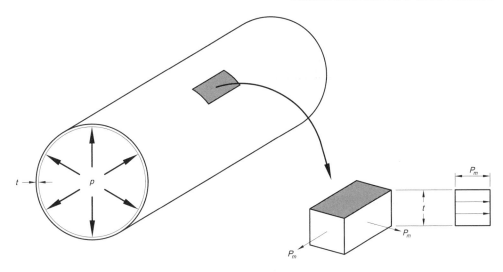

Fig. 3-2. P_m, shell internal pressure

$2.5\sqrt{Rt}$, where R is defined as $(R_1 + R_2)/2$, and t is defined as $(t_1 + t_2)/2$ where t_1 and t_2 are the minimum thickness at each of the regions considered, and R_1 and R_2 are the midsurface radii of curvature measured normal to the surface from the axis of rotation at these regions where the membrane stress exceeds 1.1 S_m. Discrete regions of local primary membrane stress, such as those resulting from concentrated loads acting on brackets, where the membrane stress exceeds 1.1 S_m shall be spaced so that there is no overlapping of the areas in which the membrane stresses exceed 1.1 S_m.

An example of a local primary membrane stress is the membrane stress in a shell produced by external loads acting on a permanent support or a nozzle connection. Another example is the membrane stress intensity at the junction of a cylindrical to conical section. P_L must not exceed 1.5 KS.

3.4.11.3 Primary Bending Stress (P_b) in Shells and Plates   The primary bending stress (P_b) as used in this manual applies to shells and plates in which the bending is across the thickness of the shell or plate and is primary in the sense that no redistribution of the load occurs as a result of yielding. An example of bending is *SI* in the central portion of a flat head or cover caused by pressure loading. P_b must not exceed 1.5 KS. The sum of P_L and P_b likewise must not exceed 1.5 KS.

3.4.11.4 Secondary Stress (Q)   A secondary stress (Q) is a normal stress, either bending or axial, or a shear stress developed by the constraint of adjacent parts or by self-constraint of a structure. The basic characteristic of a secondary stress is that it is self-limiting. Local yielding and minor distortions can satisfy the conditions that caused the stress to occur. Failure from one application of the stress is not to be expected.

Examples of secondary stress include (1) thermal stresses generally (2) bending stress intensities across the thickness of a shell or plate at a gross structural discontinuity and (3) rim bending stresses at ring girder supports.

There is no need to limit Q-type stress intensities acting alone. The combination of Q with primary stress intensities (P_L, P_m, P_b), however, should be limited to the lesser of minimum tensile strength (f_t) or 3S. The limit ensures shakedown to elastic action after a few repetitions of the load producing them and is a limit on the *SI* range. K factors (other than 1) should not be applied to the 3S limit.

3.4.12   Representative SI Classifications

Table 3-2 lists the *SI* classifications for parts typically found in penstocks and tunnel liners.

3.4.13   Equivalent Stresses

For a three-dimensional state of stress in which the three orthogonal principle stresses are S_1, S_2, and S_3, the equivalent stress (S_e) based on the distortion energy theory (the octahedral shearing stress theory) is given by

$$S_e = \{1/2\,[(S_1 - S_2)^2 + (S_2 - S_3)^2 + (S_3 - S_1)^2]\}^{1/2} \quad (3\text{-}2)$$

For a two-dimensional state of stress, the equation reduces to

$$S_e = \{S_1{}^2 - S_1 S_2 + S_2{}^2\}^{1/2} \qquad (3\text{-}3)$$

The equivalent stress (S_e) based on the distortion energy theory is equivalent to stress intensity (*SI*) based on the maximum shear stress theory.

In applying the distortion energy theory, it is suggested that the two-dimensional equation be applied using the two principal stresses that result in the greatest algebraic difference.

Table 3-2 *SI* Classifications for Typical Parts Found in Penstocks and Tunnel Liners

Case	*SI* category
Ring girder supports and stiffener rings	
Rim bending	Q
Local membrane hoop stress in the penstock in the circumferential direction	P_L
Bending stresses in ring girder and/or stiffeners acting across the depth of the girder	P_m
Bending + axial *SI* in ring	P_m
Rim bending in the penstock shell plus direct/or and axial load	$(Q + P_m)$
Saddles (penstock shell immediately above horn or tip of saddle)	
Horn bending (bending at tip of saddle)	Q
Horn hoop membrane load (hoop membrane load at tip of saddle)	P_L
Shear in shell adjacent to horn	$0.6S$
Bearing or compression from bearing loads	P_m
Bending in base plates and saddle plates across thickness of element	P_b
Loaded attachments and nozzles	
Membrane *SI* in shell of penstock caused by the loads	P_L
Bending *SI* in shell of penstock caused by the loads	Q
Nozzle necks	
Nozzle necks outside limits of reinforcement from external loads and moments, and pressure	P_L
Nozzle necks within limits of reinforcement from external loads and moments, and pressure	P_m
Bifurcations	
Membrane stress intensity away from reinforcing girders and away from changes in direction	P_m
Membrane stress intensity adjacent to reinforcing girders and changes in direction	P_L
Bending stress across thickness of shell	Q
Maximum stress intensity in reinforcing girders	$P_b + P_L$
Miscellaneous	
Membrane hoop stress from internal pressure in cylindrical and conical penstock sections and for formed shells away from discontinuities	P_m
Membrane *SI* in the vicinity of attachments under load, horn area of saddle supports, conical-to-cylindrical intersections, and penstock shells adjacent to nozzles	P_L
Bending *SI* in flat heads and covers	P_b
Temperature *SI*	Q

Note: See Fig. 3-2 through Fig. 3-6.

3.4.14 Equivalent Stress Calculations (Example)

Assume the case of a 15-ft-diameter exposed penstock supported by ring girders spaced at 60 ft in which it is necessary to determine stress intensity and equivalent stress at the top of the penstock at midspan. It is assumed that the expansion joints in the line keep longitudinal pressure stresses in the penstock at negligible values.

The weld joint reduction factor in tension is 0.85 (see Section 3.5.1), and in compression is 1.0.

The following stresses are given:

Pressure: 100 psi
Circumferential membrane stress: 15 ksi (tension)
Longitudinal membrane stress: 10 ksi (compression)

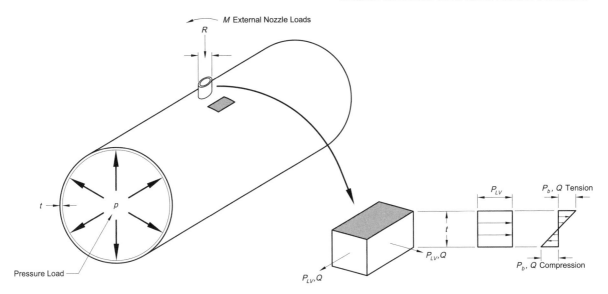

Fig. 3-3. P_{LV}, P_b, Q, shell at nozzle p and nozzle loads M, R (Q is caused by nozzle piping restraint loads only)

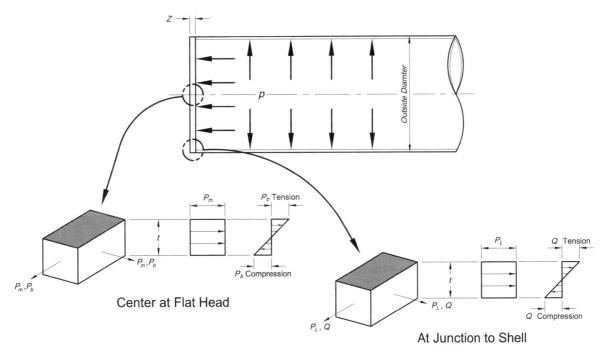

Fig. 3-4. P_m, P_b, P_L, Q, flat plate, pressure loads, p

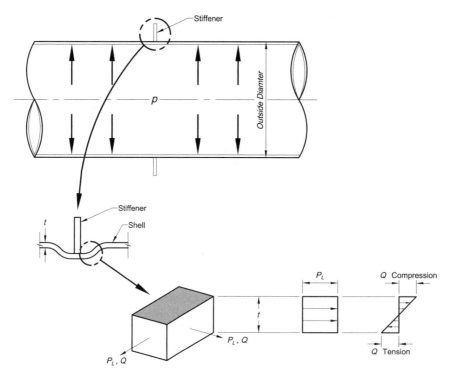

Fig. 3-5. P_L, Q, shell at stiffener/discontinuity, pressure loads, p

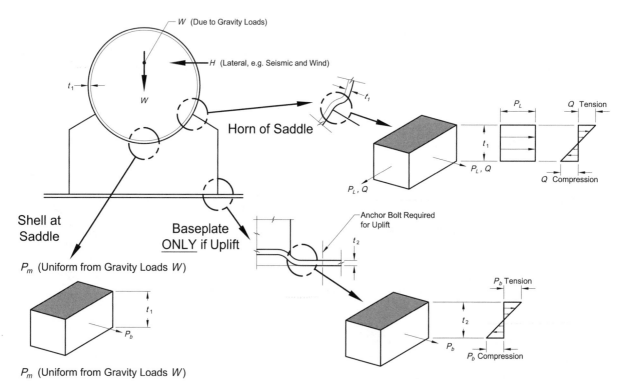

Fig. 3-6. P_L, P_m, Q, T - shell baseplate at saddle, gravity and lateral

Note: The pressure produces compression stress in the penstock shell in the radial direction. This compression ranges from −100 psi at the inner surface to zero at the outside and is assumed to vary linearly for points between the inner and outer penstock surfaces.

The weld joint reduction factors are applied to the calculated stress.

The principal stresses are

$S_1 = 15 \text{ ksi}/0.85 = 17.65 \text{ ksi}$
$S_2 = -10 \text{ ksi}/1.0 = -10 \text{ ksi}$
$S_3 = -0.05 \text{ ksi}$

Stress intensity (*SI*) is calculated as

$SI = 17.65 \text{ ksi} - (-10 \text{ ksi}) = 27.65 \text{ ksi}$

The equivalent stress (S_e) is calculated using the two-dimensional equation (Eq. 3-3) as

$S_e = [17.65^2 - 17.65(-10) + (-10)^2]^{1/2} = 24.25 \text{ ksi}$

The equivalent stress (S_e) is calculated using the three-dimensional equation (Eq. 3-2) as

$S_e = \{1/2[(17.65 - (-10))^2 + ((-10) - 0.05)^2 + 0.05 - 17.65)^2]\}^{1/2} = 24.26 \text{ ksi}$

Note that the stress intensity method is the simplest and most conservative of the three methods.

3.5 ALLOWABLE STRESS INTENSITIES

The allowable stress intensities must be adjusted by the following weld joint reduction factors and the allowable stress increase factors.

3.5.1 Weld Joint Reduction Factor (E)

Table 3-3 provides weld joint reduction factors as described in Section 3.4.5.

Table 3-3. Weld Joint Reduction Factors

Category	Type	100%RT or UT	Spot RT[a]	No RT or UT[b]
Exposed penstocks				
Longitudinal butt welds, circumferential butt welds, and spiral (helical) butt welds	Double-welded butt joints	1.00	.85	.70
	Single-welded butt joints with backing strips	.90	.80	.65
	Single-welded butt joint without backing strips (circumferential only, < ⅝ in. thick and <24 In. OD)	NA[c]	NA[c]	.60
Double full-fillet lap		NA[c]	NA[c]	.55
		NA[c]	NA[c]	.55
Single full-fillet lap joints with plug welds	Circumferential only, for <24 In. OD $T_{max} = ½$ in.	NA[c]	NA[c]	.50
Tunnel liners and buried penstocks				
Longitudinal butt welds, circumferential butt welds, and spiral (helical) butt welds	Double-welded butt joints	1.00	.85	.70
	Single-welded butt joints with backing strips	.90	.80	.65
	Single-welded butt joints without backing strips (circumferential only, < ⅝ in. thick and <24 In. OD)	NA[c]	NA[c]	.60
Double full-fillet lap		NA[c]	NA[c]	.55
		NA[c]	NA[c]	.55
Single full-fillet lap joints with plug welds	Circumferential only, for <24 in. OD, $T_{max} = ⅝$ in.	NA[c]	NA[c]	.50
Single full-fillet lap		NA[c]	NA[c]	.45
Bifurcations				
Longitudinal and circumferential butt joints in shell plates (double-welded butt joints)		1.00	.85	.70
Sickle plate splices[d] (double-welded butt joints)		1.00	.85	.70
Shell plate to sickle tee welds		1.00	NA[c]	NA[c]

Note: See Section 11.45 for a more detailed discussion of joint types and their usage.
[a] Spot RT is defined in the ASME BPVC, Section VIII, Division 1, Paragraph UW-52, ASME (2010a)
[b] A minimum of 20% MT is required for these joints
[c] NA = Not allowed
[d] Includes butt welds in webs and flanges of C-clamps

Table 3-4. Allowable Stress Increase Factors

Ch	K Factor
Normal Operating Conditions	1.00
Intermittent Conditions	1.33
Emergency Conditions	1.5
Exceptional Conditions	2.5
Construction and Hydrotest Conditions	1.33

3.5.2   Allowable Stress Increase Factor (K)

Table 3-4 provides the allowable stress increase factors.

3.5.3   Allowable Stress Intensities

The basic allowable stress intensity (S) is the smaller of tensile strength/2.4 or yield strength/1.5 as indicated in Section 3.4. Table 3-5 gives allowable stress intensities.

3.5.4   Allowable Shear Stresses (All Loading Conditions)

For the design of fillet welds and partial penetration welds, the following formula gives the allowable shear stress for the weld metal:

$$0.3K \times \text{(tensile strength of welding material)} \quad (3\text{-}4)$$

where K = stress increase factor (Table 3-4).

The allowable shear stress for the base metal is given by

$$0.6SK \quad (3\text{-}5)$$

The allowable peak shear stress for the base metal is given by

$$0.8SK \quad (3\text{-}6)$$

3.5.5   Buckling Factors of Safety

The buckling factor of safety for a dewatered and empty tunnel liner is 1.5.

The buckling factor of safety for an exposed penstock is 2.4 (based on a theoretical factor of safety of 3.0 and a knockdown factor of 0.8).

REFERENCES

AISC. (2010). "Specification for structural steel buildings." *ANSI/ AISC 360-10,* Chicago.

ASCE. (2010). "Minimum design loads for buildings and other structures." *ASCE 7-10,* Reston, VA.

ASME. (2010a). "Boiler and pressure vessel code, section VIII, division 1: Rules for construction of pressure vessels." New York.

ASME. (2010b). "Boiler and pressure vessel code, section VIII, division 2: Alternative rules." New York.

USBR. (1986). "Welded steel penstocks." Engineering monograph no. 3, U.S. Bureau of Reclamation, Denver.

Table 3-5.   Allowable Stress Intensities

K factor or SI category	Normal conditions	Intermittent conditions	Emergency conditions	Exceptional conditions	Construction and hydrostatic test conditions
Increase Factor, K	1.0	1.33	1.5	2.5	1.33
General Primary Membrane, P_m	1.0S	1.33S^3	1.50S^2	2.50S^2	1.33S^3
Local Primary Membrane, P_l	1.5S	2.0S^3	2.25S^2	2.5S^2	2.0S^3
Primary Bending, P_b	1.5S	2.0S^3	2.25S^2	2.5S^2	2.0S^3
Local Primary Membrane Plus Primary Bending ($P_l + P_b$)	1.5S	2.0S^3	2.25S^2	2.5S^2	2.0S^3
Secondary ($P_l + P_b + Q$)	The lesser of f_t or 3S	The lesser of f_t or 3S	[1]	[1]	[1]

[1]Secondary stress intensities need not be calculated except for emergency conditions when fatigue evaluation is required.
[2]Not to exceed 0.9 times tensile strength.
[3]Not to exceed minimum specified yield strength.

FOLLOW THE FOOT NOTES

CHAPTER 4

Exposed Penstocks

Source: Photograph courtesy of Denver Water; reproduced with permission.

Exposed steel penstocks constitute one of the most commonly used configurations found in the industry. Their popularity comes not only from the apparent simplicity of installation but also from the ease of inspection and maintenance while in service. Meanwhile, a number of challenges face the engineer caused by the large number of loading conditions and extremes that can be experienced by a pipeline exposed to the elements. The following section discusses the parameters to be considered in the design and analysis of exposed penstocks.

4.1 PENSTOCK SHELL DESIGN AND ANALYSIS

The principal factors that govern the required shell thickness are shipment and handling and resistance of imposed loads, considering the appropriate allowable stresses. Additional

factors in determining shell thickness include mill and fabrication tolerances and criteria for corrosion allowance, if elected in lieu of coating and lining.

4.1.1 *Minimum Shell Thickness*

Although the predominant stress-causing load on a penstock section (unrestrained and free from internal or external appurtenances) is the stress from internal pressure, it is important to recognize that other stress conditions can exist and should be considered in determining the required shell thickness. These stresses can result from beam action, differential temperatures, and loads caused by end closures free to move in a longitudinal direction. As such, and in the majority of cases, the stress analyses for determining the shell thickness end in a biaxial state of stress and are resolved by application of either the Huber–Hencky–von Mises theory or the stress intensity approach discussed in Section 3.4.

Considering that the shell thickness is to be specified within recommended mill and fabrication tolerances, the minimum thickness of the penstock shell should be established based on the following generalized criteria:

1. Thickness required to resist the loads imposed by internal pressure, weight of penstock contents, or other applied forces;
2. Thickness required to ensure adequate strength during shipment and handling, as required by this manual, and to resist excessive vertical deflection between supports;
3. Thickness required to resist longitudinal loads; and
4. Thickness required to avoid pipe buckling caused by internal vacuum.

The concept of minimum thickness based on these guidelines is not absolute because the engineer needs to consider load combinations for various conditions, then combine the stresses in the shell that are developed at the same location. The thickness is then modified in an iterative manner to ensure that the maximum stresses do not exceed the allowable.

4.1.1.1 Shell Thickness to Resist Applied Loads   The basic applied loads that drive penstock shell thickness generally consist of those resulting from internal pressure, circumferential bending attributable to the weight of the penstock and its contents, and shear stresses at supports.

4.1.1.1.1. Internal Pressure   The thickness of the penstock shell is computed on the basis of limiting the hoop tensile stress in the steel, S_H, to a defined maximum value for the internal pressure that is being analyzed. As shown in Chapter 1, hoop stress $S_H = Pr/t$, from which the required thickness of the penstock can be defined as

$$t = \frac{Pr}{S_H} \qquad (4\text{-}1)$$

where t = penstock shell thickness required to resist the design pressure P (in.);

   P = internal pressure (at centerline of pipe) (psi); and

   r = inside penstock radius (in.).

Depending on the method used to determine the penstock wall thickness, the definition of S_H in Eq. (4-1) can vary. For analysis in accordance with ASME-based procedures, S_H becomes SE; where S is the basic allowable stress intensity for the design load condition resulting in pressure P, and E is the weld joint reduction factor, as given in Section 3.5.1. For analysis in accordance with AWWA-based procedures, S_H equals the allowable stress for the given design condition resulting in pressure P.

Many penstocks are fabricated as spiral-wound pipe where the welded seam is a helix and therefore has different internal pressure principal stresses than straight seam pipe. Table 1 in Sundberg (2009) presents formulas for calculat-

ing the tensile stresses normal to both the spiral seam and the coil splice.

4.1.1.1.2. Circumferential Bending   Circumferential bending moments in a penstock shell occur whenever the penstock is partially filled. Shell design for a partly filled penstock has been treated in several articles, none of which gives a complete analysis. Good treatments of this complex analysis are given by Roark and Young (1975), Schorer (1933), and USBR (1940). Calculations using the formulas presented by Roark and Young must be done with great precision because the expression for the bending moment represents the algebraic sum of large numbers of nearly equal terms. Circumferential bending moments for a completely filled penstock (zero pressure) exist only for a penstock supported on saddles.

4.1.1.1.3. Shear Stresses at Supports   The shearing stresses developed in a transverse penstock section are caused by the external loads, including the weight of the shell and the water. When the shell is held to a cylindrical shape by a stiffener ring, for example, the developed shear is tangent to the shell at all points and varies from zero at the top to zero at the bottom, with a maximum at mid-depth twice the average value over the entire section. When the shell is free to deform, as with a saddle support, the tangential shear stresses act on a reduced effective cross section and the maximum stress occurs at the horn of the saddle. There is further increase in this shear stress because a portion of the shell above the saddle is noneffective, tending to increase the shear in the effective portion.

4.1.1.2 Thickness for Shipping and Handling, and Vertical Deflection   Two traditional guidelines for determining pipe wall thickness involve the stiffness required to resist typical shipping and handling loads and the stiffness required to provide serviceability in locations where the penstock spans between supports.

4.1.1.2.1. Shipping and Handling   The minimum thickness, t, of the penstock shell for shipping and handling can be calculated using the formula of Pacific Gas and Electric Company (PG&E) or of the U.S. Bureau of Reclamation (USBR).

1. PG&E formula:

$$t = \frac{D}{288} \qquad (4\text{-}2)$$

2. USBR formula:

$$t = \frac{D + 20}{400} \qquad (4\text{-}3)$$

where t = minimum thickness (in.) and

   D = nominal penstock diameter (in.).

The USBR formula yields a slightly smaller minimum wall thickness for pipe diameters above about 51 in. and a slightly larger minimum wall thickness for pipe diameters below about 51 in. This manual accepts either minimum wall thickness as satisfactory but advises the engineer that special design, manufacturing, shipping, or installation conditions may be necessary for the smaller minimum thickness.

The handling thickness must be checked to determine if the shell can support itself adequately at a point load, as if it were resting on a flat surface and loaded by its own weight. Roark and Young (1975) show that the maximum bending stress (S_{max}) for this loading condition is located at the point of contact with the supporting surface and can be calculated as follows:

$$S_{max} = \frac{9R^2W}{t} \tag{4-4}$$

where t = shell thickness (in.);

W = unit weight of the shell material (lb/in.^{3}); and
R = radius of the middle surface of the pipe shell (in.).

The engineer should give consideration to the use of temporary internal braces to strengthen the pipe during handling and installation if it is found that the handling thickness governs the design.

4.1.1.2.2.   Vertical Deflection   Traditional practice has successfully used a support spacing that limits the maximum deflection of the water-filled pipe, acting as a beam between supports, to 1/360 of the span. AISC (2010) only specifies that deflection not impair the serviceability of the structure. The commentary to this specification acknowledges that deflection limits of 1/360 of the span for floors and 1/240 of the span for roof members have been traditionally used but notes that engineers evaluating stiffness and deflections in buildings need to consider numerous serviceability factors.

For aboveground penstocks, the primary nonstrength serviceability considerations are to prevent deflections that can damage a penstock's coatings or linings and also provide sufficient stiffness to prevent excessive penstock vibration. Modern coatings and linings typically are not damaged by deflections of 1/240 of the span. Therefore, a deflection of 1/240 of the span is acceptable, as long as the engineer has confirmed that the proposed penstock configuration has sufficient stiffness not to develop excessive vibrations. See Chapter 1 for a discussion of penstock vibration analysis.

4.1.1.3   *Longitudinal Stresses*   Longitudinal stresses are imposed on a penstock shell from several loading conditions. These stresses are generally categorized by the following action conditions: (1) beam action, (2) stiffener ring restraint at rim, (3) buckling caused by axial compression, (4) longitudinal strain and stress (Poisson's effect), (5) temperature-related effects, and (6) thrust loads resulting from internal pressure.

4.1.1.3.1.   Beam Action   When a penstock rests on its supports, it acts as a beam. The beam load consists of the weight of the pipe, the contained water, and any external live loads, such as ice, snow, wind, or earthquake. If the penstock is to function as a beam, deformation of the shell at the supports must be limited by the use of properly designed stiffener rings, ring girder supports, or saddle supports. On the assumption that large shell deformations can be prevented, the beam stresses can be computed using the theory of flexure. For a cylinder section, the resulting longitudinal stress intensity (S_L) is given by

$$S_L = \frac{M_B}{\pi r^2 t} \tag{4-5}$$

where S_L = resulting longitudinal stress intensity (psi);

M_B = bending moment caused by beam action (in.-lb);
r = inside penstock radius (in.); and
t = nominal shell thickness (in.).

4.1.1.3.2.   Stiffener Ring Restraint at Rim   Because of restraint imposed on the shell by a rigid ring girder or stiffener ring, secondary longitudinal bending stresses are developed in the pipe shell adjacent to the ring girder or stiffener ring. Although this is a local stress, which decreases rapidly with an increase in distance away from the ring, it should be considered in designing the plate for longitudinal stresses. These secondary stresses are flexural stresses caused by the bending deformation of the shell near the stiffener because the shell at the stiffener cannot expand radially in the same manner as the more distant shell portion. The maximum longitudinal bending stress (S_{LR}), as provided by Schorer (1933), is given by

$$S_{LR} = (1.82)\left(\frac{A_r - Ct}{A_r + 1.56t\sqrt{rt}}\right)\left(\frac{Pr}{t}\right) \tag{4-6}$$

where S_{LR} = resulting longitudinal bending stress intensity caused by ring girder action (psi);

r = inside penstock radius (in.); and
A_r = area of girder ring(s), plus shell area under and between rings (in.^{2});
C = length of penstock shell measured between the outside vertical faces of girder rings at a support (in.);
t = nominal shell thickness (in.);
r = inside penstock radius (in.); and
P = value of internal pressure at centerline of shell (psi).

The longitudinal bending stress (S_{LRX}) in an axial direction, x distance from the ring girder edge, can be found by

$$S_{LRX} = S_{LR}\left(e^{-x/z}\right)\left(-\sqrt{2}\right)\cos\left(\frac{x}{2} + \frac{\pi}{4}\right) \tag{4-7}$$

where S_{LR} = resulting longitudinal bending stress intensity caused by ring girder action (psi);

x = distance from the outside face of the ring girder at which point the ring restraining bending stress is being calculated (in.); and

z = an empirical constant based on attributes of the cylindrical shell (in.), determined by

$$z = \frac{\sqrt{rt}}{\sqrt[4]{3(1-v^2)}}$$

(for steel with $v = 0.3$, $z = 0.78\sqrt{rt}$).

In current practice, if the maximum longitudinal restraining stresses are excessive, the shell thickness is increased on each side of the stiffener ring for a minimum length of $LR = 2.33\sqrt{rt}$. These secondary longitudinal restraining stresses in the shell should be combined with other stresses that may exist simultaneously at this location, and the result used in shell thickness computations, based on stress intensification factors shown in Chapter 3.

These formulas for stresses caused by ring girder action are for a penstock under pressure. A good treatment for calculating stress conditions with a half-full penstock is given by Schorer (1933). Because it may control the design, this loading condition should be reviewed in combination with other applicable loads.

Shell restraint stresses may also develop in the vicinity of concrete encasements, such as anchor blocks or other structures in contact with the penstock. In these cases, the engineer must take into consideration not only the radial restraint but also the restraint that prevents rotation of the penstock around horizontal and vertical axes. Some indication of the magnitude of these stresses may be realized by the use of finite element analyses.

4.1.1.3.3. Buckling Caused by Axial Compression When a thin shell is subjected to excessive axial compression, the acting longitudinal stresses may cause direct buckling or wrinkling failure of the shell. Axial compression may be caused by beam bending action, by temperature expansion in a longitudinally restrained penstock, by forces developed because of resistance against sliding, by seismic loads, and by the compressive force developed from the weight of an inclined penstock with bottom anchorage. Several investigations have been carried out on axial buckling. A treatment for determining the allowable buckling stress in a thin shell, considering the effects of imperfections caused by fabrication and eccentric loading, is given by Donnell and Wan (1950). Experimental tests show that the following allowable compressive stresses can be safely carried without buckling failure by wrinkling.

For general buckling

$$S_{\text{allow}} = 1.5 \times 10^6 \left(\frac{t}{r}\right) \text{ but } \leq S_g \qquad (4\text{-}8)$$

For local buckling

$$S_{\text{allow}} = 1.8 \times 10^6 \left(\frac{t}{r}\right) \text{ but } \leq S_l \qquad (4\text{-}9)$$

where S_{allow} = allowable compressive stress (psi);

S_g = allowable general membrane stress as defined in Chapter 3 (psi);

S_l = allowable local membrane stress as defined in Chapter 3 (psi);

t = nominal shell thickness (in.); and

r = inside penstock radius (in.).

Eq. (4-9), for local buckling, can be applied to a general primary membrane stress field with a circumferential width not to exceed $2.4\sqrt{rt}$, as given by Reed (1968). Such fields can occur in the penstock shell near appurtenances or other locations that may result in similar local deformation. Eq. (4-8) should be applied to larger stress fields, as might be typical at the midpoint between supports.

A more refined treatment of penstock buckling under pressure is given by Baker et al. (1972).

4.1.1.3.4. Longitudinal Strain and Stress (Poisson's Effect) Radial expansion caused by internal pressure on an axially restrained shell causes longitudinal contraction (Poisson's effect) with a corresponding longitudinal tensile stress equal to

$$S_{LP} = v\, S_H \qquad (4\text{-}10)$$

where S_{LP} = longitudinal stress from Poisson's effect (psi);

v = Poisson's ratio (0.3 for steel); and

S_H = hoop stress caused by internal pressure (psi).

This stress occurs only if the pipe is axially restrained, and it should be combined algebraically with other longitudinal stresses that may occur concurrently.

4.1.1.3.5. Temperature-Related Effects The conditions under which thermal stresses occur can be distinguished in two ways:

a. The temperature and shell conditions are such that there would be no stresses caused by temperature except for the constraint from external forces and/ or restraints. In this case, the stresses are calculated by determining the shape and dimensions the shell body would take if unrestrained and then finding the forces required to bring it back to its restrained shape and dimension. Having determined these "restoring" forces, the stresses in the shell are calculated using applicable formulas.

b. The form of the body and thermal conditions are such that stresses are produced in the absence of external constraints, solely because of the incompatibility of the natural expansions and contractions of the different parts of the body.

Where the position of ring girder footings, valve flanges, and other external attachments have to be fixed, thermal stresses and resulting displacements are important when the penstock is empty and when setting the penstock in place.

Temperature differentials must be determined for the installation during both its construction and its operation. In addition, the position of the sun should be considered when determining thermal stresses and displacements, and the resultant stresses and displacements should be resolved into components for use in design.

If a thin-walled shell of a given length with both ends fixed is subjected to an outside temperature (T) on one side and an outside temperature ($T + \Delta T$) on the opposite side, and the temperature gradient between the two sides is assumed to be linear, then the fixed-end moments (M_T) that develop at the end of the shell are given by

$$M_T = \frac{E_Y I \Delta T \alpha}{D} \qquad (4\text{-}11)$$

and the maximum resulting bending stress (S_{TB}) is given by

$$S_{TB} = \frac{MC}{I} = \left(\frac{E_Y I \Delta T \alpha}{D} \right) \frac{D}{2I} = \frac{E_Y \Delta T \alpha}{2} \qquad (4\text{-}12)$$

where M_T = fixed-end moments (in.-lb);

S_{TB} = maximum resulting bending stress (psi);
E_Y = Young's modulus for steel (lb/in.^{2});
ΔT = differential temperature or change in temperature (°F);
D = nominal penstock diameter (in.);
α = temperature coefficient of expansion (in./in./°F);
M = any moment;
C = distance to the most extreme fiber of shell subject to bending (in.); and
I = moment of inertia of a section (in.^{4}).

Expansion joints can be used to minimize longitudinal thermal effects by removing the restraint imposed at the ends of a penstock. With the restraint removed, the load caused by temperature is the resistance to sliding between shell and support and between shell and connecting joint (expansion or coupling). For resistance to sliding forces of expansion joints see Section 9.1.1.

4.1.1.3.6. Thrust Loads Resulting from Internal Pressure

A longitudinal thrust force is generated at locations in a penstock where the longitudinal flow is changed. If the penstock is free to move at these locations, the longitudinal force can generate a longitudinal stress in the penstock adjacent to the fitting or appurtenance. The maximum longitudinal stress caused by internal pressure is equal to half of the circumferential stress generated by internal pressure at that location in the penstock. Applying this result to Eq. (4-1) results in the following equation for wall thickness based on longitudinal stress caused by pressure:

$$t = \frac{Pr}{2S_H} \qquad (4\text{-}13)$$

As discussed previously for internal pressure design, depending on the analysis method chosen, the definition of S_H will vary.

4.1.1.4 Thickness Required to Resist Buckling Caused by Vacuum

When penstocks are subject to vacuum, buckling instability can occur at stresses well below the yield strength of the material. Design for buckling is paramount for the engineer because failure of this type is preceded by little if any warning. Chapter 6 contains specific guidelines regarding design for buckling caused by vacuum.

4.1.2 Loading Combinations

Given the above potentially active stress conditions, the engineer should take into consideration probable combinations of loadings that may result in higher principal stresses. The stresses considered under normal conditions are those between supports and at supports:

1. Between supports
   a. Longitudinal stresses caused by beam bending;
   b. Longitudinal stresses caused by longitudinal movement under temperature changes and internal pressure;
   c. Circumferential (hoop) stress caused by internal pressure; and
   d. Equivalent stress based on the Huber–Hencky–von Mises theory of failure or the stress intensities approach.
2. At supports
   a. Circumferential stresses in the supporting ring girder or saddle caused by bending and direct stresses and tensile stress caused by internal pressure;
   b. Longitudinal stress in the shell at the supports caused by beam action and longitudinal movement from temperature changes and internal pressure;
   c. Bending stresses in the shell imposed by the support (ring girder or saddle); and
   d. Equivalent stress based on the Huber–Hencky–von Mises theory of failure or the stress intensities approach.

Also, it may be important to consider the shear stress because this stress may occur at or near supports and throughout the shell structure. It is important to keep in mind the secondary tension and compression stresses that can occur at an element of the shell and how the resulting Huber–Hencky–von Mises stress or stress intensities may govern design when these secondary stresses are combined with the primary hoop tension and beam bending longitudinal stresses.

Loading combinations are also influenced by the operational state of the penstock. For example, stresses imposed in the pipe in the dewatered state may be much greater than those in the pressurized condition because of the temperature extremes that can be experienced by an empty penstock.

4.1.3 Tolerances

Penstock material plate or prefabricated pipe should be ordered with a thickness equal to or greater than the minimum thickness for handling or the thickness calculated for the design. No additional adjustment needs to be made to the shell thickness if the specified mill tolerance is not greater than the smaller of 0.01 in. or 6% of the nominal thickness. If greater tolerances are allowed, the design plate thickness should be increased to account for the larger undertolerance. Tolerances for steel plates and/or shapes are given by ASTM A6 (ASTM 2010a) and ASTM A20 (ASTM 2010b).

Also, for an acceptable design thickness of the shell plate without any upward adjustment, it is important to apply minimum/maximum acceptance tolerances for shop and field weldment alignments at weld joints. The requirements of ASME BPVC, Section VIII, Division 1 (ASME 2010a) should be specified.

4.2 CONCRETE PIERS

4.2.1 General

Concrete piers support the penstock at intermediate points between the anchor blocks. Pier spacing is determined by optimizing the penstock span length after all combined hoop and longitudinal shell stresses are determined. Pier spacing usually ranges from 20 to 100 ft. Spacing also is influenced by other factors in the pipe configuration, such as locations of expansion joints and/or coupled joints. The two concrete pier configurations commonly used are ring girder supports (Fig. 4-1) and saddle supports (Fig. 4-2). Where steel sad-

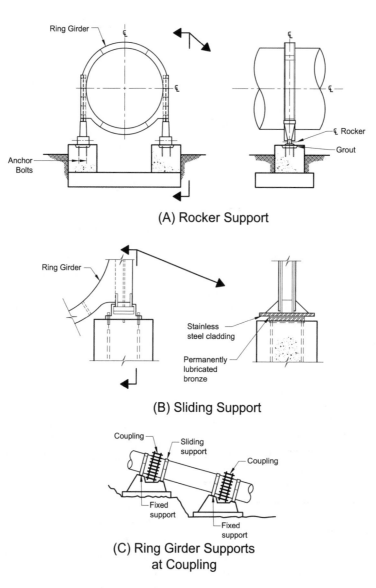

(A) Rocker Support

(B) Sliding Support

(C) Ring Girder Supports
at Coupling

Fig. 4-1.   Typical ring girder supports

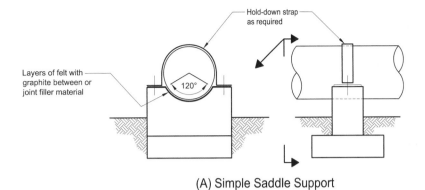

(A) Simple Saddle Support

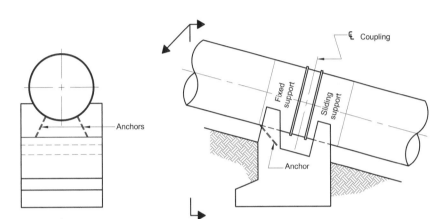

(B) Pair of Saddles at Sleeve Coupling

Fig. 4-2. Typical saddle supports

dles are used, the concrete piers are similar to those used at ring girders.

Ring girder supports are used for larger pipe diameters where the reaction load at the support is too large to be reasonably accommodated by a plain pipe shell. The pipe reaction loads are transmitted from the ring girder to the concrete piers by sliding bearing plates, rollers, or rocker assemblies. These details are developed to prevent lateral movement of the pipe but allow longitudinal movement or rotation. Where it is desired to prevent longitudinal movement of the pipe, the piers can also be detailed with anchor bolts through the ring girder base plates to form a fixed support point (Fig. 4-1).

Concrete saddles generally are used for the smaller pipe diameters and shorter spans (usually 20 to 40 ft) where the pipe span end reaction at the support does not cause overstress of the pipe shell and thus require a ring stiffener. To provide the most ideal condition for pipe and concrete saddle stress distribution at the reaction point, the saddle contact angle is usually 120 to 180 degrees, with the smaller angle giving the higher steel stress but the lower lateral load concrete stress at the edges. Two pieces of fabric or ultra-high-molecular-weight (UHMW) pad should be placed between

the pipe and concrete surface and secured accordingly. Alternately, a stainless steel slide plate can be attached to the penstock with a piece of UHMW pad placed between the slide plate and the concrete surface and secured accordingly. Also, wooden wedges or other resilient materials, such as rubber, are often placed in the top edge of the saddle (saddle horn) to reduce stress concentration. Saddles can be detailed to allow the pipe to slide or to anchor the pipe as illustrated in Fig. 4-2.

The final overall dimensions of saddles, piers, and their base mats are determined by the underlying soil or rock properties and by the stability evaluation described in Chapter 8. In most cases, the concrete reinforcement steel required is based on the minimum temperature steel required. However, specific details must be checked, such as the narrow cross section of the concrete near the top of the saddles (saddle horns) subjected to lateral reaction loads plus radial pipe expansion. These same lateral loads require checking of the reinforcing across the center of the saddle under the pipe.

Because concrete piers need to be cast in place, they typically require considerable costly field labor. Therefore, except where the piers of an existing penstock are in good condition and can be reused when replacing a steel penstock shell, more recent experience has been that ring girders can be more cost effective. This situation occurs because ring girders reduce the amount of costly field labor by allowing much longer spacing between supports and require only minimal cast-in-place concrete footings. Ring girders also offer several advantages over concrete saddles that extend the service life of the steel penstock shell.

- Ring girders eliminate the interface between the steel shell and the concrete saddle, which is a common location for accelerated shell corrosion because this space retains moisture and the steel surface exteriors are inaccessible for maintenance painting.
- Ring girders eliminate the high-stress concentration at the saddle horn that results in local shell deformation dimpling and cracking that is commonly the initial result from overstress as concrete saddles age. This situation is particularly problematic if the saddles support less than the bottom half of the steel shell.
- Ring girders provide an alternate to concrete saddles, which historically have presented structural issues at the tips of the saddle. When concrete saddles are formed higher around the sides of the steel shell to reduce the local horn stresses, the upper sections of concrete saddles are subjected to these shell reactions that historically have resulted in the saddles experiencing structural shear cracking. Properly reinforcing the upper horn sections of a concrete saddle for the penstock's horn and lateral forces can be difficult because the forming offers limited space for rebar reinforcement.

4.2.2   Loads and Combinations

Concrete supports must be designed for the dead load of the pipe and contained water (and ice and snow if applicable), longitudinal forces resulting from frictional resistance caused by longitudinal strain (Poisson's ratio effect) and temperature movements, and lateral forces caused by wind and earthquake forces. In addition to the loads that can act on the pipe and therefore on the concrete support, the support also may experience additional loads, including buoyancy caused by groundwater and prestressing load if rock anchors are used.

The load combinations and service level categories used in the concrete support design must be consistent with those of the pipe design. The most critical combinations must be determined separately for the overall stability evaluation (see Chapter 8) and then for the detailed reinforced concrete design. The loading conditions can be categorized as follows:

Construction conditions: Dead load with pipe empty or full, buoyancy, wind, prestress load, and temperature;

Test conditions: Dead load, buoyancy, wind, prestress load, internal test pressure, and temperature; and

Operating conditions (static and dynamic cases): Dead load, buoyancy, wind, seismic, prestress load, internal pressure (steady state and transient), and temperature.

For each condition, the worst combinations of values of each component and associated friction effects should be evaluated. Also, if no expansion joints are used, Poisson's ratio effects must be included in the loading.

4.3   RING GIRDERS

Ring girders normally are used to support long-span exposed steel penstocks. The purpose of the ring girder is to support the exposed penstock, its contents, and all live and dead loads, as defined in Section 3.2. Also, ring girders stiffen the penstock shell and maintain the pipe section's roundness, thus allowing the penstock to be self-supporting, acting as either a simple beam or a continuous beam when the penstock is supported at more than two locations. Fig. 4-3 shows a typical ring girder supporting a large-diameter penstock. See Section 3.4 for further definitions of types of stresses to be considered for ring girders.

4.3.1   Analysis

Detailed ring girder stress analysis includes combining circumferential and longitudinal stresses in the penstock shell at the ring girder junction in accordance with Section 3.4.

Added to the longitudinal beam stresses are longitudinal stress caused by pressure on the exposed pipe end at the expansion joint, longitudinal stress caused by frictional force at the supports, longitudinal stress caused by frictional force at expansion joints or sleeve-type couplings, longitudinal stress caused by gravity force (if the penstock is sloping), and localized bending stress in the shell caused by ring restraint.

Away from supports, the penstock shell is also designed for combined longitudinal and circumferential stresses using the same procedure; however, bending stress from localized ring restraint is neglected. For exposed penstocks, it is common to thicken the shell in the vicinity of the ring girder. A detailed analysis method for ring girders has been published by USBR (1940). Also, an abbreviated version has been published (USBR 1986).

4.3.1.1   Vertical Loads   The basic procedure for ring girder design is to first locate the supports and then determine the reaction at the supports, assuming that the penstock acts as a continuous beam. A trial ring geometry is selected and the centerline of the support legs is located such that the column centerline is approximately collinear with the centroid of the ring plus shell section. To minimize the ring bending moment, the centerline of the support legs should be located approximately $0.04r$ in. outside the centroid of the ring plus shell section. The shell length (L_l)

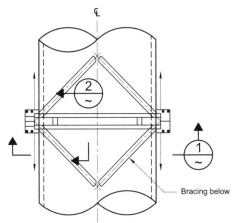

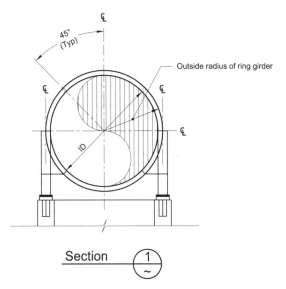

Fig. 4-3a. Typical ring girder supporting a large-diameter penstock

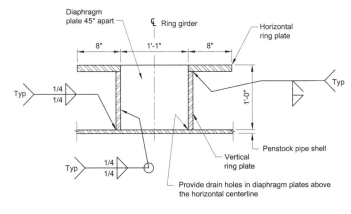

Fig. 4-3b. Typical ring girder cross-section

that is assumed to be effective on either side of the ring is given by

$$L_1 = 0.78\sqrt{rt} \qquad (4\text{-}14)$$

where r = inside radius of the shell (in.) and

t = shell thickness (in.).

For ring girders that use more than one ring, the maximum shell length (L_2) that is assumed to be effective between the rings must not exceed $1.56\sqrt{rt}$.

Next, section properties for the ring and the portion of the penstock shell that will participate in composite action with the ring are determined. Then the maximum ring and shell stresses for the vertical loads are calculated by combining the direct stress, bending stress, and pressure stress. Both the inside shell stress and the outside ring tip stress then are compared with allowable stresses as defined in Section 3.4. The ring geometry is revised, and the above analysis is repeated until stresses are acceptable.

4.3.1.2 Lateral Loads Ring girders also must support lateral forces caused by wind or seismic conditions. Lateral forces should be evaluated according to Sections 1.5 and 1.6. The equivalent static lateral force must be not less than 15% of the vertical load.

Horizontal seismic loading of ring girders produces a maximum bending at the horizontal spring line of the penstock (where the support legs are attached to the ring). Vertical load stresses are combined with the seismic load stresses, and the area near the support leg attachments should be investigated to determine the stress magnitude and maximum stress location. The stresses then are compared with the allowable stresses defined in Chapter 3 to determine if the ring girder is adequate. If not, the ring geometry is revised, and the above analysis is repeated until stresses are acceptable.

Also, the engineer must check stresses that occur in ring girders when the penstock is half full. These stresses are compared with the material and allowable stresses indicated in Section 3.4. The ring geometry must be revised until the stresses are acceptable.

4.3.2 Slide Bearings

Slide bearings provide low resistance to longitudinal forces acting on the penstock support as the result of temperature changes, pressure, and gravity loads on inclined penstocks. Fig. 4-4 shows a typical slide bearing detail. Low resistance to friction can also be obtained by using slide bearings made of virgin polytetrafluorethylene (Teflon), high-density polyethylene (HDPE), or other products manufactured for the purpose of supporting heavy loads. These materials can reduce the coefficient of friction to 0.05 or less. The upper surface of the bearing should be slightly larger than the lower surface to prevent debris from contaminating the contact surface of the bearing.

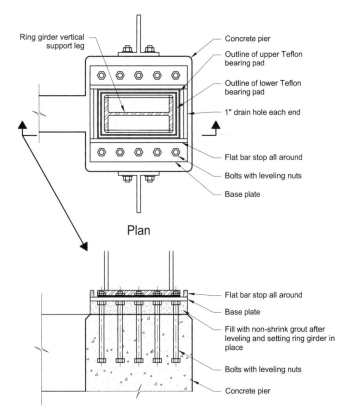

Ring girder vertical support leg

Concrete pier

Outline of upper Teflon bearing pad

Outline of lower Teflon bearing pad

1" drain hole each end

Flat bar stop all around

Bolts with leveling nuts

Base plate

Plan

Flat bar stop all around

Base plate

Fill with non-shrink grout after leveling and setting ring girder in place

Bolts with leveling nuts

Concrete pier

Fig. 4-4.   Ring girder slide bearing

An alternative to slide bearings are rocker- or roller-type bearings. Rocker- or roller-type bearings can be advantageous in some installations, especially those where the long-term functionality of slide bearings may be a concern.

4.4   SADDLES

Saddles are a type of support for exposed penstocks. The support engages less than the full perimeter of the penstock, generally between 120 and 180 degrees of arc. Saddles are simpler to construct than full-perimeter ring girder supports but generally are spaced closer together than ring girders. The closer spacing is necessary because saddles do not stiffen the penstock shell against radial deformations to the same extent as ring girder supports.

Saddles, serving the same functions as ring girders, act as supports to carry water and penstock dead (material) loads or as construction supports. Saddles may be designed to act as anchors to resist pressure forces at bends in the penstock and also loads directed along the length of the penstock. The loads acting along the length of the penstock are from friction forces generated at supports as a penstock expands or contracts because of temperature changes or pressure surges, and from the axial component of gravity loads on inclined penstocks. Saddles may be of steel or reinforced concrete.

Steel saddles generally are fabricated from structural grades of steel (ASTM A36 is the most common) and are of welded construction. Welded-on steel saddle design is detailed in Section 4.4.4 and is based on the guidelines presented in ASME BPVC, Section VIII, Division 2 (ASME 2010b). Fig. 4-5 illustrates several typical saddle configurations.

4.4.1   Reinforcing Plates

Reinforcing plates sometimes are used between the saddle and penstock shell to stiffen the shell and limit local stresses in the shell immediately around the saddle. To be effective in reducing shell stresses, reinforcing plates must extend in each direction beyond the saddle they bear upon by an amount, in the longitudinal direction, of at least $0.78\sqrt{R_m t}$ plus the differential growth, and in the circumferential direction by $\theta/24$ degrees of arc. Reinforcing plates should be attached to the penstock shell by a continuous fillet weld. Since reinforcing plates generally are welded to the penstock, they must be of the same material as the penstock shell or of a compatible material with the same nominal chemical composition and mechanical properties as the penstock shell material. Also, the reinforcing plate material must be heat-treated to similar notch toughness as the shell and must have good weldability to the penstock. See Section 2.4 for specific material requirements for reinforcing plates. Corners of the reinforcing plate should be cut with a radius to reduce stress concentrations.

4.4.2   Stiffener Rings

Full-circumference stiffener rings can be used on either side of the saddle to stiffen the penstock shell or can be placed in the same cross section as the saddle itself and made integral with the saddle. Stiffener rings make possible span lengths approaching those permitted by ring girder supports. Stiffener rings placed directly over the saddle generally are welded to the penstock. When stiffener rings are welded to the penstock, they must be of the same material as the penstock shell or of a compatible material with the same nominal chemical composition and mechanical properties as the penstock shell material. Also, welded-on stiffener rings must be heat-treated to similar notch toughness as the shell and must have good weldability to the penstock. See Section 2.4 for specific material requirements for stiffener rings.

4.4.3   Expansion Provisions

If not fixed, saddles generally are designed to permit sliding relative to the penstock, either at the saddle-to-penstock interface or at the base of the saddle. If sliding is at the base of the saddle, keeper bars are required to prevent the penstock from moving in the transverse direction.

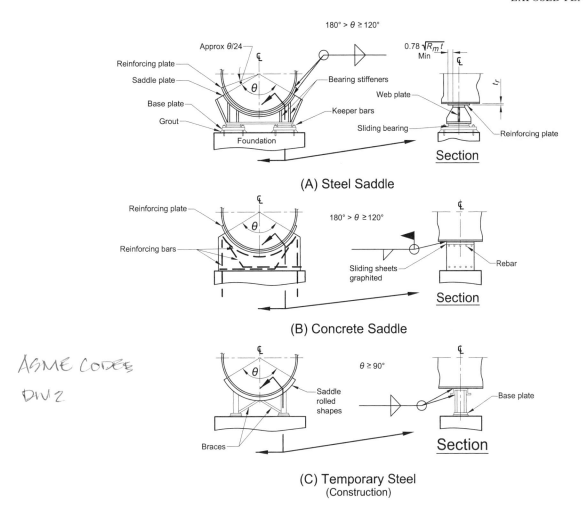

Fig. 4-5.   Saddle configurations

4.4.4   Welded-on Saddle Design

4.4.4.1   Saddle Geometry   The geometry of the saddle support is detailed in Fig. 4-6. Saddles should be in continuous contact with the penstock shell for a span, θ, of at least 120 degrees, but not more than 180 degrees. Reinforcing plates can be used to reduce the stresses in the penstock shell. The required width of a reinforcing plate, b_1, is

$$b_1 = b + 1.56\sqrt{R_m t} \qquad (4\text{-}15)$$

where b = longitudinal width of the contact surface between the saddle support and the penstock shell (in.);

R_m = mean radius of the penstock shell (in.); and
t = penstock shell thickness (in.).

According to Zick (1985), the saddle width at the interface with the penstock does not control proportioning the design. A minimum saddle width (dimension b in Fig. 4-6) of 12 in. for steel saddles and 15 in. for concrete saddles is recommended.

The minimum circumferential span of the reinforcing plate, $θ_1$, is

$$θ_1 = θ + \frac{θ}{12} \qquad (4\text{-}16)$$

where θ = the circumferential span of the saddle support contact with the penstock shell (radians).

Stiffening rings may be attached to the penstock at the saddle to reduce local stresses. The rings may be placed in the plane of the saddle, similar to a ring girder, or one ring may be placed on each side of the saddle, equidistant from the saddle centerline. When one stiffening ring is placed on each side of the saddle, as in Fig. 4-7(b), the maximum inside face to inside face distance between the rings, h, is R_m. If both rings are placed such that h is less than $1.56\sqrt{R_m t}$, they must be considered as a single stiffening ring located in the plane of the saddle in the stress calculations.

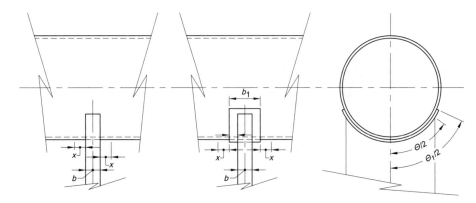

Fig. 4-6. Welded-on saddle geometry
Source: Reprinted from ASME (2010b), by permission of the American Society of Mechanical Engineers. All rights reserved.

4.4.4.2 Shear Stresses For a penstock with a stiffening ring in the plane of the saddle support, the maximum shear stress in the shell is located at the spring line. The shear stress at this location is given by

$$\tau_1 = \frac{T}{\pi R_m t} \qquad (4\text{-}17)$$

where T = maximum shear force at the saddle support (lb).

For a penstock with one stiffening ring on each side of the support, or no stiffening rings, the shear stress in the shell is a maximum at a circumferential distance $\theta/2 + \beta/20$ from the invert (Fig. 4-8). The shear stress at this location is given by

$$\tau_2 = \frac{K_1 T}{R_m t} \qquad (4\text{-}18)$$

where K_1 = coefficient (Fig. 4-9).

The absolute value of the calculated stresses τ_1 or τ_2 must not exceed $0.8S$.

4.4.4.3 Circumferential Stresses The distribution of circumferential bending moment at the saddle support is dependent on the use of stiffening rings at that location, as shown in Fig. 4-10.

1. For a penstock without a stiffening ring, or with a stiffening ring in the plane of the saddle, the maximum circumferential bending moment is located at the horn of the saddle (a circumferential distance of $\theta/2$ from the invert), as shown in Fig. 4-10(a), and is given by

$$M_\beta = K_3 Q R_m \qquad (4\text{-}19)$$

where K_3 = coefficient (Fig. 4-9) and

Q = maximum reaction at saddle support from all loads (lb).

For a penstock with a stiffening ring on each side of the saddle, the maximum circumferential bending moment is located at a circumferential distance ρ from the top of the penstock, as shown in Fig. 4-10(b), and is given by

$$M_\beta = K_6 Q R_m \qquad (4\text{-}20)$$

where K_6 = coefficient (Fig. 4-9).

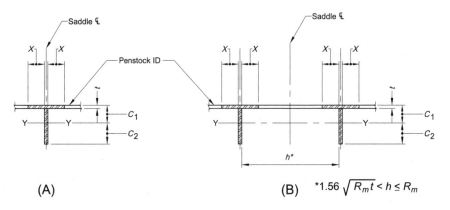

(A)

(B) $^*1.56\sqrt{R_m t} < h \le R_m$

Fig. 4-7. Stiffening ring orientation
Source: Reprinted from ASME (2010b), by permission of the American Society of Mechanical Engineers. All rights reserved.

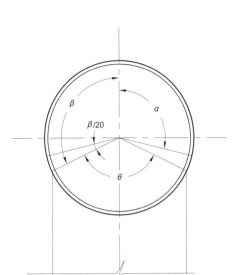

Fig. 4-8. Angular location of shear stresses in penstock shell
Source: Reprinted from ASME (2010b), by permission of the American Society of Mechanical Engineers. All rights reserved.

2. The maximum width of penstock shell, x (Fig. 4-7), that contributes to the strength of the shell at the saddle location is given by

$$x = 0.78\sqrt{R_m t} \qquad (4\text{-}21)$$

3. The circumferential stresses in the penstock shell when stiffening rings are not used can be calculated as follows:
   a. The maximum circumferential compressive stress in the shell at the base of the saddle support is given by

$$\sigma_6 = \frac{-K_2 Q k}{t(b + 2x)} \qquad (4\text{-}22)$$

   where K_2 = coefficient (Fig. 4-9).
   b. The circumferential compressive stress in the shell at the horn of the saddle (a circumferential distance $\theta/2$ from the invert) is given by
   (1) At supports adjacent to a free pipe end (a free pipe end is one connecting to a coupling, expansion joint, etc.) where the free end is a distance, d, from the centerline of the support, and $d \le L/2$

$$\sigma_7 = \frac{-Q}{4t(b + 2x)} - \frac{3K_3 Q}{2t^2}\left(\frac{4R_m}{K_{A1}}\right) \qquad (4\text{-}23)$$

   where $K_{A1} = \min\left[4R_m,\ d + \dfrac{L}{2}\right]$ and

   L = centerline distance between supports (in.).

(2) At other supports

$$\sigma_7^{\,*} = \frac{-Q}{4t(b + 2x)} - \frac{3K_3 Q}{2t^2}\left(\frac{4R_m}{K_{A2}}\right) \qquad (4\text{-}24)$$

where $K_{A2} = \min\left|\,4R_m,\ L\,\right|$.

Note that if support spacing is not equal on each side of a given support, then the average of the support spacing on each side of that support should be used for L.

   c. Circumferential bending stresses σ_6, σ_7, and $\sigma_7^{\,*}$ can be reduced by the addition of a reinforcement plate at the saddle location. These plates must be welded to the penstock shell and meet the requirements given in Section 4.4.4.1. If reinforcement plates are used, the circumferential compressive stresses are given by

$$\sigma_{6,r} = \frac{-K_2 Q k}{b_1(t + \eta t_r)} \qquad (4\text{-}25)$$

$$\sigma_{7,r} = \frac{-Q}{4(t + \eta t_r) b_1} - \frac{3K_3 Q}{2(t + \eta t_r)^2}\left(\frac{4R_m}{K_{A1}}\right) \qquad (4\text{-}26)$$

$$\sigma_{7,r}^{\,*} = \frac{-Q}{4(t + \eta t_r) b_1} - \frac{3K_3 Q}{2(t + \eta t_r)^2}\left(\frac{4R_m}{K_{A2}}\right) \qquad (4\text{-}27)$$

where $k = 0.1$ for support welded to penstock;

η = shell to reinforcing plate strength reduction factor, $= S_r/S$, but not more than 1.0;
S = allowable stress intensity for shell (psi);
S_r = allowable stress intensity for reinforcing plate (psi); and
t_r = reinforcing plate thickness (in.).

   d. If $t_r > 2t$, the compressive stress plus bending stress at the ends of the reinforcing plate should be computed using the following equations. In these equations, coefficient $K_{3,1}$ is computed using the equation for K_3 found in Fig. 4-9, but substituting θ_1 for θ. See Eq. (4-16) for values of θ_1.

4. At supports adjacent to a free pipe end, where free end is a distance, d, from the centerline of the support, and $d \le L/2$

$$\sigma_{7,1} = \frac{-Q}{4t(b + 2x)} - \frac{3K_{3,1} Q}{2t^2}\left(\frac{4R_m}{K_{A1}}\right) \qquad (4\text{-}28)$$

5. At other supports

$$\sigma_{7,1}^{\,*} = \frac{-Q}{4t(b + 2x)} - \frac{3K_{3,1} Q}{2t^2}\left(\frac{4R_m}{K_{A2}}\right) \qquad (4\text{-}29)$$

$$K_1 = \frac{\sin\alpha}{\pi - \alpha + \sin\alpha\cos\alpha}$$

$$K_2 = \frac{1 + \cos\alpha}{\pi - \alpha + \sin\alpha\cos\alpha}$$

$$K_3 = \frac{\dfrac{3\cos\beta}{4}\left(\dfrac{\sin\beta}{\beta}\right)^2 - \dfrac{5\sin\beta\cos^2\beta}{4\beta} + \dfrac{\cos^3\beta}{2} - \dfrac{\sin\beta}{4\beta} + \dfrac{\cos\beta}{4} - \beta\sin\beta\left[\left(\dfrac{\sin\beta}{\beta}\right)^2 - \dfrac{1}{2} - \dfrac{\sin 2\beta}{4\beta}\right]}{2\pi\left[\left(\dfrac{\sin\beta}{\beta}\right)^2 - \dfrac{1}{2} - \dfrac{\sin 2\beta}{4\beta}\right]}$$

$$K_4 = \frac{\cos\beta\left[1 - \dfrac{\cos 2\beta}{4} + \dfrac{9\sin\beta\cos\beta}{4\beta} - 3\left(\dfrac{\sin\beta}{\beta}\right)^2\right]}{2\pi\left[\left(\dfrac{\sin\beta}{\beta}\right)^2 - \dfrac{1}{2} - \dfrac{\sin 2\beta}{4\beta}\right]} + \frac{\beta\sin\beta}{2\pi}$$

$$K_5 = \frac{1}{2\pi}\left\{\left[-\frac{1}{2} + (\pi - \beta)\cot\beta\right]\cos\rho + \rho\sin\rho\right\}$$

$$K_6 = \frac{1}{2\pi}\left\{\rho\sin\rho + \cos\rho\left[\frac{3}{2} + (\pi - \beta)\cot\beta\right] - \frac{(\pi - \beta)}{\sin\beta}\right\}$$

Notes:

1. $\alpha = 0.95\left(\pi - \dfrac{\theta}{2}\right)$

2. The relationship between ρ and θ is given by $0 = (\tan\rho)\left[0.5 + \dfrac{\theta}{2}\cot\left(\pi - \dfrac{\theta}{2}\right)\right] - \rho$. *Values for ρ*

   for a specified θ are shown in the table below.

Relationship between ρ and θ							
θ	120°	130°	140°	150°	160°	170°	180°
ρ	93.661°	91.046°	87.935°	84.201°	79.654°	74.007°	66.782°

3. $\beta = \pi - \dfrac{\theta}{2}$

4. The angles θ, β, and ρ are in radians in the calculations.

5. See Table 4-1 for Coefficients and Related Angular Values for Various Angles of θ.

Fig. 4-9. Coefficient equations for saddle supports
Source: Reprinted from ASME (2010b), by permission of the American Society of Mechanical Engineers. All rights reserved.

Table 4-1. Coefficients and Related Angular Values for Various Angles of θ

Coefficient	Angular value						
θ	120°	130°	140°	150°	160°	170°	180°
$θ_1$	130°	140.8°	151.7°	162.5°	173.3°	184.2°	195°
α	114°	109.3°	104.5°	99.8°	95°	90.3°	85.5°
β	120°	115°	110°	105°	100°	95°	90°
$β_1$	115°	109.6°	104.2°	98.8°	93.3°	87.9°	82.5°
ρ	93.661°	91.046°	87.935°	84.201°	79.654°	74.007°	66.782°
K_1	1.17069	1.02223	0.90034	0.79885	0.71325	0.64017	0.57707
K_2	0.76026	0.72579	0.69711	0.67329	0.65357	0.63738	0.62427
K_3	0.05285	0.04492	0.03788	0.03168	0.02624	0.02151	0.01744
$K_{3,1}$	0.04492	0.03733	0.03072	0.02499	0.02008	0.01592	0.01242
K_4	0.34047	0.32962	0.31674	0.30209	0.28592	0.26847	0.25000
K_5	0.27087	0.25585	0.23869	0.21901	0.19633	0.17000	0.13911
K_6	0.05809	0.05082	0.04323	0.03549	0.02785	0.02069	0.01460

6. The circumferential stresses in the penstock with an external stiffening ring in the plane of the saddle support are given by the following equations.

   a. The maximum compressive stress in the penstock shell is calculated as follows:

$$\sigma_6{}^* = \frac{-K_2 Q k}{A} \qquad (4\text{-}30)$$

where A = cross-sectional area of the stiffening ring and associated penstock shell (in.²).

   b. The circumferential compressive plus bending stress at the horns of the saddle support is calculated as follows:

$$\sigma_8{}^* = \frac{-K_4 Q}{A} + \frac{K_3 Q R_m c_1}{I} \quad \text{(shell)} \qquad (4\text{-}31)$$

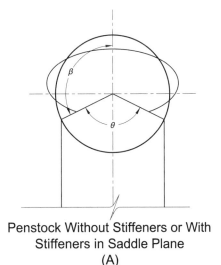

Penstock Without Stiffeners or With
Stiffeners in Saddle Plane
(A)

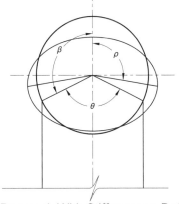

Penstock With Stiffeners on Both
Sides of Saddle Plane
(B)

Fig. 4-10. Angular location of maximum circumferential bending moment in penstock shell
Source: Reprinted from ASME (2010b), by permission of the American Society of Mechanical Engineers. All rights reserved.

$$\sigma_9^* = \frac{-K_4 Q}{A} - \frac{K_3 Q R_m c_2}{I} \quad \text{(stiffening ring)} \quad (4\text{-}32)$$

where σ_8^* = stress in penstock shell (psi);

K_4 = coefficient (Fig. 4-9);
K_3 = coefficient (Fig. 4-9);
c_1 = distance between the extreme element of the penstock-stiffening ring cross section to the neutral axis of the penstock-stiffening ring cross section (in.) (Fig. 4-7);
I = moment of inertia of cross-sectional area, A (in.) (Young and Budynas 2002);
σ_9^* = stress in stiffening ring (psi); and
c_2 = distance between the extreme element of the penstock-stiffening ring cross section to the neutral axis of the penstock-stiffening ring cross section (in.) (Fig. 4-7).

7. The circumferential compressive plus bending stress at the horns of the saddle support when one external stiffening ring is located on each side of the saddle is calculated by the following equations.
   a. The maximum compressive stress in the shell is calculated as follows:

$$\sigma_6 = \frac{-K_2 Q k}{t(b + 2x)} \quad (4\text{-}33)$$

   b. The circumferential compressive plus bending stress at a circumferential distance ρ from the top of the penstock is given by

$$\sigma_{10}^* = \frac{-K_5 Q}{A} - \frac{K_6 Q R_m c_1}{I} \quad \text{(shell)} \quad (4\text{-}34)$$

$$\sigma_{11}^* = \frac{-K_5 Q}{A} + \frac{K_6 Q R_m c_2}{I} \quad \text{(stiffening ring)} \quad (4\text{-}35)$$

where σ_{10}^* = stress in penstock shell (psi);

K_5 = coefficient (Fig. 4-9); and
σ_{11}^* = stress in stiffening ring (psi).

8. The acceptance criteria for the circumferential stresses are as follows:
   a. The absolute value of σ_6, $\sigma_{6,r}$, σ_7, σ_7^*, σ_{7r}, σ_{7r}^*, $\sigma_{7,1}$, $\sigma_{7,1}^*$, σ_8^*, and σ_{10}^*, as applicable, must not exceed the allowable stress intensity of the shell material.
   b. The absolute value of σ_6^*, as applicable, must not exceed the smaller of the allowable stress intensity of the shell or the stiffening ring materials.
   c. The absolute value of σ_9^*, and σ_{11}^*, as applicable, must not exceed the allowable stress intensity of the stiffening ring material.

4.4.4.4 Saddle Support The saddle must be designed to resist the horizontal force, F_h, at the minimum section at the low point of the saddle. The horizontal force, F_h, is be calculated as follows:

$$F_h = Q \left(\frac{1 + \cos \beta - 0.5 \sin^2 \beta}{\pi - \beta + \sin \beta \cos \beta} \right) \quad (4\text{-}36)$$

4.4.5 Detailing

Various configurations have been used successfully in penstock construction. Details illustrated in Fig. 4-5 are typical of those used and should not be construed as the only acceptable details.

The sliding surface at the base of the steel saddle shown in Fig. 4-5(A) also could be located at the saddle plate–wear plate interface. In that case, the sliding bearing at the base of the saddle would not be necessary, and the base of the saddle would be anchored to the foundation with anchor bolts.

4.5 STIFFENERS TO RESIST EXTERNAL PRESSURE

4.5.1 Circumferential Stiffening

Full circumferential stiffener rings should be provided on exposed penstocks when required to resist external pressure such as vacuum.

4.5.2 Stiffener Spacing

To determine spacing of stiffener rings and required shell thickness, the procedure in UG-28 of ASME BPVC, Section VIII, Division 1 (ASME 2010a) must be followed.

4.5.3 Moment-of-Inertia Requirements

The size of stiffener rings must meet the moment-of-inertia requirements of UG-29 of the ASME BPVC, Section VIII, Division 1 (ASME 2010a).

Alternatively, the moment of inertia (I) of intermediate stiffener rings must satisfy the formula

$$I \geq \frac{pL D_o^3}{77,300,000 \left(N^2 - 1\right)} \quad (4\text{-}37)$$

where I = moment of inertia of the composite stiffener ring and participating part of the penstock shell (the shell length must not exceed $1.1 \, (D_o t)^{1/2}$ (in.^{4});

p = external pressure (psi);
L = stiffener spacing (in.);
D_o = outside diameter (in.);

N = number of complete waves into which a circular ring will buckle;

$$N^2 = \frac{0.663}{\left(\dfrac{H}{D_o}\right)\left(\dfrac{t}{D_o}\right)^{1/2}} \leq 100 \; ;$$

H = distance between ring girder supports or end stiffeners (in.); and

t = penstock shell thickness (in.).

Welds attaching stiffener rings to the penstock must be in accordance with UG-30 of the ASME BPVC, Section VIII, Division 1 (ASME 2010a), which gives acceptable sections and structural shapes for use as stiffener rings and methods of attachment.

Stiffener ring splices should be complete joint penetration butt joints designed to develop the full section of the stiffener ring.

4.5.4 Line of Support

Line of support for purposes of determining stiffener spacing may be a cone–cylinder junction, provided the moment of inertia of the junction meets the rules for stiffener rings specified in UG-29 of the ASME BPVC, Section VIII, Division 1 (ASME 2010a). Ring girder supports are often adequate to serve also as stiffeners to resist external pressure but must meet Section 4.5.3 moment-of-inertia requirements.

4.5.5 Factor of Safety

Although the ASME Code rules are based on a theoretical factor of safety of 3 for external pressure, the tolerances were established to limit the buckling pressure to not less than 80% of that for a perfectly circular vessel. Implicitly, the true factor of safety is $3 \times 0.8 = 2.4$. This factor of safety is not considered overly conservative for exposed penstocks constructed to normal fabrication tolerances. A factor of safety of 3 is used in the moment-of-inertia equation for stiffener rings in Section 4.5.3.

4.5.6 Attachment of Stiffener Rings

Fillet-welded attachment of stiffener rings is permitted. The maximum size of fillet welds used to attach stiffener rings to a penstock made of heat-treated material, such as ASTM A517 (ASTM 2010c), which is quenched and tempered, should be limited to 3/8 in., and the welds must be continu-

ous. Stitch welding is not recommended for any material. With the exception of stitch welding, the rules specified in UG-30 of the ASME BPVC, Section VIII, Division 1 (ASME 2010a) are recommended.

REFERENCES

AISC. (2010). "Specification for structural steel buildings." *AISC 360-10,* Chicago.

ASME. (2010a). "Boiler and pressure vessel code, section VIII, division 1: Rules for construction of pressure vessels." New York.

ASME. (2010b). "Boiler and pressure vessel code, section VIII, division 2: Alternative rules." New York.

ASTM. (2008). "Standard specification for carbon structural steel." *ASTM A36/A36M,* West Conshohocken, PA.

ASTM. (2010a). "Standard specification for general requirements for rolled structural steel bars, plates, shapes, and steel piling." *ASTM A6/A6M,* West Conshohocken, PA.

ASTM. (2010b). "Standard specification for general requirements for steel plates and pressure vessels." *ASTM A20/A20M,* West Conshohocken, PA.

ASTM. (2010c). "Standard specification for pressed vessel plates, alloy steel, high-strength, quenched and tempered." *ASTM A517/A517M,* West Conshohocken, PA.

Baker, E. H., Kovalevsky, L., and Rish, F. L. (1972). *Structural analysis of shells,* McGraw-Hill, New York.

Donnell, L. H., and Wan, C. C. (1950). "Effects of imperfections on buckling of thin cylinders and columns under axial compression." *J. Applied Mechanics,* 17, 73–83.

Reed, Robert E., Jr. (1968). "Remarks on imperfections of axially loaded cylinders," NASA TM-1552, National Aeronautics and Space Administration, Springfield, VA.

Roark, R. J., and Young, W. C. (1975). *Formulas for stress and strain,* McGraw-Hill, New York.

Schorer, H. (1933). "Design of large pipelines." *Transactions of the American Society of Civil Engineers* 1993, Vol. 98, 101, ASCE, New York.

Sundberg, C. (2009). "Do the math—Spiral welded steel pipe is a good bet." *Pipelines 2009: Infrastructure's Hidden Assets: Proc., Pipelines 2009 Conf.,* ASCE, Reston, VA.

USBR. (1940). "Penstock analysis and stiffener ring design." Bulletin No. 5, Part V, Boulder Canyon Project Final Design Report, U.S. Bureau of Reclamation, Denver.

USBR. (1986). "Welded steel penstocks." Engineering monograph no. 3, U.S. Bureau of Reclamation, Denver.

Young, Warren C., and Budynas, Richard G. (2002). *Roark's formulas for stress and strain,* 7th Ed., McGraw-Hill, New York.

Zick, L. P. (1985). "Useful information on the design of plate structures," *Steel plate engineering data*—vol. 2, American Iron and Steel Institute, Washington, DC.

CHAPTER 5

Buried Penstocks

Source: Photograph courtesy of National Welding Corp.; reproduced with permission.

5.1 DESIGN OF BURIED PENSTOCK SHELLS

When a steel penstock installation requires the pipeline to be buried below ground or within fill material, the penstock shell must be analyzed to resist not only internal pressure and other hydrodynamic loads but also external loads caused by earthfill, loads resulting from excessive deflections, potential movements of the ground, and numerous live loads.

Shell design for internal pressure and other hydraulic and hydrodynamic conditions is performed in the same manner as for an exposed penstock. Shell design for external soil

and other loads must take into consideration the flexibility of the pipe and the shell thickness needed to resist the external loads acting on the pipe before pressurization. For lower pressure, lower velocity, or lower risk penstocks (as discussed in the Preface), pipe may be designed as a waterline following AWWA Manual M11 (AWWA 2004).

External dead and live loads on flexible pipe produce compressive stresses in the pipe wall, which tend to reduce the hoop stress caused by internal pressure. The two key elements for shell design are the flexibility of the pipe and the type of soil and its characteristics.

5.1.1 Pipe Flexibility

Steel pipe, normally considered for penstock applications, usually functions as a flexible conduit within the flexibility limits imposed by the requirement that there be no damage to the lining and coatings used in the applications.

Pipes with D/t ratios, where D = nominal pipe diameter and t = pipe wall thickness, of less than 288 do not display excessive deflections if the backfill is compacted to a density greater than that of the zone of critical void ratio. The critical void ratio is the density (or void ratio) at which no volume change occurs. A standard density of 80% is considered above the critical void ratio. Stulling and increased compaction efforts must be applied if the D/t ratio is greater than 288.

Acceptable deflections of the pipe shell, based as a percentage of the outside pipe diameter for buried pipe, are as follows:

1. Flexible lined and coated pipe: 5%;
2. Flexible coated and mortar-lined pipe: 3%; and
3. Mortar-lined and coated pipe: 2%.

5.1.2 Load Types

Two types of earth loads are applicable to buried pipe design:

1. Trench loading, in which the pipe is laid in an excavated trench and backfilled and
2. Fill earth loading, in which the pipe is laid on a graded or prepared ground surface and fill is placed around and over the pipe.

5.1.3 Analyses for Buried Pipe

Excellent and detailed analyses and sample examples for the analyses of buried pipe are presented in AWWA Manual M11 (AWWA 2004), *Steel Plate Engineering Data,* Vol. 3 (AISI 1989), *Structural Mechanics of Buried Pipes* (Watkins and Anderson 2000), and ASCE Manual of Practice 119 (ASCE 2009).

5.1.4 Special Considerations

For an underground penstock installation, the following must be considered:

1. Protection against corrosion by coating;
2. Cathodic protection;
3. Vacuum design;
4. Flotation and drainage;
5. Specialty construction relating to trenching, pipe laying, and backfilling;
6. Welding and construction requirements; and
7. Inspection, hydrotesting, and maintenance requirements.

5.2 EXTERNAL PRESSURE

Generally, buried penstocks do not require stiffener rings to resist vacuum or earth pressures. The design for external pressures and backfill forces is dependent on the diameter-to-thickness ratio, backfill material and density, and built-in elongation of the vertical penstock diameter. The penstock wall and embedment must be designed to resist external pressure, external backfill forces, and vacuum.

5.2.1 Method

External loadings that can contribute to buckling include external water pressure, soil pressure surcharge (using the Marston theory above the penstock), live loads on top of the surcharge (where the effect on the penstock is determined by the Boussinesq method), and vacuum or net external pressure. Design methods to address external loadings are covered in the references listed at the end of this chapter.

5.2.2 Increased Resistance to Collapse

If external pressure exceeds the factored allowable external pressure permitted by the method shown in AWWA Manual M11 (AWWA 2004), the embedment quality or the ring stiffness must be increased. Increasing ring stiffness by increasing the wall thickness is a possibility. However, one of the following methods should be considered and is often more effective.

5.2.2.1 Importing Embedment By using imported backfill rather than native backfill, the soil support may be increased so additional pipe stiffness is not required. With a better grade of backfill, the soil modulus is increased and a better compactive effort can also be obtained

5.2.2.2 Using Enhanced Soil Using soil cement offers an option where placement and compaction under the penstock haunches is difficult or the availability or cost of quality backfill is prohibitive. An increasingly popular procedure for bedding and backfill is "flowable fill," which is a slurry that is poured into the trench. Because the fill is flowable, the penstock is in contact and is supported by its embedment. The problem of compacting soil under the haunches is eliminated. Flat spots at the invert are avoided. The slurry is not limited to select soil with less than 10% fines; native soil is often used. Very little portland cement is needed (one sack per cubic yard). Unconfined compressive strength should be in the range of 50 psi—no greater than 100 psi—to avoid stress concentrations. Average flow consistency must be 12 in. when determined in accordance with ASTM D6103-04 (ASTM 2004). Pipe stress and ring deflection are reduced significantly. It is essential that the flowable fill not shrink more than the allowable ring deflection.

5.2.2.3 Using Stiffener Rings Stiffener rings may be used instead of increasing the penstock shell thickness.

Stiffener ring design and spacing should follow the procedure given in Section 4.5.

5.3 BENDS

Bends in buried penstocks must meet the same requirements as those for bends in exposed penstocks given in Section 9.12.

5.4 BURIED JOINTS

Buried field joints, similar to exposed joints for aboveground penstocks, provide either rigidity or flexibility in the penstock, according to the design requirements.

Rigid joints that must provide longitudinal restraint along the penstock to resist axial and bending forces caused by temperature and pressure can be either full-penetration, butt-welded joints, which provide the greatest strength, or lap-welded joints, which also may provide sufficient strength and are less expensive. Single- or double-welded lap joints often are used for thinner wall thicknesses and lower pressures. Lap-welded joints provide some installation flexibility and are generally less costly than butt-welded joints. When lap-welded joints are used, longitudinal stresses at the joint must be checked carefully. For allowable stresses on welded joints, see Chapter 3. For field welding criteria, see Chapter 11.

Flanged and bolted joints and mechanically tied joints may be used to join buried penstocks. These types of joints are more commonly used on exposed penstocks. Such joints generally are placed in vaults. Criteria for flanges and bolted joints are found in Section 9.8.

Because thermal expansion and contraction of a buried penstock is less severe than that for an exposed penstock, buried joints that must provide penstock flexibility may be either flexible couplings or joints with rubber gaskets. Flexible joints allow angular joint deflection but cannot transfer appreciable longitudinal stresses. Rubber gasketed O-ring joints are generally used for nominal diameters up to 72 in. and for normal operating conditions up to 250 psi. Such joints may be the least costly joints and may help reduce installation costs. Buried flexible joints must be restrained from penstock thrust forces with suitable thrust ties or anchor blocks. Alternatively, joints can be welded to make up lengths sufficiently long to resist thrust forces through soil–pipe friction.

The lap-welded joints and O-ring rubber gasket joints are typically used when the penstock is designed as a waterline, as mentioned in Section 5.1.

Flexible couplings also are used for closures, roll-out sections, and other points where articulating joints are required. Expansion joints for buried penstocks typically are not recommended. When provided, couplings are generally placed in vaults for accessibility. Flexible joints are installed in pairs with a spool piece between them to accommodate differential settlement. Proper installation procedures must be observed.

During installation and before backfilling, buried penstocks are exposed to the same atmospheric and solar conditions as aboveground penstocks. Care must be taken to maintain alignment, and consideration must be given to expansion and contraction caused by temperature changes. Some precautions include shading penstock sections already joined together and completing the backfilling on completed sections before closure welds are made.

Closure welds often are made during the coolest part of the day to minimize temperature-induced stresses. These welds are made on a special closure lap joint at 400-ft to 500-ft intervals. The special closure joint is stabbed deeper than the normal closed position; all joints are welded except the closure; partial backfill is placed over all pipe except the closure to aid in cooling and contraction of the pipe; and then the weld is made.

REFERENCES

AISI. (1989). "Welded steel pipe." *Steel plate engineering data*, vol. 3, Revised Ed., Washington, DC.

ASCE. (2009). "Buried flexible steel pipe: Design and structural analysis." *Manual of Practice No. 119,* Reston, VA.

ASTM. (2004). "Standard test method for flow consistency of controlled low strength material (CLSM)." *ASTM D6103-04,* West Conshohocken, PA.

AWWA. (2004). "Steel water pipe: A guide for design and installation," 4th Ed. *Manual M11,* Denver.

Watkins, R. K., Anderson, L. R. (2000). *Structural mechanics of buried pipes*, CRC Press, Boca Raton, FL.

PIPE STIFFNESS normally < 10% of total STIFFNESS

CHAPTER 6

Steel Tunnel Liners

Source: Photograph courtesy of National Welding Corp.; reproduced with permission.

Steel tunnel liners often are required to prevent the migration of tunnel leakage caused by unfavorable geologic conditions in the surrounding rock mass. When a steel liner is installed in a tunnel, it must be analyzed to determine that it can resist internal operating pressures as well as resist buckling from external water pressures when the tunnel is dewatered infrequently for inspection and maintenance.

This chapter discusses procedures for the analysis and design of steel tunnel liners for

1. Internal pressure;
2. External pressure (steel tunnel liners without stiffeners);

3. External pressure (shell between stiffeners when stiffeners are used); and
4. External pressure (stiffener and contributing contiguous portion of shell).

The installer and engineer must pay particular attention to buckling and buoyancy issues during the backfill or grout installation process. Liners are sometimes internally or externally braced and/or externally anchored to resist buckling or buoyant forces.

Further considerations for the magnitude and conditions of internal design pressure loading and the sources and magnitudes of external pressure loading when the tunnel is dewatered are presented within this chapter.

6.1   DESIGN OF STEEL LINERS FOR INTERNAL PRESSURE

A pressurized steel tunnel liner interacts elastically with the contiguous concrete backfill and surrounding rock continuum to share its load. In the interest of obtaining an economic design, the steel tunnel liner should be analyzed for only that portion of the internal pressure that it alone will be required to carry. The depth of rock cover necessary to consider load sharing for various rock conditions is considered in Fig. 3-22 of ASCE (1989). When rock cover and surrounding rock conditions are inadequate for load sharing, the steel liner must carry the full internal pressure. Rock moduli for various rock types and graphs showing how rock modulus affects load sharing also are presented in ASCE (1989).

When rock cover and rock conditions are favorable, the ratio of the pressure carried by the steel liner to the total internal pressure can be determined in accordance with an elastic analysis, where the radial displacement (Δ_s) of the steel tunnel liner shell can be determined per ASCE (1989). A more detailed development is presented in Appendix B of ASCE (1989). In Fig. 6-1, the radial temperature gap (Δ_K) results from concrete shrinkage plus the temperature differences between the temperature reached during erection, including the effect of the temperature rise during hydration of cement and the lowest operating water temperature; Δ_c is the radial deformation in the concrete backfill behind the steel liner; Δ_D is the radial deformation in the cylinder of any distressed, fissured rock around the tunnel excavation; and Δ_E is the radial deformation at the inner radius of the unfissured or more competent rock surrounding the tunnel.

In tropical and semitropical zones, Δ_K is defined as zero because the tunnels and construction materials reach ambient air temperature after they are open to forced ventilation for a long period, and the average water temperature in the reservoir is at or near the average air temperature. In addition, the final concrete temperatures can be expected to be on the order of the same magnitude if properly controlled during placement. In northern climates, consideration should be given to the possible existence of a gap and its being properly accounted for.

Concrete or cellular backfill is typically used to fill the majority of the void between the tunnel excavation and the steel liner, while grouting or contact grouting is used to seal any remaining annular space when a high resistance to buckling is required. The concrete backfill is assumed to be radially fractured during loading so that strain in the radial direction is not affected by Poisson's ratio for homogeneous, continuous material. Similarly, blasting effects during excavation usually are the cause for a distressed, fissured zone of rock around the excavated tunnel periphery, so that strain in the radial direction in that material also is unaffected by

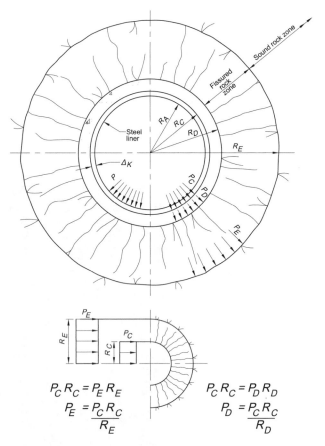

Fig. 6-1.   Elastic interaction among steel liner, concrete backfill, grout, and surrounding rock continuum
Source: Moore (1990), with permission from PennWell Corp.

Poisson's ratio. In Fig. 6-1, the pressure transmitted to the face of the concrete (P_C) can be assumed to be distributed radially outward through the fissured materials inversely proportional to the radius. The thickness of the zone of distressed, fissured material can vary from little or nothing in a machine-bored tunnel to several feet in a conventional drilled and blasted tunnel in jointed rock.

The moduli of rock deformation should reflect the effects of overlying rock mass at tunnel depth. Investigations should be carried out accordingly. Geophysical measurements can be made in the tunnel to determine the depth of fissured rock and its relative modulus of deformation with respect to the undisturbed, nondistressed parent rock.

The elastic interaction method presented here is valid only when the vertical and horizontal in situ stresses are approximately equal. A computer program is presented in Fig. 2 of Moore (1990), which can be used to determine the portion of the internal pressure carried by the steel liner. When significant differences exist between the vertical and horizontal in situ stresses, a three-dimensional finite element analysis (FEA) should be used.

In some cases, the steel liner must be designed and fabricated for installation before completing the excavation of the tunnel. The moduli of deformation would need to be determined conservatively on the basis of tests performed in exploratory drill holes and/or exploratory adits in the various rock units encountered. The values of the moduli can then be considered representative. However, during the actual tunnel excavation, zones of material may be encountered locally over short reaches that have a lower modulus than that determined for use in the design. One way to resolve this problem, short of substituting a heavier plate, is to provide circumferential hoop reinforcement on the outside of the steel liner throughout the local zone of lower resistant rock. The use of circumferential reinforcing, however, is not very effective, especially when the liner has an appreciable thickness. In addition, if a temperature reduction will exist between the liner installation and the subsequent operation conditions, a gap could exist behind the liner when the tunnel is filled and the liner subjected to internal pressure. Circumferential reinforcement can be significantly less beneficial under such circumstances. Closely spaced multilayer reinforcement can then only account for small reductions in the value of the rock modulus of deformation.

6.2 DESIGN OF STEEL LINERS FOR EXTERNAL PRESSURE

Experience has shown that buckling occurs in a single lobe. Amstutz (1970), Jacobsen (1974), and Montel (1960) developed their analyses in a similar manner based on the theory that a single lobe is formed.

The critical external buckling pressure for an unstiffened steel liner is determined by considering a gap between the steel liner and the concrete backfill surround caused by concrete shrinkage and a temperature difference. The gap can realistically vary from 0 to 0.001 times the radius.

The radial gap results from concrete shrinkage plus the temperature differences between the material temperatures during construction. This shrinkage includes the effects of the temperature rise during hydration of cement and the rise in ambient temperatures caused by forced ventilation into the tunnel using warmer outside air over a long period during construction.

The steel liner is the form for the concrete backfill. Under certain conditions, the steel liner might reach temperatures near 80°F when the concrete sets because of a combination of ambient air temperature and hydration effects. If the tunnel is later dewatered during winter when the water temperature is 34°F, a maximum 46°F temperature difference could exist for a short duration. A 46°F temperature difference produces a gap equal to $0.0003 \times R_A$ (Fig. 6-1).

Permissible tolerances during fabrication and erection allow an out-of-roundness, creating an elliptical shape with a 1% difference between the measured maximum and minimum diameters. This difference should not be considered in arriving at the critical radial design gap. A simple analysis demonstrates that the equivalent increase in curvature on the flattest curvature created for permitted ellipticity is only 1.5% larger than the originally specified circular radius [see Appendix C, Chapter 3, of ASCE (1989)]. The development of a lobe in an unstiffened liner usually involves a sector of the shell subtended by an angle of less than 80 degrees. The indented lobe would be expected to develop at the weakest point, which would be at the flattest curvature produced by the ellipticity of the penstock shell.

During concrete placement, the concrete backfill is introduced behind the steel tunnel liner from a pump line at the top of the tunnel. The concrete flows down around the steel liner. Because of a combination of frictional drag along the lower external surface of the liner, poor venting, and settlement occurring during consolidation, a void can be created under the bottom invert of the steel liner. This void can subtend an angle of approximately 40 to 60 degrees (0.70 to 1.05 radians). The void is usually shallow, and little grout is needed to fill it. When correctly executed, contact grouting between the steel liner and the concrete backfill eliminates these voids. The determination of the critical radial gap can assume that contact grouting has been properly executed when careful inspection and quality control have been exercised during construction. Therefore, temperature and shrinkage effects alone should govern the selection of an appropriate radial gap for purposes of analysis and design.

6.2.1 Amstutz Formulation

Amstutz assumes that a single lobe or indent will form at one particular spot (1970). A considerable extent of the indent always occurs parallel to the axis of the pipe because only the small resistance of the plate to bending has to be overcome. The Amstutz analysis, therefore, has been limited to a circumferential ring of unit width. A new mean radius is developed for the indent or lobe, with two outward and one inward half-waves forming around the mean arc line subtended by this new mean radius (Fig. 6-2). It is common practice to use the Amstutz analysis to determine a reasonable value for the tunnel liner thickness, and then this value can be used as a starting value for the more accurate Jacobsen analysis presented in Section 6.2.2.

Amstutz developed his buckling theory for the forces and displacements on the pipe wall element represented by the mean arc line subtended by a corresponding new mean radius. The Amstutz formulas are presented in his (1970) paper, "Buckling of Pressure-Shaft and Tunnel Linings." The Amstutz formulas for determining the stress condition in unstiffened cylindrical steel pipe are

$$\frac{\sigma_N - \sigma_V}{\sigma^*_y - \sigma_N} \left[\left(\frac{r}{i} \right) \sqrt{\frac{\sigma_N}{E^*}} \right]^3 \cong 1.73 \left(\frac{r}{e} \right) \left[1 - 0.225 \left(\frac{r}{e} \right) \frac{\sigma^*_y - \sigma_N}{E^*} \right]$$

(6-1)

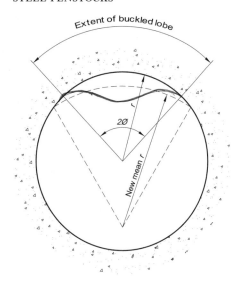

Fig. 6-2. Amstutz buckling pattern
Source: Moore (1990), with permission from PennWell Corp.

$$P_{cr} \cong \left(\frac{F}{r}\right)\sigma_N \left[1 - 0.175\left(\frac{r}{e}\right)\frac{\sigma_y^* - \sigma_N}{E^*}\right]$$

(6-2)

where $i = \dfrac{t}{\sqrt{12}}, e = \dfrac{t}{2}, r = \dfrac{D}{2}, F = t$;

$$\sigma_V = -\left(\frac{\Delta}{r}\right)E^* ;$$

$\Delta/r =$ gap ratio, for gap ratio between steel and concrete
= γ (for Figs. 6-3 and 6-4);

$r =$ tunnel liner radius;

$D =$ tunnel liner diameter;

$t =$ plate thickness;

$E =$ modulus of elasticity of liner;

$E^* = E/(1 - \upsilon^2)$;

$\sigma_y =$ yield strength;

$\sigma_N =$ circumferential axial stress in plate liner ring;

$\mu = 1.5 - 0.5[1/(1 - 0.002E/\sigma_y)]^2 =$ supporting effect coefficient (which can be set equal to 1 to allow for shape irregularities);

$\sigma_y^* = u\sigma_y/(1 - \upsilon + \upsilon^2)^{1/2}$;

$\upsilon =$ Poisson's ratio = 0.3; and

$P_{cr} =$ critical external buckling pressure.

The calculated P_{cr} can then be divided by an appropriate factor of safety (typically 1.5, in accordance with Section 3.5.5) to calculate a maximum allowable external pressure (P_{all}).

Amstutz's formulas have been reworked by E. T. Moore (Moore 1990) for ease of solution by computer. Fig. 6 of Moore's paper presents a computer program to solve Amstutz's equations for the critical buckling pressure on a cylindrical pipe without stiffeners.

For very small gap ratios and relatively thick liners, the value of σ_N approaches σ_y at the critical buckling pressure, and some caution must be exercised. When the axial stress in the liner approaches the yield stress from a position of low stress, the strain is proportional to stress and the ratio is linear because of the constancy of E, the modulus of elasticity. However, as the stress approaches the yield stress, the linearity no longer holds, and strain begins to increase faster than the stress increases, until finally when yield stress is reached, the strain increases, with no further increase in stress. This phenomenon, of course, can occur only with a decreasing value of E. E has been assumed constant in the Amstutz formulation. If the axial stress were to exceed about 80% of the yield stress (σ_y), the Amstutz formulation must be modified to reflect the phenomenon of a decreasing E by adopting a reducing modulus. However, in addition to complicating the computational process, integrating in a formula for reducing the modulus that is conservatively dimensioned shows such a significant reduction in buckling pressure when $\sigma_N \geq 0.8\sigma_y$ that the result is unacceptable and only results for $\sigma_N < 0.8\sigma_y$ have utility.

Another limitation also must be observed. When Amstutz developed his equations, personal computers were not available. To make his formulation tractable and reduce the number of unknown variables to be solved, he introduced a number of coefficients. He determined that within certain limits some variables could be represented by constants (i.e., coefficients) without affecting the results. ε is an expression in the differential equation for the inward deformation of the shell at any point. Amstutz (1970) indicated that an acceptable range for ε is $5 < \varepsilon < 20$. However, inspection of Fig. 6-5 shows that a more acceptable range would be $10 < \varepsilon < 20$. Substituting the actual values of F, W, and Y of Fig. 6-5 when $\varepsilon < 10$ results in critical buckling pressures that are actually higher than the buckling pressures computed using the respective values of 1.73, 0.175, and 0.225 recommended by Amstutz for use in his equations. However, in very few design cases is $\varepsilon \geq 10$.

When computing the critical buckling pressure according to the Amstutz equations, the axial stress (σ_N) in the steel liner must be determined along with the corresponding value of σ. The results may be considered satisfactory when $\varepsilon > 5$, according to Amstutz's recommendation, subject to the limitation that $\sigma_N < 0.8\sigma_y$.

Figs. 6-3 and 6-4 show curves for unstiffened liners for yield stresses of 38,000 psi and 50,000 psi, respectively, based on the Amstutz equations. The upper left-hand portion of the curves defined by dashed lines is where $\varepsilon < 5$ or $\sigma_N > 0.8\sigma_y$. Results within the solid line reach of the curves can be considered satisfactory. If the Amstutz results were further limited to $\varepsilon > 10$, only a small portion of the curves could be considered to provide satisfactory results. The dashed line for $2\sigma_y/(D/t)$ indicates pressures at which the liner fails in general yielding.

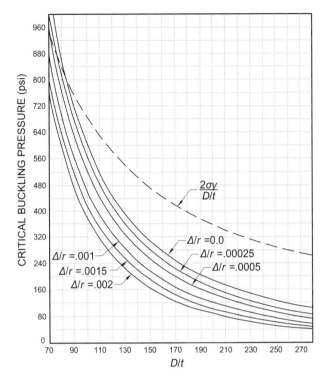

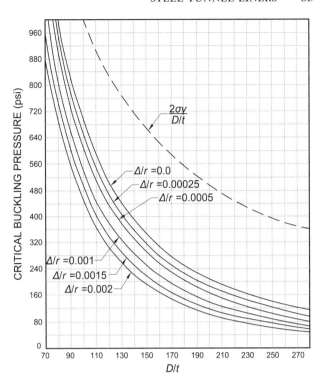

Fig. 6-3. Amstutz curves for unstiffened liners (yield stress = 38,000 psi)
Source: Moore (1990), with permission from PennWell Corp.

Fig. 6-4. Amstutz curves for unstiffened liners (yield stress = 50,000 psi)
Source: Moore (1990), with permission from PennWell Corp.

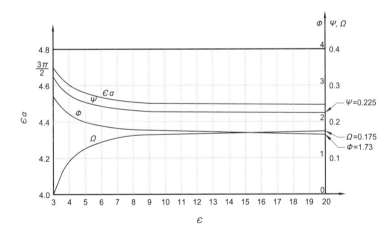

Note: At ϵ = 2, α = 180° (ϵ = 360°) and Φ and $\Psi \longrightarrow \infty$

ϵ	$\epsilon\alpha°$	$\alpha°$	$\dfrac{\tan}{\epsilon\alpha}$	$\tan\alpha$	$\epsilon\tan\alpha$	$\cos(\epsilon\alpha)$	$\sin(\epsilon\alpha)$	$\sin\alpha$	$\epsilon\alpha$	β [25]	γ [29]	∂ [35]	Φ [39]	Ψ [40]	Ω [48]
3	270°00'	90°00'	∞	∞	∞	0	−1.00000	−1.0000	4.71239	−2.6667	28.3	8.00	2.88	0.331	0
4	236°37', 2	65°54', 3	8.9446	2.2360	8.9440	−0.11112	−0.99381	0.91287	4.60104	−1.8095	32.7	16.67	2.21	0.271	0.100
5	261°11', 6	52°14', 3	6.4550	1.2910	6.4550	−0.15310	−0.98821	0.79056	4.55868	−1.3933	38.7	27.67	2.00	0.251	0.133
10	258°19', 17	25°50'	4.8409	0.48413	4.8413	−0.20231	−0.97932	0.43575	4.50868	−0.6650	71.4	119.03	1.78	0.226	0.168
20	257°40', 2	12°53'	4.5749	0.22873	4.5746	−0.21357	−0.97693	0.22297	4.49719	−0.3286	143.4	484.2	1.73	0.225	0.175

Fig. 6-5. Amstutz acceptable range for ϵ
Source: Moore (1990), with permission from PennWell Corp.

It is possible to use the Amstutz design method without making the noted simplifying assumptions in order to closely approximate the results of Jacobsen's procedure, but most engineers who have access to nonlinear analysis software simply choose to use Jacobsen's procedure.

6.2.2 Jacobsen Formulation

Jacobsen (1974) gives equations for determining critical buckling pressure for cylindrical pipe without stiffeners.

The three equations that must be solved simultaneously for the three unknowns α, β, and P_{cr} are

$$r/t = \sqrt{\left\{ \frac{\left[\left(9\pi^2/4\beta^2\right)-1\right]\left[\pi-\alpha+\beta\left(\sin\alpha/\sin\beta\right)^2\right]}{12\left(\sin\alpha/\sin\beta\right)^3\left[\alpha-\left(\pi\Delta/r\right)-\beta\left(\sin\alpha/\sin\beta\right)\left[1+\tan^2\left(\alpha-\beta\right)/4\right]\right]} \right\}} \tag{6-3}$$

$$P_{cr}/E^* = \frac{\left(9/4\right)\left(\pi/\beta\right)^2-1}{12\left(r/t\right)^3\left(\sin\alpha/\sin\beta\right)^3} \tag{6-4}$$

$$\sigma_y/E^* = \left(t/2r\right)\left[1-\left(\sin\beta/\sin\alpha\right)\right]+\left(P_{cr}r\sin\alpha/E^*t\sin\beta\right)\left[1+\frac{4\beta r\sin\alpha\tan\left(\alpha-\beta\right)}{\pi t\sin\beta}\right] \tag{6-5}$$

where α = one-half the angle subtended to the center of the cylindrical shell by the buckled lobe (radians);

β = one-half the angle subtended by the new mean radius through the half waves of the buckled lobe (radians);

P_{cr} = critical external buckling pressure (psi);

Δ/r = gap ratio, for gap between steel and concrete;

r = tunnel liner internal radius (in.);

σ_y = yield strength of liner (psi);

t = liner thickness (in.);

E^* = modified modulus of elasticity of steel liner (psi) = $E/(1-v^2)$;

v = Poisson's ratio for steel = 0.3;

E = modulus of elasticity of liner (psi); and

D = tunnel liner diameter (in.).

The calculated P_{cr} can then be divided by an appropriate factor of safety (typically 1.5, in accordance with Section 3.5.5) to calculate a maximum allowable external pressure (P_{all}).

Jacobsen's formulas have been reworked by Moore (1990) for ease of solution by computer. The critical pressure for a particular D/t ratio and gap ratio is obtained by solving the three simultaneous nonlinear equations using numerical methods.

Figs. 6-6 and 6-7 show curves for unstiffened liners for yield stresses of 38,000 psi and 50,000 psi, respectively, based on Jacobsen's equations. Within the practical range of D/t ratios from 70 to 280, the axial stress at buckling does not exceed 80% of the material's yield strength, so it is not nec-

essary to consider a reducing modulus of elasticity to replace E in the Jacobsen analysis. The axial stress at buckling is

$$\sigma_N = \frac{P_{cr}}{2} \times \frac{D}{t} \tag{6-6}$$

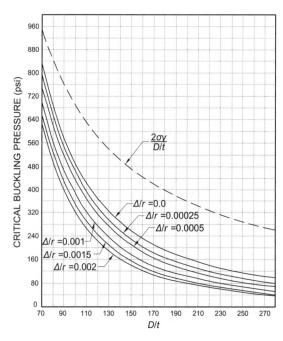

Fig. 6-6. Jacobsen curves for unstiffened liners (yield stress = 38,000 psi)

Source: Moore (1990), with permission from PennWell Corp.

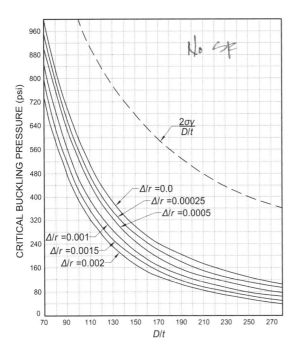

Fig. 6-7. Jacobsen curves for unstiffened liners (yield stress = 50,000 psi)
Source: Moore (1990), with permission from PennWell Corp.

The Jacobsen method gives results approximately 20% lower than the Amstutz method; however, because it does not require the consideration and application of limitations in its use, it is recommended that only the Jacobsen formulation be used for the analysis and design of cylindrical pipe to resist external buckling pressures.

6.2.3 Montel Formulation

As an alternative to the Amstutz and Jacobsen formulations, work by Montel (1960) provides a simplified formula for calculating the critical external pressure in rigidly confined cylinders. The main assumption is that the critical external pressure capacity can be estimated as the pressure that causes first yielding of the cylinder wall. Montel, using thin-ring deflection theory (Timoshenko and Gere 1961), as well as experimental results, developed a semiempirical formula for the buckling pressure of cylinders embedded in a stiff cavity.

The simplified equation is as follows

$$P_{cr} = \frac{14.1\sigma_y}{(D/t)^{1.5}[1+1.2(\delta_0 + 2\Delta)/t]} \tag{6-7}$$

where P_{cr} = critical external pressure (psi);

σ_y = yield strength of steel liner (psi);
D = tunnel liner diameter (in.);
t = liner thickness (in.);
δ_0 = amplitude of the initial localized out-of-round-ness (in.); and

Δ = gap between steel and concrete (in.).

valid for $60 < D/t < 340$; $36\text{ ksi} \le \sigma_y \le 72\text{ ksi}$; $t/10 < \delta_0 < t/2$; $\Delta/t < 0.25$; and $\Delta/D < 0.00125$.

Clearly, the critical external pressure P_{cr} of this equation is a decreasing function of both δ_0 and Δ. Furthermore, it can be readily shown that for D/t between 100 and 250, the value of P_{cr} is well below the plastic pressure p_y of the cylinder (i.e., the pressure that causes full plastification of the cylinder wall). This plastic pressure p_y can be readily calculated from the following equation:

$$p_y = 2\,\sigma_y\left(\frac{t}{D}\right) \tag{6-8}$$

where σ_y = yield strength of steel liner (psi);

D = tunnel liner diameter (in.); and
t = liner thickness (in.).

Figs. 6-8 and 6-9 show the ultimate pressure predicted by Montel's formula for steel cylinders with D/t values ranging from 70 to 275, with yield strengths of 38 ksi and 50 ksi, respectively. Comparison of those results with those in Figs. 6-3 through 6-7 from Amstutz and Jacobsen indicates that predictions from Montel's formula are accurate, despite their simplicity. Furthermore, the accuracy of Montel's formula has been confirmed through comparisons with finite element analysis, as described in the recent works of Vasilikis and Karamanos (2008 and 2009). Results from this finite element simulation are depicted in Figs. 6-10 and 6-11.

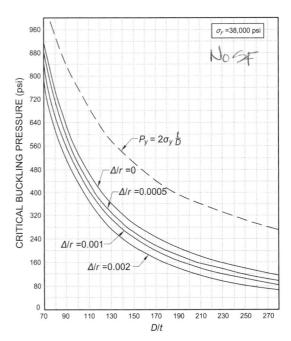

Fig. 6-8. Montel curves for unstiffened liners (yield stress = 38,000 psi)
Source: Figure courtesy of Spyros A. Karamanos; reproduced with permission.

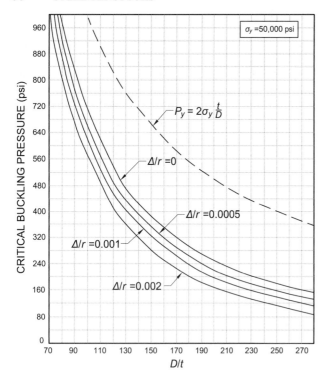

Fig. 6-9. Montel curves for unstiffened liners (yield stress = 50,000 psi)
Source: Figure courtesy of Spyros A. Karamanos; reproduced with permission.

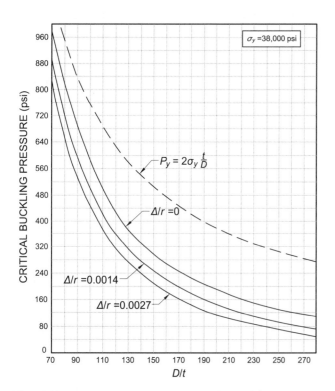

Fig. 6-10. FEA curves for unstiffened liners (yield stress = 38,000 psi)
Source: Figure courtesy of Spyros A. Karamanos; reproduced with permission.

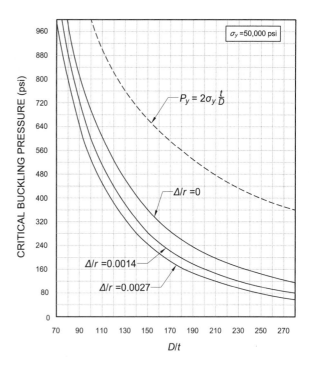

Fig. 6-11. FEA curves for unstiffened liners (yield stress = 50,000 psi)
Source: Figure courtesy of Spyros A. Karamanos; reproduced with permission.

6.3 DESIGN OF STEEL TUNNEL LINERS WITH STIFFENERS

6.3.1 Introduction

An engineer must consider using external circumferential ring stiffeners when the external pressure analysis indicates that the critical external pressure requires an unstiffened liner thickness greater than the thickness required for the internal design pressure. Economic considerations determine whether the liner shell thickness should be increased, circumferential ring stiffeners should be added, or a combination of increasing the shell thickness and adding ring stiffeners should be adopted.

Stiffener dimensions have been designed and dimensioned in accordance with several different procedures. They include the rotary symmetric formulations, the U.S. Bureau of Reclamation general yielding formulation, and the single lobe theory.

6.3.1.1 Rotary Symmetric Formulations   Rotary symmetric theories assume that the circumferential stiffener ring buckles into an even number of half waves around the circumference. Comparison of these formulations with those of Amstutz, for example, shows that the rotary symmetric theories give higher critical buckling pressures. Failures occur at external pressures below those predicted by the theories. Also, failures initiate in practice in one single lobe, as depicted in Fig. 6-2.

6.3.1.2 *General Yielding Formulation* The U.S. Bureau of Reclamation's procedure for the design and dimensioning of stiffeners has been used for relatively low external pressures. The collapse mechanism considers a nonembedded pipe shell. The stiffener ring is designed to carry the entire external load over the length (*L*) between rings. The minimum yield strength and the area of the external ring, plus the area of the effective flange, control the design. Bureau of Reclamation criteria treat the combined section stiffener ring similarly to a short column. The formulation was developed for relatively low external pressure considerations. It allows relatively wide stiffener spacing and is conservative.

6.3.1.3 *Single Lobe Formulation* As previously discussed, both Amstutz and Jacobsen have derived their formulations considering the formation of a single lobe and the confinement offered by the backfill concrete. In actual practice, stiffeners have been observed to fail by buckling in a single lobe. Buckling failure of a stiffener begins when the stress in the extreme, outside fiber of the external stiffener ring element reaches yield stress. This buckling occurs because of a combination of both axial compression and a compressive bending stress at the middle of the indent, as shown in Fig. 6-2.

6.3.1.4 *Shell Between Stiffeners* When ring stiffeners are provided at spaced intervals, the portion of the pipe between stiffeners can sustain the external pressure acting on it as long as the ring stiffeners maintain the circularity of the pipe shell at their selected locations. The pipe shell between the stiffeners can then be considered a free tube that can buckle inward, free of any constraint provided by the surrounding backfill concrete. The buckling resistance of the free tube between external ring stiffeners is treated in a separate analysis. When the external pressure is less than the critical buckling pressure of the shell, the total reactive force of the shell or free tube on the stiffener is only a small portion of the total external radial pressure on the shell between stiffeners. The equivalent force per unit length of the circumference is less than the acting external pressure multiplied by the carrying width of the pipe shell at the stiffener.

6.3.2 Design of Shell as Free Tube Between Stiffeners

The critical external pressure on the shell between stiffeners can be determined by the R. von Mises corrected instability formula for tubes loaded with radial pressure only Windenburg and Trilling (1934).

$$P_{cr} = \frac{E\left(t/r\right)}{1-v^2}\left[\frac{1-v^2}{\left(n^2-1\right)\left(\frac{n^2L^2}{\pi^2r^2}+1\right)^2}\right] + \frac{E\left(t/r\right)^3}{12\left(1-v^2\right)}\left(n^2-1+\frac{2n^2-1-v}{\frac{n^2L^2}{\pi^2r^2}-1}\right) \qquad (6-9)$$

where *r* = radius to neutral axis of shell in original formulation but for practical purposes can also be taken as the radius to outside of shell (in.);

 L = length of tube between stiffeners, i.e., center-to-center spacing of stiffeners (in.);

 t = thickness of shell (in.);

 P_{cr} = collapsing pressure (psi);

 E = modulus of elasticity (psi);

 v = Poisson's ratio; and

 n = number of lobes or waves in the complete circumference at collapse.

The calculated P_{cr} can then be divided by an appropriate factor of safety (typically 1.5, in accordance with Section 3.5.5) in order to calculate a maximum allowable external pressure (P_{all}).

The resulting graph for collapse of a free tube (from the von Mises instability formula) is illustrated in Fig. 6-12. This figure provides assistance in determining the collapse, or buckling, of a free tube. Similar formulas and graphic pre-

sentations to determine buckling of a free tube have been developed by Timoshenko (1936) and Flügge (1960).

n is the integer number for which P_{cr} is a minimum and is not an independent variable. *n* can be determined entirely by trial and error substitution; however, it is usually faster to estimate an initial *n* that is close to or equal to the final *n* by using available graphs and formulas. For most practical design cases, $6 \leq n \leq 12$. The following formula from Windenburg and Trilling (1934) can be used to estimate *n*:

$$n = \sqrt[4]{\frac{\left(\frac{3}{4}\right)\left(\pi^2\right)/4\left(1-v^2\right)^{1/2}}{\left(L/D\right)^2\left(t/D\right)}} \qquad (6-10)$$

where *v* = Poisson's ratio = 0.3;

 L = length of tube between stiffeners (in.);

 D = tunnel liner diameter (in.); and

 t = thickness of shell (in.).

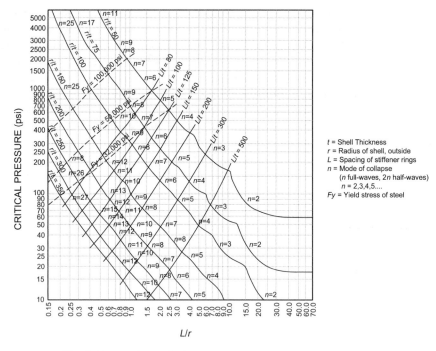

Fig. 6-12. Collapse of a free tube (von Mises)
Source: Moore (1990), with permission from PennWell Corp.

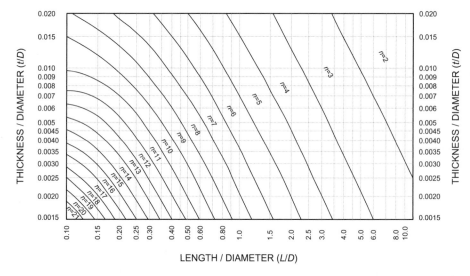

Fig. 6-13. Estimation of *n*
Source: Moore (1990), with permission from PennWell Corp.

For steel penstocks with $v = 0.3$ the formula reduces to

$$n = \sqrt[4]{\frac{7.061}{\left(L/D\right)^2\left(t/D\right)}} \qquad (6\text{-}11)$$

Fig. 6-13 can also be used to estimate *n*. When the value of *n* is calculated, it is rounded to the nearest whole integer.

The shell between stiffeners must be dimensioned to have a critical buckling pressure slightly in excess of the stiffeners. The same safety factor should be used for both the shell between stiffeners and for the stiffeners.

6.3.3 Design of Stiffeners

Circumferential ring stiffeners as analyzed by both Amstutz and Jacobsen provide the required circularity at the stiffener locations to prevent buckling of the shell. The Amstutz procedure for analyzing stiffeners is similar to the procedure he developed for unstiffened cylindrical shells. However, the value of ε at the critical external pressure is always determined to be well below the value of 3 at which 2α is greater than 180 degrees (Fig. 6-2) and Amstutz's equations are no longer valid (Fig. 6-5). The Jacobsen equations do not have

this limitation, and a reliable answer is obtained with $2\alpha > 180$ degrees. For this reason, it is recommended that the Jacobsen procedure be used to analyze and design stiffeners.

Jacobsen (1974) gives equations for determining critical buckling pressure on stiffeners. The three equations that must be solved simultaneously for the three unknowns α, β, and P_{cr} are

$$\frac{Fr_{NA}^2}{I_F} = \frac{\left[\left(9\pi^2/4\beta^2\right)-1\right]\left[\pi-\alpha+\beta\left(\sin\alpha/\sin\beta\right)^2\right]}{\left(\sin\alpha/\sin\beta\right)^3\left[\alpha-\left(\pi\Delta/r_{NA}\right)-\beta\left(\sin\alpha/\sin\beta\right)\left[1+\tan^2\left(\alpha-\beta\right)/4\right]\right]} \tag{6-12}$$

$$P_{cr}/E = \left[\left(9\pi^2/4\beta^2\right)-1\right]\left[\frac{\left(I_F\right)\sin^3\beta}{r_{NA}^3\sin^3\alpha}\right] \tag{6-13}$$

$$\sigma_y/E = \frac{h}{r_{NA}}\left(1-\frac{\sin\beta}{\sin\alpha}\right)+\frac{P_{cr}r_{NA}\sin\alpha}{EF\sin\beta}\left[1+\frac{8\beta hr_{NA}F\sin\alpha\tan\left(\alpha-\beta\right)}{12\pi I_F\sin\beta}\right] \tag{6-14}$$

where α = one-half the angle subtended to the center of the cylindrical shell by the buckled lobe (radians);

β = one-half the angle subtended by the new mean radius through the half waves of the buckled lobe (radians);

P_{cr} = critical radial external buckling pressure (psi);

I_F = moment of inertia of external stiffener and contributing portion of shell (in.) (Jacobsen 1974);

F = cross-sectional area of external circumferential ring stiffener and pipe shell between stiffener rings (in.) (Moore 1990);

h = distance from neutral axis of stiffener to outer edge of ring stiffener element (in.);

r_{NA} = radius to neutral axis of ring stiffener (in.);

σ_y = yield strength of liner and stiffener material, whichever is less (psi);

E = modulus of elasticity of liner and stiffener material (psi); and

Δ/r = gap ratio, i.e., gap–liner radius.

The contributing width of the shell is given by $1.57\sqrt{(rt)} + t_s$ where

r = radius of liner shell (in.);

t = thickness of liner shell (in.); and

t_s = thickness at base of external ring stiffener (in.).

Jacobsen's formulas can be reworked to obtain three simultaneous, nonlinear equations as functions of α, β, and

P_{cr}, which are set equal to zero for solution by numerical methods [see Moore (1990), which also presents a computer program for solving Jacobsen's equations for the critical buckling pressure on circumferential ring stiffeners].

The calculated P_{cr} can then be divided by an appropriate factor of safety (typically 1.5, in accordance with Section 3.5.5) to calculate a maximum allowable external pressure (P_{all}).

Some general rules or guidelines can be followed to help an engineer initiate selection of an appropriate stiffener configuration. Fig. 6-14 shows one such configuration. Sample guidelines are presented at the end of Moore (1990).

6.3.4 Design of Weld between Shell and External Ring Stiffener

The welds connecting the external stiffeners to the tunnel liner shell must be designed to carry the shear at the section. The following expression from Jacobsen (1983) is used to determine the maximum total shear (V) carried by the stiffener:

$$V = E\left(I_F\right)\left(\frac{\pi a}{l}\right)\left(\frac{1}{\rho^2}-\frac{\pi^2}{l^2}\right) \tag{6-15}$$

where V = maximum total shear (psi);

E = modulus of elasticity of liner and stiffener material (psi);

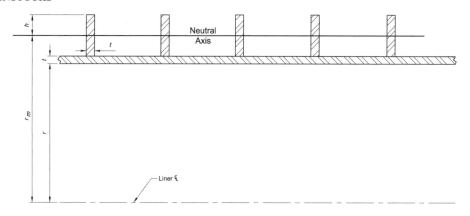

Fig. 6-14. Ring stiffener and thin-shell configuration

I_F = moment of inertia of external ring stiffener and contributing portion of shell (in.) (Jacobsen 1974);

$$a = \left(\frac{l}{\pi}\right)\tan(\alpha - \beta) \text{ (in.)};$$

α and β are determined by the computer program of Fig. 13 in Moore (1990);

r = internal radius of tunnel liner (in.);

$$\rho = r\frac{\sin\alpha}{\sin\beta} = \text{new mean radius of lobe (in.); and}$$

$l = \dfrac{\rho\beta}{1.5}$, where β is in radians and l is in inches.

The unit shear for designing the welds at the base of the external stiffeners is then determined from the expression

$$v = \frac{VQ}{\left(I_F\right)t_s} \tag{6-16}$$

where v = unit horizontal shear at the base of external circumferential ring stiffener where it is welded to the shell (psi/in.);

V = maximum total shear (psi);

Q = static moment of the area of external portion of stiffener alone about the axis of the total stiffener, which includes a contributing portion of shell (in.) (Jacobsen 1974);

I_F = moment of inertia of external ring stiffener and contributing portion of shell (in.) (Jacobsen 1974); and

t_s = width of external stiffener at connection to shell (in.).

It is recommended that the weld size adopted be capable of developing the full tensile strength of the connecting leg if angles or tee sections are used as external ring stiffeners.

6.4   TUNNEL LINER BENDS

Bends for tunnel liners have the same requirements as for exposed penstocks, as discussed in Section 9.12.

6.5   SEEPAGE RINGS

Seepage rings are used primarily to minimize seepage behind the steel liner section and secondarily to create a longer seepage path for the potential reduction of external pressure (Fig. 6-15).

Seepage rings usually are installed at or near the upstream end of the tunnel liner. When deemed appropriate by the engineer, they can be used at other locations, such as near the powerhouse where the penstock enters the plant, along with separate seepage drainage below the penstock.

Although the primary purpose of the rings is to minimize potential seepage, it is recommended that consideration be given to the ring acting in conjunction with the shell and that a ring–shell design be done. Allowance should be made for higher acceptable stresses up to 90% of yield, based on the liner design condition. Low or intermediate strength material can be used, unless compatibility of welding material with the liner is a requirement.

One, two, or three seepage rings may be used, depending on the engineer's preference. It is suggested that at least one ring be used for static head pressures up to 400 ft, two rings for pressures up to 800 ft, and three fo r pressures above 800 ft.

The radial depth of the stiffener ring must be about one-third the nominal thickness of the concrete encasing the steel liner, but not less than 4 in. or more than 8 in. Ring thickness depends on the ring–shell design analyses but should not be less than 3/8 in.

Seepage rings can be fabricated from plate sectors, bent structural angle sections, or bent structural tee sections. Fillet welds customarily are used to attach the ring to the shell. However, complete joint penetration welds should be considered for attaching rings formed from plate sections.

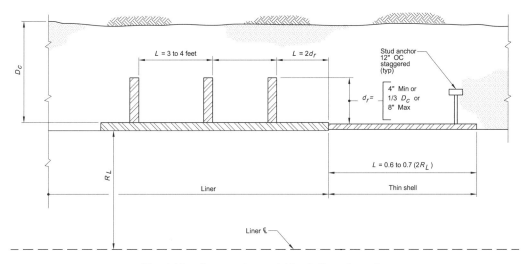

Fig. 6-15. Seepage ring and thin-shell configuration

Ring location and spacing may be dictated by the ring–shell analyses. As a rule of thumb, the most upstream ring is located approximately two times the radial height of the ring from the end of the steel liner. The second and subsequent rings are located downstream approximately three to four feet away from the first ring and each other.

Recent steel liner designs for tunnels consider the installation of a thin liner shell as the most upstream shell. This thin shell is welded to the liner and provides the end flexibility needed in the liner to prevent the creation of a leakage or pressure gap at the liner-to-concrete interface because of potential radial movement of the concrete liner caused by internal pressure. This shell, with an axial length approximately 0.60 to 0.70 times the nominal liner diameter, is designed to act as an impervious membrane to the concrete lining (just upstream of the steel liner). The thin shell is made of the same material as the steel liner and is attached to the liner with a butt weld. The thickness of the thin shell is a matter of the engineer's preference and can range from 3/8 in. to 1/2 in. Because it is thin, the shell must be anchored extensively to its encasing concrete using studs, hooked bars, U-bars, or spirals. This anchoring ensures against a buckling failure from external pressure and also ensures that the shell will perform as a thin, watertight membrane to the concrete lining as it expands radially outward because of the internal pressure. If a thin-shell concept is incorporated in the design, the seepage rings for the liner still should be located at the last liner section and not on the thin shell.

6.6 TUNNEL LINER JOINTS

6.6.1 Welded Joints

Welded joints for tunnel liners are subject to the same requirements as for exposed penstocks, as discussed in Chapter 11. In some instances, particularly where scheduling is tight, it may prove beneficial to wrap the penstock liner sections to reduce installation time by reducing the required number of field longitudinal welds from three to one. The method entails fabrication of a liner section with only one remaining longitudinal joint, then overlapping the joint and compressing the shell to a smaller diameter. Care is needed so as not to have the shell yield. The smaller diameter is retained by straps and allows the liner section to be installed in tunnels of almost the same diameter as the final shell diameter. Once installed, the straps are released and the shell springs back to the original intended diameter.

6.6.2 Butt Joint with Backing

Where clearance between the liner and tunnel wall is too restrictive to permit adequate working access for welders and nondestructive examination (NDE) technicians, the use of a backed-up weld detail is permissible (see Section 11.3.2).

6.6.3 Grout Connections

Most liners require connections or penetrations for rock consolidation grouting and for contact grouting between rock and concrete and between concrete and liner. These penetrations are sealed off after completion of grouting operations. The actual details used vary considerably (Figs. 6-16, 6-17, and 6-18). A 2 3/8-in. diameter hole in the penstock wall, however, is the maximum that can be considered as self reinforced (ASME 2010). After the grouting process is completed, the plug at the grouting port is typically seal-welded to provide a permanent leakproof seal.

6.7 TUNNEL LINER TRANSITIONS

Transitions for tunnel liners are subject to the same requirements as for exposed penstocks, as discussed in Chapter 4.

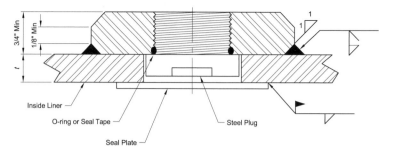

Fig. 6-16. Grout connection: threaded grout pad

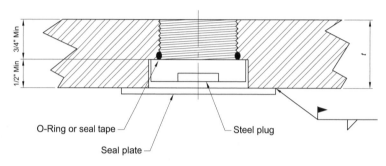

Fig. 6-17. Grout connection: direct threading of thick-walled pipe

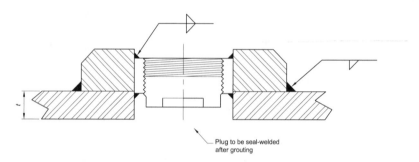

Fig. 6-18. Grout connection: half coupling grout pad

REFERENCES

Amstutz, E. (1970). "Buckling of pressure-shaft and tunnel linings." *International Water Power and Dam Construction,* 22(11), 391–399.

ASCE. (1989). "Tunnels and shafts." *Civil Engineering Guidelines for Planning and Designing Hydroelectric Developments*, vol. 2, Reston, VA.

ASME. (2010). "Boiler and pressure vessel code, section VIII, division 1: Rules for construction of pressure vessels." New York.

Flügge, W. (1960). *Stresses in shells,* Springer-Verlag, Berlin.

Jacobsen, S. (1974). "Buckling of circular rings and cylindrical tubes under external pressure." *International Water Power and Dam Construction,* 26(12), 400–407.

Jacobsen, S. (1983). "Steel linings for hydro tunnels." *International Water Power and Dam Construction*, 35(June), 23–25.

Montel, R. (1960). "Formule semi-empirique pour la détermination de la pression extérieure limite d'instabilité des conduits métalliques lisses noyées dans du béton." *La Houille Blanche,* 15(5), 560–568.

Moore, E. T. (1990). "Designing steel tunnel liners for hydro plants." *Hydro Review*, 9(October), 72–90.

Timoshenko, S. (1936). *Theory of elastic stability,* McGraw-Hill, New York.

Timoshenko, S., and Gere, J. M. (1961). *Theory of elastic stability,* McGraw-Hill, New York.

Vasilikis, D., and Karamanos, S. A. (2008). "Buckling of unconfined and confined thin-walled steel cylinders under external pressure." *Pipelines 2008—Pipeline Asset Management: Maximizing Performance of Our Pipeline Infrastructure,* Reston, VA.

Vasilikis, D., and Karamanos, S. A. (2009). "Stability of confined thin-walled steel cylinders under external pressure." *International J. Mechanical Sciences,* 51(1), 21–32.

Windenburg, D. F., and Trilling, C. (1934). "Collapse by instability of thin cylindrical shells under external pressure." *Transactions of ASME*, 56(11), 819–825.

CHAPTER 7

Wye Branches and Branch Outlets

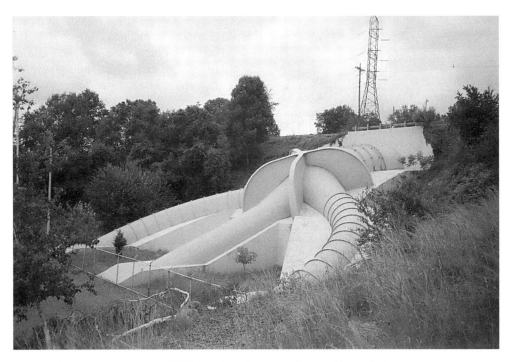

Source: Photograph courtesy of MWH Americas, Inc.; reproduced with permission.

Penstock works often involve branching the tunnel or penstock into two or more penstocks for multiple hydroelectric power turbines or discharge outlets. Such branches, commonly referred to as wye branches or bifurcations, require special design that is unique from project to project.

For underground facilities, wye branches are constructed of steel or reinforced concrete cast against a solid rock mass where the internal loading is transferred to the rock. For outdoor powerhouses with surface penstocks, wye branches usually are fabricated of steel and encased in a concrete anchor block to transfer the hydraulic thrusts to the surrounding foundation. Only steel wye branches are discussed in this section. When underground penstocks are steel lined, it is preferable to locate the steel wye branches so that they can be encased in concrete. Penstock shells are typically subject to internal static and dynamic water pressures. In certain conditions, the penstock may be subject to external water pressures or grouting pressures in the dewatered condition. Steel liners and branches where a portion of the design pressure is taken by rock embedment or concrete encasement are not covered in this manual. Only the design of steel wye branches is presented.

In general, pipe wyes or branches interrupt the hoop behavior of the pipe. Additional pipe reinforcement or stiffeners are typically required as additional support for the steel shell. Typically, the reinforcement should be sufficient to make the pipe equal in strength to the pipe without the opening. Wye branches must be suitably reinforced so that no substantial stress concentrations or deformations occur. In addition to structural strength, geometry considerations to achieve proper streamlining are required to reduce hydraulic losses for hydropower facilities.

Design methodologies vary in the design of wyes and branches. Before digital computers were common, wye and branch design was performed using simplified curved-beam procedures or design aids such as nomographs. Curved-beam procedures were incorporated by USBR (1986). In 1955, Swanson et al. (1955) published a procedure to design wyes and branches using nomographs. Current finite-element programs enable the engineer to design with more precision and detail. The simplified methods, design aids, and the finite element method are all used in the industry and are addressed in this manual.

7.1   LAYOUTS

In this manual, a branch outlet is defined as an outlet where the main penstock continues past the outlet pipe. When several branches occur in succession from the main penstock, the group of branches can be described as a manifold. When the main pipe terminates into smaller conduits, it is defined to be a wye. A wye usually branches the penstock into two (bifurcation) smaller conduits. Though less common, in some cases the penstock branches into three (trifurcation) smaller conduits. Several layouts for wyes and branch outlets and are presented in Figs. 7-1 and 7-2.

The symmetrical wye may be a single symmetric bifurcation or a series of bifurcating pipes in which the branch pipes are parallel to the direction of the main pipe. Generally, when the bifurcating pipe is the straight symmetrical wye type (Fig. 7-2), the internal angle between the two branching pipes should range between 60 and 90 degrees.

Nonsymmetrical wyes distribute several branch pipes in the same direction from the straight main pipe, as shown in Fig. 7-2. To reduce the head loss, it is advantageous to decrease the bifurcating angle. As the bifurcating angle decreases, more reinforcing material is required at the bifurcating point.

When pressures become high, external reinforcement may become restrictive and uneconomical. Other types of wyes that may be used in high-pressure conditions are spherical wyes and internally reinforced wyes.

7.1.1   Externally Reinforced Wyes and Branches

Figs. 7-1 and 7-2 present several methods of supporting wyes and branches. Because the hoop action of the pipes is interrupted by the openings, pipe reinforcement is required. Unsupported pipe areas are highlighted in the figures with shading, and these unsupported loads must be redistributed. Using simplified methods, the loads required to be redistributed are shown graphically.

Branch outlets of the type shown in Figs. 7-1(f) and 7-1(g) are not desirable when the diameter of the branch pipe is in excess of, say, three-fourths of the diameter of the

penstock, since the curvature of the reinforcement becomes very sharp. In such cases, the type of reinforcement shown in Fig. 7-1(e) is preferable. Branch outlets and wyes are usually designed so that the header and the branches are in the same plane.

Branch outlets as shown in Figs. 7-1(a), 7-1(b), and 7-1(c) may be reinforced with a simple curved plate designed to meet the requirements of ASME (2010). Branch outlets and wyes as shown in Figs. 7-1(d) to 7-1(g), inclusive, and in Figs. 7-2(b) to 7-2(f), inclusive, may be reinforced with one or more girders. The type and size of reinforcement required depends on the pressure, the extent of the unsupported areas, and clearance restrictions.

For branch outlets that intersect in a manner as shown in Figs. 7-1(d) and 7-1(e) and for wyes as shown in Fig. 7-2(f), three or four exterior horseshoe girders may be used, the ends of which are joined by welding. The welding of the ends of the horseshoes can sometimes be facilitated by use of a round bar at the junction to reduce fabrication difficulties and reduce stress concentrations. For trifurcations, a better practice is to join the two ring stiffeners by butt-welding and then butt-weld the two horseshoes to the ring. This practice avoids the possibility of the round bar tearing vertically from the top.

7.1.2   Spherical Wyes and Branches

One alternative for high-pressure penstocks is the spherical-type wye. A spherical header system is shown in Fig. 7-3. Circumferential reinforcing rings are required at the joining part of the branch and the spherical shell. The reinforcement of the opening must be designed under ASME (2010) in the section on requirements for large-diameter openings. Limits of reinforcement and compact reinforcement rules also must meet the same requirements.

A common variation of the spherical header incorporates conical and knuckle transitions between the main pipe and the spherical header, as well as similar transitions between the branch pipes and the spherical header. This configuration eases the discontinuity stresses at the openings in the spherical element by use of transition members. This shape generally carries most of the stresses in membrane tension and substantially reduces surface bending stresses. An example of this configuration is shown in Fig. 7-3. Finite element analysis (FEA) techniques are recommended to aid in the design of this type of configuration.

7.1.3   Internally Reinforced Wyes

One commonly used internally reinforced wye is the internal single-member splitter in the form of a crescent. The crescent must be shaped so that it is in tension only (Fig. 7-4). The centroid of the adjacent load must be coincident with the centroid of the crescent plate to justify the use of the tensile-stress-only concept.

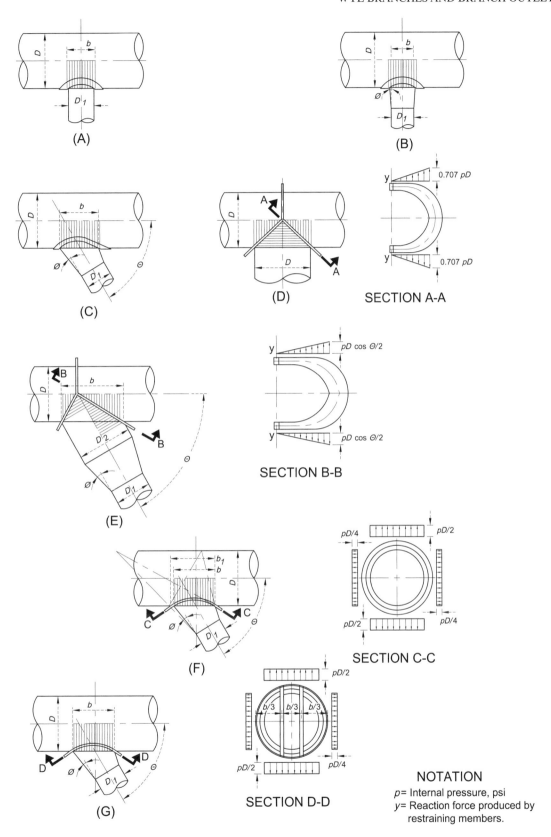

Fig. 7-1. Branch outlet reinforcement and loading diagrams
Source: Original figure appears in USBR (1986); reproduced courtesy of U.S. Bureau of Reclamation.

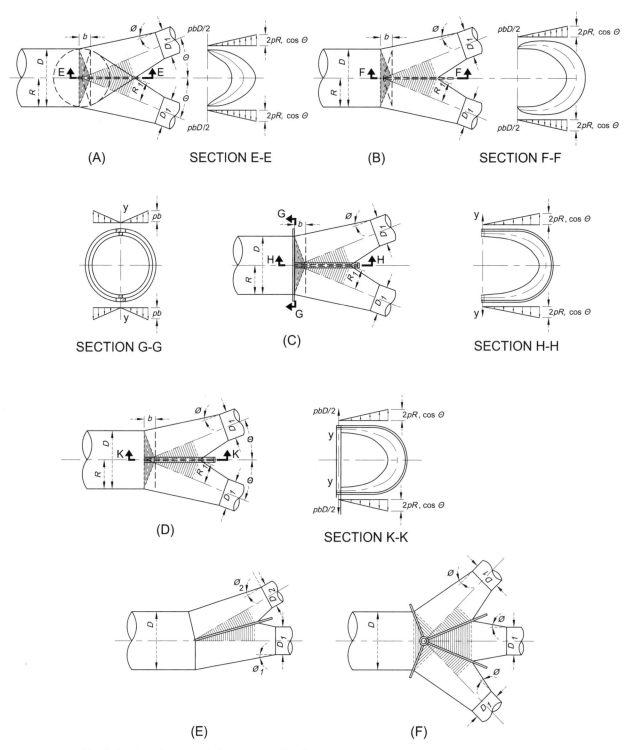

Fig. 7-2. Branch outlet reinforcement and loading diagrams
Source: Original figure appears in USBR (1986); reproduced courtesy of U.S. Bureau of Reclamation.

The inward projection of the splitter plate allows the engineer to minimize the eccentricity between the centroid of the load and the centroid of the reinforcing crescent plate. Ideally, this eccentricity will be reduced to zero, resulting in a tensile-stress-only condition in the splitter plate because the bending moment has been eliminated. In most cases, the eccentricity is not completely eliminated and some bending occurs. FEA techniques are recommended for this type of configuration.

In some cases, a combination of girders and tie rods may be used, but such arrangements are not recommended because of hydraulic issues. In particular, tie rods produce

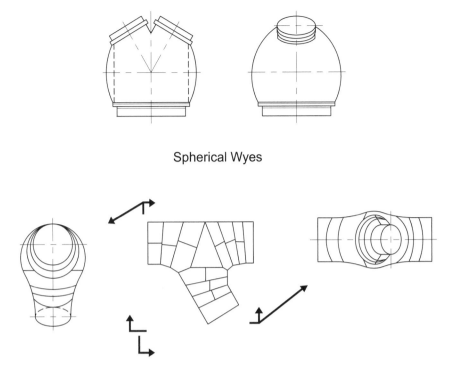

Spherical Wyes

Fig. 7-3. Spherical wyes with knuckle and conical transitions

vortex shedding downstream of the rod, which tends to induce vibrations.

7.2 HYDRAULIC CONSIDERATIONS

Wye branches must be designed for smooth hydraulic flow to avoid excessive head loss, vibration, and cavitation. They must be geometrically detailed to evenly proportion the flow distribution, eliminate acceleration or deceleration of flow in the adjoining branches, and thus minimize head loss.

For power-generating facilities, hydraulic efficiencies must be considered in the design and project economics. Head losses in the penstock, including losses in wyes and

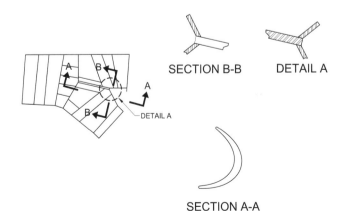

SECTION B-B   DETAIL A

SECTION A-A

Fig. 7-4. Internally reinforced wyes

branches, contribute to inefficiencies in the power generation system and may result in lost generating revenue or, in the case of pumped-storage projects, additional pumping costs. The additional cost of more complex wyes and branches with conical shapes may be more than offset by additional hydraulic efficiencies over the life of the project. Sizes and shapes of wyes and branches may also be limited by constructability issues. For example, wyes with a small deflection angle may be more efficient hydraulically but may be more difficult to fabricate. An example of a bifurcation with a small deflection angle is shown in Fig. 7-2(e).

Right-angle cylindrical branch or outlet tees shown in Figs. 7-1(a) and 7-1(d) are hydraulically inefficient and should be avoided whenever possible. The use of a frustum of a cone with convergence of 6 to 8 degrees, as shown in Fig. 7-1(b), reduces the branch loss to approximately one-third that of a cylindrical branch or outlet.

Branch losses may also be reduced by joining the branch pipe to the main pipe at an angle less than 90 degrees, as shown in Fig. 7-1(c). Ordinarily, branch outlet deflection angles vary from 30 to 75 degrees. The hydraulic efficiency increases as the deflection angle decreases. However, difficulties may be encountered in reinforcing branch outlets and wyes with deflection angles less than 45 degrees.

For wyes used in penstocks designed for low-velocity flow, an internal horseshoe girder, sometimes called a splitter, as shown in Figs. 7-2(a) and 7-4, may be used. Although the splitter is structurally efficient, it causes disturbances if the flow is not equally divided between the branches. To compensate for this undesirable flow constriction, the main pipe may be flared to increase the cross-sectional area in the region of the splitter. The use of conical sections in the main pipe and conical reducers extending to the branch pipes ensures minimal splitter plate design coupled with favorable flow and head-loss characteristics.

For a spherical wye branch system, flow-rectifying plates are provided inside the spherical shell interior to reduce the head loss resulting from the suddenly increased inner volume at the sphere part. The flow-rectifying plates are provided with sufficiently large water-pressure equalizing holes to transmit the impact of flowing water pressure to their exterior. The flow-rectifying plates should not be rigidly welded to the spherical shell because such welding would restrict the deformation of the sphere caused by water pressure. The head loss at the spherical wye depends upon the ratio of cross-sectional area of main and branch pipes, branching angle, ratio of flow distribution, flow, and other factors.

Hydraulic considerations must be included in the layout and design of power-generating facilities. Hydraulic design should consider the penstock layouts, including wyes and branches. Several different branches should be detailed, initial costs should be estimated, and the value of energy caused by head loss should be calculated. The wye branch with the lowest total cost should be evaluated in light of hydraulic adequacy. Physical or numerical hydraulic modeling may be performed as part of the overall design.

7.3 STRESS LIMITS

Branch and wye connections are complex structural systems. Often simplifying assumptions are made in the design. Exact stress analysis of these connections is at times difficult to perform. Even advanced FEA method results may be subject to modeling assumptions used. Allowable stresses used in design are dictated by the method used. Depending on approximations used in design, allowable stress levels must be adjusted with respect to the design method used.

7.3.1 Nomograph Method

The nomograph method is based on minimum yield strength of 30 ksi. In the procedure, the design pressure was kept to 1.5 times the working pressure to approximate an allowable stress of 20 ksi. Adjustments are required for different maximum stresses and allowable stresses based on steel type and grade used. Because of uncertainties in this procedure, the engineer may choose to use lower allowable stresses, as defined in Section 7.3.2.

7.3.2 Simplified Beam Method

It is prudent to design these connections for a lower allowable stress than for simple pipe shells. This practice is a result of the inability to determine critical stresses in complex structures with the same degree of accuracy as simpler penstock structures. Consequently allowable stresses are limited to

$$S_e \leq 0.5 f_y \text{ and}$$
$$S_e \leq 0.25 f_u$$

where S_e, f_y, and f_u are allowable, yield, and ultimate stress values, respectively.

7.3.3 *Finite Element Method*

For well-defined models consisting of plate and shell elements, the shell of the wye and the reinforcing members should be designed for general membrane stress levels equal to the lesser of the following to account for uncertainty in the action of the curved beams and stress concentrations:

$$S_e \leq 0.67 f_y \text{ and}$$
$$S_e \leq 0.33 f_u.$$

Surface stress allowables at points of geometric discontinuity are limited to 3 times the allowable membrane stress. Local membrane stresses may be 1.5 times the allowable membrane stress provided they attenuate within a distance of $\sqrt{rt}$ (where r is the inside radius of curvature of the wye at the point being analyzed and t is the thickness of the wye at the point being analyzed) to a value equal to 1.1 times the general membrane allowable stress.

7.4 NOMOGRAPH DESIGN METHOD

A design procedure using nomographs was presented by Swanson et al. (1955). This method has advantages, such as it provides a simplified procedure and minimum design time. The design is based on the experience of past test data compiled by Swanson. It proportions the size of the stiffener by estimating the maximum stress at the most critical point. Location of critical point is not provided.

The nomograph method is limited to branch outlets in which the main pipe continues past the outlet pipe as shown in Fig. 7-9. The nomograph method is not applicable to bifurcations or trifurcations. The nomograph method is also limited to stiffener plates with rectangular cross section only. More structurally efficient shapes, such as stiffeners with outer flanges, are not incorporated in the procedure. The depth of stiffener plate is limited to 30 times the thickness. Also, the depth of plate is restricted to be less than the diameter of the pipe.

Swanson et al.'s nomograph method (1955) is based on a minimum yield strength stress of 30 ksi. In the procedure, the design pressure was kept to 1.5 times the working pressure to approximate an allowable stress of 20 ksi. Adjustments are required for different maximum stresses and allowable stresses, based on steel type and grade used.

Because the nomograph method sizes the stiffener by estimating the maximum stress at a critical point, it cannot provide the stress distribution along the stiffener. Also, it cannot provide axial loads, moments, and shears along the stiffener supports. In addition, local stress intensities at stiffener–pipe shell connections and stresses in the pipe shell are not provided.

It is recommended that this method be used for initial design and that more detailed design methods, such as FEA methods, be used for final design.

7.4.1 *One- or Two-Plate Design*

The nomograph method, based on design working pressure plus surge allowance, includes a safety factor that keeps stresses well below the yield point of steel. The minimum yield strength of steel that is the basis of this nomograph is 30,000 psi. The design pressure used in the nomograph was kept to 1.5 times the working pressure to approximate an allowable stress of 20,000 psi. The procedure is as follows:

Step 1. Lay a straightedge across the nomograph (Fig. 7-5) through the appropriate points on the pipe diameter (see Step 2b) and internal-pressure scales; read off the depth of plate from its scale. This reading is the crotch depth for 1-in.-thick plate for a two-plate, 90°, and wye-branch pipe.

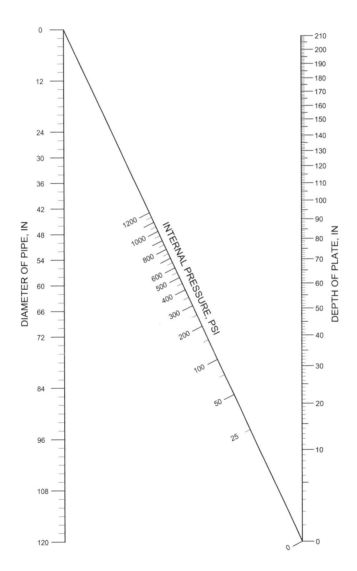

Plate thickness, 1 in, deflection angle, 90°.

Fig. 7-5. Nomograph for selecting reinforcement plate depths of equal-diameter pipe
Source: Swanson et al. (1955). Reprinted from *Journal AWWA,* by permission. ©1955 American Water Works Association.

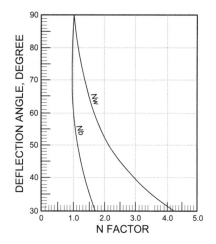

Fig. 7-6. *N*-factor curves
Source: Swanson et al. (1955). Reprinted from *Journal AWWA,* by permission. ©1955 American Water Works Association.

Step 2a. If the wye branch deflection angle is other than 90°, use the *N*-factor curve (Fig. 7-6) to get the factors which, when multiplied by the depth of plate found in Step 1, will give the wye depth d_w and the base depth d_b for the new wye branch.

Step 2b. If the wye branch has unequal-diameter pipe, the larger diameter pipe will have been used in Steps 1 and 2a, and these results should be multiplied by the *Q*-factors found on the single-plate stiffener curves (Fig. 7-7) to give d'_w and d'_b. These factors vary with the ratio of the radius of the small pipe to the radius of the large pipe.

Step 3. If the wye depth d_w found so far is greater than 30 times the thickness of the plate (1 in.), then d_w and

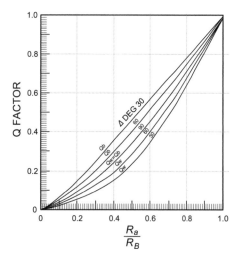

Fig. 7-7. *Q*-factor curves
Source: Swanson et al. (1955). Reprinted from *Journal AWWA,* by permission. ©1955 American Water Works Association.

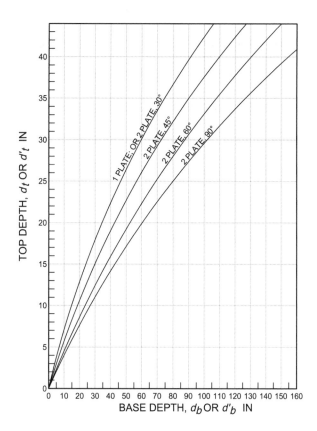

d'_t and d'_b are one-plate design dimensions: d_t and d_b are two-plate design dimensions.

Fig. 7-8. Selection of top depth
Source: Swanson et al. (1955). Reprinted from *Journal AWWA,* by permission. ©1955 American Water Works Association.

d_b should be converted to conform to a greater thickness, *t*, by use of the general equation

$$d = d_1 \left(\frac{t_1}{t} \right)^{\left(0.917 - \frac{\Delta}{360} \right)}$$

where d_1 = existing depth of plate;

t_1 = existing thickness of plate;
d = new depth of plate;
t = new thickness of plate selected; and
Δ = deflection angle of the wye branch.

Step 4. To find the top depth d_t or d'_t, use Fig. 7-8, in which d_t or d'_t is plotted against d_b or d'_b. This dimension gives the top and bottom depths of plate at 90° from the crotch depths.

Step 5. The interior curves follow the cut of the pipe, but the outside crotch radius in both crotches should equal d_t plus the radius of the pipe, or in the single-plate design, d_t plus the radius of the smaller pipe. Tangents connected between these curves complete the outer shape.

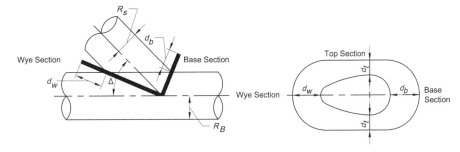

Fig. 7-9. Wye branch plan and layout
Source: Swanson et al. (1955). Reprinted from *Journal AWWA,* by permission. ©1955 American Water Works Association.

The important depths of the reinforcement plates, d_w, d_b, and d_t (Fig. 7-9), can be found from the nomograph. If a curved exterior is desired, a radius equal to the inside pipe radius plus d_t can be used, both for the outside curve of the wye section and for the outside curve of the base section.

7.4.2 *Three-Plate Design*

Section 7.4.1 covers the design of one- and two-plate wye branches without touching on a three-plate design because of its similarity to the two-plate design. The function of the third plate is to act like a clamp in holding down the deflection of the two main plates. In doing so, it accepts part of the stresses of the other plates and permits a smaller design. This decrease in the depths of the two main plates is small enough to make it practical by simply adding a third plate to a two-plate design. The additional plate should be considered a means of reducing the deflection at the junction of the plates. The two factors that dictate the use of a third plate are diameter of pipe and internal pressure. When the inside diameter is greater than 60 in. and the internal pressure is greater than 300 psi, a ring plate can be advantageous. If either of these factors is below the limit, the engineer should be allowed to choose a third plate.

If a third plate is desired as an addition to the two-plate design, its size should be dictated by the top depth, d_t. Because the other two plates are flush with the inside surface of the pipe, however, the shell plate thickness plus clearance should be subtracted from the top depth. This dimension should be constant throughout, and the plate should be placed at right angles to the axis of the pipe, giving it a half-ring shape. Its thickness should equal the smaller of the main plates.

7.5 CURVED-BEAM DESIGN METHOD

A penstock wye or branch connection usually has several stiffening beams to resist the loads applied by the shell of the pipe and at times may have internal tension members

called tie rods, although they are undesirable because of hydraulic considerations. Historically, the standard design procedure developed by USBR (1986) presented a systematic design procedure using numerical integration and solution of deflection and stress equations. The method incorporated rib shortening, shear deformation of the stiffener beams, and variable flange width. Although valid, this systematic design process is not efficient considering available modern computing methods. However, a spreadsheet can be used to drastically cut down the time involved with this design method.

In this manual, the USBR procedure and design assumptions are maintained, but computer beam and frame analysis methods are recommended for deflection and force computations.

To analyze the wye and branch connections using beams, many simplifications and approximations are used. The localized effect of structural discontinuities, restraints of the stiffening beams, foundation support, and dead load of the water-filled pipe are neglected. End-load effects and conicity of the outlet pipes are also neglected and considered to be small in comparison to the vertical load on the beams.

This analysis method provides an approximate design that offers an improvement over the nomograph method, particularly if the process can be automated using a spreadsheet. However, it should be noted that localized stress intensities at connections are not computed and shell stresses by the stiffeners are not determined. Because of these limitations, a reduction in allowable stresses, as discussed in Section 7.3, is recommended.

Structural analysis of a wye and branch consists of five general steps:

1. Determination of the stiffeners and, if applicable, tie rods resisting the unbalanced load;
2. Determination of the load on the resisting members;
3. Determination of the shell portion acting with stiffeners;
4. Analysis of the loaded structure; and
5. Interpretation of analysis results.

The beam design method has not been typically used in the design of internal stiffeners such as an internal sickle plate. It is recommended that more advanced FEA be used for internal stiffeners.

7.5.1 Determination of a Structural System Resisting an Unbalanced Load

The parts of the branch connection resisting the unbalanced pressure load consist of one or more of the following: crotch plates or C-clamps, stiffening beams, stiffening rings, or internal tie rods. A portion of the pipe shell adjacent to the stiffener is assumed to act integrally as an effective flange. Several stiffener layouts are presented in Figs. 7-1 and 7-2. Stiffener sizes are determined through an iterative process. Initial sizes are estimated, and geometry and size are adjusted until stress levels are satisfied. The strength of this design procedure with the aid of beam analysis programs is that geometry and beam sizes can be modified efficiently.

A bifurcation is presented in Fig. 7-10. The structural supports supporting the unbalanced loads consist of a crotch girder, a ring girder, and an internal tie rod. Tie rods are not desirable because of hydraulic considerations but are included in the example to illustrate design issues.

7.5.2 Determination of the Loads on Resisting Members

The reinforcement of a wye and branch should be proportioned to carry the unsupported loads, the areas of which are shown shaded in Figs. 7-1 and 7-2. Unsupported pipe areas

are highlighted in the figures with shading. The pressure loads on these unsupported loads are supported by stiffeners. The total load to be carried by the reinforcement is equal to the product of the internal pressure and the unsupported area projected to the plane of the fitting. The stiffener beams are assumed to carry the vertical component of the membrane girth stress resultant at the line of attachment of the shell to the stiffener. This load varies linearly from zero at the top centerline of the pipe to a maximum at the horizontal centerline of the pipe. Loads carried by the stiffeners and tie rods are shown graphically in Figs. 7-1 and 7-2.

For the bifurcation example, loads are developed using pressure, radii, and angles defined in Fig. 7-10. These loads are applied vertically on the stiffener shown in Fig. 7-11. Note that this method does not consider horizontal pressure load on the bifurcation.

7.5.3 Determination of the Shell Portion Acting with Stiffeners

The angle at which the shell intersects the beam is considered inconsequential because the flange effect is obtained by shear of the pipe walls. The shell acting with the flange may be considered oriented perpendicular to the stiffener web, as shown in Fig. 7-12.

An initial moment diagram may have to be assumed to determine the amount of shell acting with the stiffener, as shown in Fig. 7-12. The moment shape can be verified from the beam or frame analysis of the support system. From the shape of the moment diagram, an amount of the shell acting as an effective flange width may be approximated (USBR 1964). The moment diagram is divided into sections, each portion fitting a shape along the developed elastic axis, as shown in Fig. 7-11. From the moment diagram shape, an estimate of the effective flange width may be determined. The way in which the effective flange width is chosen is largely a matter of judgment and experience, but previous analyses show that some latitude may be tolerated in choosing an effective flange width without seriously affecting the final results (USBR 1964).

7.5.4 Analysis of the Loaded Structure

Once the layout of the stiffeners with initial beam depths and flange widths is defined, the mesh can be discretized into a series of beam elements. The centroid of the beam considering the effective flange width at each point is used to define the geometry. Total cross-sectional areas, shear areas, and moments of inertia of the beam elements, which include the effective flange widths, are determined for analysis.

Results must be adjusted considering the radius of curvature of the beam. For curves with large beam depth over radius ratios, the neutral axis diverges from the centroid axis, significantly affecting stress results. This effect is presented in Fig. 7-13. If the engineer models a curved beam with

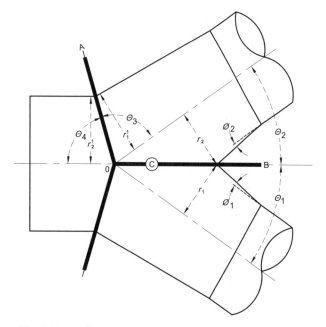

Fig. 7-10.   Bifurcation procedure: geometry
Source: Original figure appears in USBR (1964); reproduced courtesy of U.S. Bureau of Reclamation.

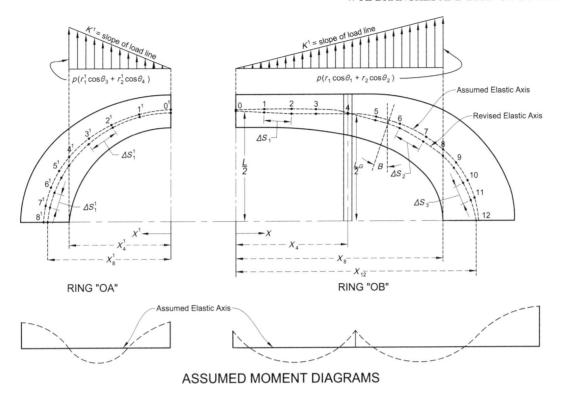

ASSUMED MOMENT DIAGRAMS

Fig. 7-11. Bifurcation procedure: loads and general behavior
Source: Original figure appears in USBR (1964); reproduced courtesy of U.S. Bureau of Reclamation.

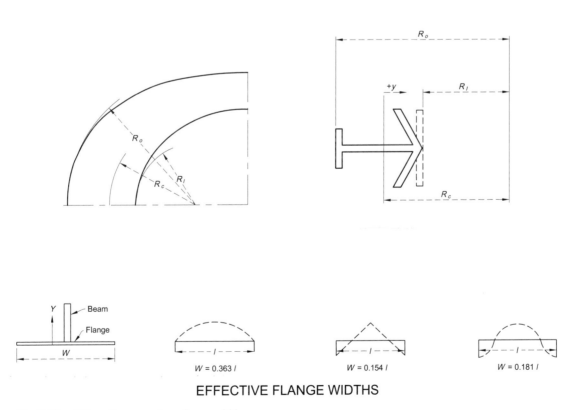

EFFECTIVE FLANGE WIDTHS

Fig. 7-12. Bifurcation procedure: flange widths
Source: Original figure appears in USBR (1964); reproduced courtesy of U.S. Bureau of Reclamation.

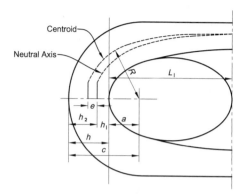

Fig. 7-13. Curved-beam effects
Source: Swanson et al. (1955). Reprinted from *Journal AWWA*, by permission. ©1955 American Water Works Association.

straight-beam segments approximating a curve, the output needs to be modified for curved-beam effects. Modeling the curvature with beam elements may not incorporate eccentricities in the stiffener at tight curves.

7.5.5 Interpretation of Analysis Results

For crotch plates or C-clamps with small radii of curvature, correction factors are required by curved-beam theory. Dimensions must be assumed initially, to be verified or adjusted in design. Axial, bending, and shear stress need to be investigated. Weld connection design is required for stiffener connection to the shell, between connecting stiffeners and, if applicable, in the outer flange to the web.

Interpretation of the stresses obtained in any structure is done by appraisal of the general acceptability of the assumptions made in the method of structural action, the applied loading, and the accuracy of the analysis. Allowable stress for this type of analysis is reduced because of the approximate nature of the methodology.

7.6 FINITE ELEMENT ANALYSIS DESIGN METHOD

The finite element analysis design method provides a more complete representation of the penstock shell–stiffener system. The nomograph method was developed from tests and provides an estimate of the maximum stress. The curved-beam method provides an improved design but uses many simplifications, as discussed in the previous section. These two design procedures have limitations on the types of wyes and stiffeners used for design. Design using FEA methods with proper modeling can determine load and stress distributions in the stiffeners and penstock shell. Loads are proportionally distributed to the shell and stiffeners based on the relative stiffness of the various members. The FEA method can also provide much more detailed stress information, deformations, and localized stress concentrations. The FEA method is becoming the standard design tool for penstock wye design in the hydropower industry.

7.6.1 Geometry and Modeling

Defining the geometry of a wye with stiffeners in 3-D space can be complex because each point in the mesh is located using *x*, *y*, and *z* coordinates. Polar coordinates may be used, but the definition of two or more branches with stiffeners would still be complex. In addition to the coordinates, the connectivity of the shell elements must be defined. Geometry definition can be developed with the aid of software such as AutoCAD. Also, some finite element programs, such as Abaqus and ANSYS, have highly developed mesh-generation capabilities. Solid models may also be generated using such programs as AutoCAD Inventor.

Typically, shell elements are used in FEA of wyes. Shells and plates need to model membrane (axial), in-plane, and out-of-plane bending behavior. Most FEA models are evaluated using linear elastic analysis. Loads are typically applied as internal pressure on the inside face of the shell elements. Modeling should be such that the entire pressure load is self-contained in the shell–stiffener system. External pressure could be modeled, but the engineer needs to be aware that buckling analysis is more complex and is not supported by linear-elastic methods.

Modeling proper boundary conditions in the FEA model is important to obtain realistic results. External boundary conditions should be applied in such a manner that they would not result in an artificial support or restraint of the wye. At times, a support is provided to avoid numerical instability of the FEA model. If such a restraint is provided, the designer needs to verify that the reaction on that support is zero. At the ends of the model, the pipes actually continue beyond the model along the penstock. To account for this phenomenon, the pipe ends must be modeled at a sufficient distance away from the pipe intersection, and the boundary condition at the ends should be such that the pipes can deform outward in the radial direction. Boundary conditions can also be used to reduce the model size. If the model is symmetric, half the FEA mesh can be generated as long as proper boundary conditions are applied to model symmetry. An example of an FEA half model is presented in Fig. 7-14.

It is recommended that FEA be performed with linear-elastic material modeling. In general, stiffened wyes should resist loads in the linear elastic range. In practice, it is possible that localized yielding may occur at joints. Nonlinear design is not recommended. Once model behavior is in the nonlinear range, there may be different possible path-dependent solutions, depending on which areas yield first. Also, load cycles would then have to be investigated. For example, if there is yielding under a pressure load and the pipe is dewatered and then repressurized, additional yielding potential would have to be reevaluated.

7.6.2 Model Validation

There is no design manual that provides specific steps in developing and analyzing an FEA wye branch model. FEA modeling can be considered an art based on the engineer's

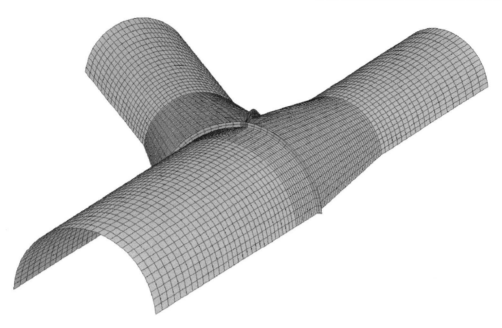

Fig. 7-14.   FEA half model
Source: MWH Americas, Inc.

experience. Different engineers or firms may not be consistent in the modeling approach. Good judgment is required in modeling. Any assumption in modeling needs to be justifiable and defensible. Results may vary, depending on the modeling scheme, fineness of the mesh chosen, element shape, and element distribution. A finer meshed model picks up more localized stress and stress concentrations.

QA/QC for FEA modeling and analysis can be difficult because results may vary according to modeling assumptions used. In general, the reviewer is required to be experienced in FEA analysis. The reviewer should review deformations in addition to stresses to verify proper model behavior.

7.7   ADDITIONAL CONSIDERATIONS

Wye branches should be of welded construction, tested by radiographic examination (RT) where possible, and stress relieved. When RT is not possible, ultrasonic examination (UT) should be performed. Postweld heat treatment may be required depending on the material properties and thickness, especially if the reinforcing members are thicker than $1\frac{1}{2}$ in. Before acceptance, the wyes should be hydrostatically tested in accordance with Chapter 14. See Section 14.3.2 for additional testing requirements for wye branches.

REFERENCES

ASME. (2010). "Boiler and pressure vessel code, section VIII, division 1: Rules for construction of pressure vessels." New York.

Swanson, H. S., Chapton, H. J., Wilkinson, W. J., King, C. L., and Nelson, E. D. (1955). "Design of wye branches for steel pipe." *Journal American Water Works Association,* 47, 6.

USBR. (1964). "Stress analysis of wye branches." *Engineering monograph no. 32*, Denver.

USBR. (1986). "Welded steel penstocks," *Engineering monograph no. 3*, Denver.

CHAPTER 8

Anchor Blocks

Source: Photograph courtesy of Richard Stutsman, RDS Consulting; reproduced with permission.

8.1 GENERAL

8.1.1 *Types of Penstock Support*

A penstock can be supported in a variety of ways, depending on the initial design selected, existing geological conditions, and penstock profile. The penstock can be totally buried, partially buried, or supported aboveground.

There are different advantages and disadvantages among buried, partially buried, and elevated steel penstocks, and different site-specific conditions, project criteria, and owner preferences determine which approach is most advantageous for a particular application.

8.1.1.1 Buried Steel Penstocks Advantages of buried steel penstocks include the following:

1. The continual support provides structural redundancy.

2. They typically allow easier access to the penstock right-of-way.

3. If burial can be integrated with the adjacent landscape, it can offer less environmental disruption of natural drainage routes.

4. They can be more aesthetic because they are less visually obtrusive.

5. They are less prone to vandalism.

6. They are better suited for flatter topographies accessible to heavy equipment and soil depths and material easily excavated.

7. Minimized thermal gradients can reduce the need and number of expansion joints.

8. Experience has shown that in colder areas, fully buried steel penstocks typically experience less long-term internal surface corrosion. By contrast, inevitable sea-

sonal thermal expansion and contraction of elevated penstocks typically causes the internal surface layers to disband from the base material, resulting in shell pitting and material loss if the internal coating is not periodically replaced.

Disadvantages of buried steel penstocks include the following:

1. The most significant is that the shell exterior is not readily accessible for future inspection or repair.
2. Burial can be difficult or impossible in steep topography, where the ground is rock, or where the soil overburden depth is shallow. Also, buried penstocks are not amenable for penstock routes that need to traverse sharply undulating landscapes with features such as streams and gullies.
3. Buried penstocks are more susceptible to stray ground currents that can induce shell corrosion.
4. Buried penstocks are not amenable to coupling-type field connections and typically require field-welded connections.
5. Exterior corrosion can be a problem in aggressive soil conditions.

Buried penstocks benefit from the addition of soil restraint, which may supplement or replace the need for anchor blocks. Refer to Chapter 5 and AWWA (2004) for further discussion on soil restraint design and methodology.

8.1.1.2 Partially Buried Steel Penstocks Partially buried steel penstocks are typically not considered advantageous because of their susceptibility to long-term exterior corrosion, particularly at the interface between the buried and exposed portion of the penstock, where corrosion may be more acute. Proven modern exterior corrosion protection systems, such as tape coating, deteriorate and become damaged if exposed long term. Also partial burial does not allow convenient access for the necessary periodic repainting of the entire steel shell exterior.

8.1.1.3 Aboveground Elevated Steel Penstocks Advantages of elevated aboveground penstocks include the following:

1. The use of ring girder supports makes the entire steel shell available for inspection and repair.
2. The use of ring girder supports allows for long spans between supports that can reduce the number of field supports.
3. Elevated penstocks can be the only practical method to traverse sharply varying and steep terrain, such as streams and gullies.
4. Aboveground penstocks can reduce the amount of site disturbance in sensitive environmental areas, such as traversing wetlands.
5. Elevated penstocks typically significantly reduce the amount and cost of transporting and placing fill, compared to buried penstocks.

6. Elevated penstocks are less susceptible to stray ground currents that can induce shell corrosion.
7. Coupling field connections that do not require field welding can be used on elevated penstocks.

One disadvantage of elevated aboveground penstocks is that they experience higher temperature variations, which require additional provisions, such as more frequent expansion couplings or thrust restraints.

See also the advantages of buried penstocks.

8.1.2 Locations and Purposes of Anchors

Anchors are installed at bends to resist hydrostatic pressure and hydrodynamic forces at longitudinally restrained bends, longitudinal forces from thermal changes where displacements cause problems, and transverse wind and seismic loads. Anchors also help by preventing shifts in the pipeline during installation. The judicious placement of anchors can control the natural frequency of the parts of the penstock and reduce the effects of vibration. One of the most important aspects of considerations for anchor block placement and design is whether the penstock circumferential field connections are able to transmit the shell's longitudinal reactions. For example, steel penstocks with field-welded circumferential joints can typically readily transmit these forces, resulting in longer lengths of pipe, which effectively offer resistance to these longitudinal forces. By contrast, unrestrained circumferential field joints, such as couplings or gasketed bell-and-spigot field connections, frequently require more anchor restraint for resisting longitudinal loads.

Generally, expansion joints in the penstock are installed to control the pipe's longitudinal expansion and contraction caused by temperature variations and to minimize shell longitudinal stresses. If no expansion joints are used, longitudinal forces can be high and must be considered in the pipe stress, anchor design, and effect on displacement-sensitive equipment, such as hydroelectric turbines.

When expansion joints are used, a maximum spacing of 500 ft between anchors in straight sections of pipe is normally used because of the magnitude of thermal elongation of the pipe and the need for fixed points during erection.

8.1.3 Geologic Conditions

Geological considerations and the remedies for adverse conditions are outlined in Section 1.7.

Movement of the foundation at anchor blocks is not desirable and can cause excessive penstock stress. Sound, stable rock provides the best foundation for anchor blocks, and if economically feasible, the penstock alignment should be located at or near surface rock during the initial layout phase. If sound rock cannot be found or reached for the foundations, and "softer" earth materials (e.g., weak or weathered rock and soil) comprise the foundation, settlements must be analyzed and appropriate provisions must be made in the penstock design. Provisions for soil improvement or piles may be necessary to attain suitable foundation conditions.

The surface and subsurface conditions at anchor block sites must be investigated thoroughly by a geotechnical engineer to verify that the natural foundation satisfies the design requirements or to determine that foundation improvements are necessary to obtain satisfactory anchor block installations. When dealing with weathered rock, the weathering profiles can be erratic and must be carefully evaluated. Misjudgments about the weathering profile can result in over- or underexcavations for the foundations and may necessitate redesign of the supports. The bottom of the foundations should be below the frost line. The geotechnical conditions must be clearly described and portrayed in the project documents to provide the installer with all available information about the foundation.

Suitable cutoff walls or collars should be provided at the penstock inlet to minimize water leakage along the exterior of the penstock. Leakage of acidic water can dissolve carbonate and sulfate rocks, and continuous seepage can cause surface and subsurface erosion of soil materials. This situation may require underdrains to intercept and divert leakage, slope protection to prevent erosion, and a regular inspection program to check for changing conditions.

8.1.4 Foundation Stability

For sliding analysis, the factors of safety shown below are the recommended minimums. The foundation material (soil, rock) resistance values used in this analysis should be those determined by a qualified geotechnical engineer to be appropriate allowances for design use for the specific site. Higher factors of safety are required when foundation conditions warrant or when limited subsurface investigation has been performed.

8.1.4.1 Sliding These are values for the minimum factor of safety against sliding for particular conditions under working (nonfactored) loads:

1. Static pressure plus momentum force for in-service conditions: 1.5;
2. Static pressure plus momentum force and wind for in-service conditions: 1.3;
3. Test pressure: 1.3;
4. Static plus transient pressures resulting from load rejection: 1.3; and
5. Seismic loading combined with static pressure plus momentum force: 1.2.

The resistance against sliding is a function of the sliding coefficient of friction times the vertical load, shear key if used, rock or soil anchors if used, prestressed rock or soil anchors if used, and passive soil resistance if available. A shear key increases the sliding resistance by the weight of soil or rock bounded between the top and bottom of the shear key and between the front of the shear key and front of the foundation times the coefficient of friction. The added sliding resistance of rock or soil anchors shall be based on field tests of the actual size and type of anchors to be used. Anchors are normally assigned a factor of safety of 4. The resistance of rock or soil anchors shall not be combined with passive pressure resistance. Prestressed rock or soil anchors increase the sliding resistance by the value of the prestressing force minus the estimated amount of relaxation times the coefficient of sliding friction. Sliding resistance of prestressed anchors shall not be combined with passive pressure resistance.

Geotechnical investigations must be made before finalizing the value of the sliding coefficient of friction. See Section 8.1.5 for typical friction values. These investigations also should be used in determining the required embedment for soil and rock anchors and values to be used for passive resistance.

When shear keys are used, considerable care must be taken during excavation to ensure that the rock is not weakened by fracturing the rock. Once final rock excavation is complete, the rock surface should be sealed with either a gunite cover or a concrete seal slab to prevent degradation of the exposed rock. Before sealing, loose material should be removed by either an air or water jet or by prying. Once a gunite or seal slab is used, though, the coefficient of sliding resistance will be the lesser of the friction values between concrete to rock and concrete to concrete.

If passive soil resistance is considered at the toe of the anchor, the engineer must recognize that movement of the soil is required to develop passive resistance. If no movement is allowed, such as with passive or prestressed anchors, then passive soil resistance cannot be developed.

8.1.4.2 Overturning Anchors must be designed so that the resultant of the vertical loads (including the weight of piers and earth covering the base) and the longitudinal forces intersect the base within its middle third.

The following are values for the minimum factor of safety against overturning for a particular condition:

1. Static pressure in-service conditions: 1.5;
2. Dynamic conditions, transient pressure resulting from load rejection: 1.3;
3. Seismic loading combined with static pressure or test pressure: 1.2; and
4. Extreme conditions, dam safety criteria earthquake: 1.05.

The resistance against overturning is a function of vertical loads, passive pressure if applicable, and rock anchors if used. The limitations on the use of passive resistance noted above in the sliding case also apply here. When rock anchors are used, their total contribution must not exceed one-half the contribution of the gravity loads. A mass that provides at least a factor of safety of 1.0 is required to control vibrations in the penstocks.

The bearing stresses should be calculated, and the bearing capacity of the foundation material must be verified by a geotechnical investigation (see Section 8.1.5).

8.1.4.3 Hydrostatic Uplift Uplift forces commonly are not encountered or are mitigated by drainage systems. If there is uplift (e.g., anchors under water or in soil acting as a semifluid), the combination of vertical loads and anchors must yield a safety factor greater than 1.2. The calculations must be based on the actual density of the material that is acting like a fluid. The factor of safety against buoyancy should be a minimum of 1.5. These forces also must be considered when checking sliding and overturning.

8.1.4.4 Movement at Anchors Normally a penstock system is designed to have no significant foundation movement at anchors. Primarily, movements in the penstock are allowed to occur at couplings, expansion joints, or in sections of continuous pipe that are adequately flexible to absorb the movement without developing excessive stresses. The potential for movements at or settlements of the anchors should be considered in the entire penstock assembly. If the magnitude of potential movements and settlements of the anchors cannot be determined or controlled, the anchors might be detailed with bolted connections to allow future penstock realignment.

8.1.4.5 Slope Stability In addition to designing anchors, it is essential that an overall slope stability analysis be performed on the supporting medium of the anchors. Proper consideration also must be given to slope protection. A washout or saturation of the slope caused by penstock leakage or natural drainage may cause the anchor to fail. Armoring of earth backfill with concrete slope protection, gunite, riprap, or slope paving may be desirable (see Section 1.7).

8.1.5   Foundation Strength

Table 8-1 gives the estimated foundation strengths of various materials. These values must be checked by testing before final design is completed and before consultation with a qualified geotechnical engineer.

8.1.6   Codes, Allowables, and Methods

Reinforced concrete supports must be designed in accordance with ACI (2008).

8.2   ANCHORS FOR PENSTOCKS

8.2.1   General Theory

Anchorages are major components of a penstock system and are usually necessary at penstocks that have characteristics such as the following:

1. They have alignment bends where the penstock has nonrigid shell girth shell joints that cannot transfer the bend's longitudinal reactions.
2. They provide vertical support for the dead weight of steeper penstocks with features such as couplings and expansion joints that cannot transfer axial reactions.
3. Penstock movement needs to be restrained, such as at the connection to a nonembedded turbine pressure case where penstock movement would alter turbine alignment.
4. The penstock needs to resist large lateral forces, such as seismic or wind loads.
5. There is not sufficient burial length or weight to fully restrain the movement of buried pipe to acceptable limits.

Typical concerns for a buried penstock are the interface with a rigid interface, where movement would induce unacceptable shell stresses, and at lateral bends adjacent to elevated sections, where movement could cause misalignment of elevated penstock supports.

The function and dimensions of any anchor depend on how adjacent, nonrigid girth joints are distributed along the penstock barrel. Such nonrigid shell girth joints include sleeve couplings, expansion joints, gasketed bell and spigot, or other gasketed joints. These joints divide the barrel shell into separate units that are watertight but structurally discontinuous.

Table 8-1.   Foundation Strength of Material for Preliminary Design

Material	Load capacity in bearing (tons/ft^2)	Shear capacity on keys (tons/ft^2)	Static friction coefficient "f", concrete on foundation material
Soft clay[a]	—	—	—
Hard clay[b]	1.5	—	0.33
Loose sand and gravel[b]	2.0	—	0.40
Confined gravel[b]	3.0	—	0.50
Soft rock	10.0	5.0	0.60
Hard rock	40.0	20.0	0.65

[a]A qualified geotechnical engineer should be consulted for strength parameters in soft clay. Values can approach zero for certain materials.

[b]A qualified geotechnical engineer should be consulted for shear capacity recommendations for shear keys in this material.

When water is in a penstock, the transverse components of internal pressure are simply contained by hoop tension. In contrast, axial components developed by pressure, temperature, and velocity become vectors, which are described by centerline alignment, the degree of fluid pressure, and cross-sectional area. Unbalanced axial vectors result wherever any of these three factors change, as at penstock bends or at reducer sections and line valves. These unbalanced vectors can be restrained within the vessel wall if the penstock barrel is made structurally continuous either by welded rigid girth joints or bolted flanged joints. When the penstock contains nonrigid joints, the penstock must have some means to resist the longitudinal force from changes in cross-sectional area, presence of line valves, or changes in alignment for each section between nonrigid joints. This resistance is usually provided by installing penstock anchors at or near these locations.

A typical design for a penstock installed above ground uses sleeve-type couplings or single expansion joints between anchor blocks. Both designs are shown in Fig. 8-1. Nonrigid joints effectively sever the penstock into free body segments extending between the two joints. A typical bend is anchored between two straight penstock legs, and the axial force vectors intersect at the point of intersection of the bend. Their vector resultant is the major anchor load.

An alternative aboveground design might use a penstock barrel that includes a greatly reduced number of nonrigid girth joints (less than one for each bend). This design has a cost advantage in reducing the number of fixed concrete

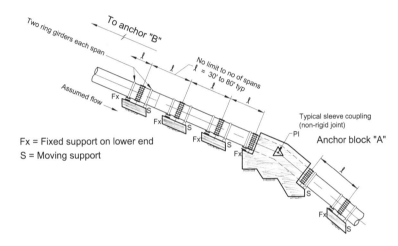

(A) Penstock with Sleeve Couplings (Profile)

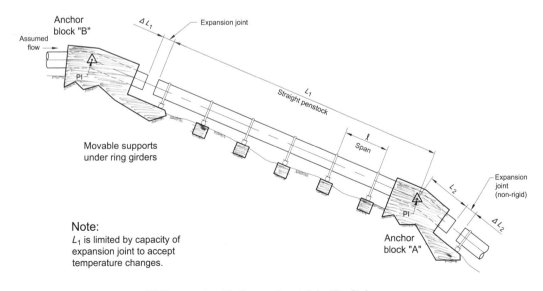

(B) Penstock with Expansion Joints (Profile)

Fig. 8-1. Anchor details of typical aboveground penstocks near anchors

anchors. Conceptually, this penstock could absorb thermal and other movements by allowing free flexure of the barrel between major anchorages. This is feasible for smaller diameters, and successful installations have been made. This theory is included in Bouchayer (1962). Situations are limited for which this alternative design may be practical.

For a buried penstock design, the field joints are often welded and do not require a specific anchorage. Nonrigid buried joints are commonly sealed by rubber O-rings or by sleeve-type couplings. Restraint of unbalanced vectors at nonrigid buried joints is equally as important as for aboveground situations. Earth friction and earth bearing capacity are used as major anchoring forces on a buried penstock. It is often possible to achieve the necessary anchorage by using rigid or harnessed joints for a predetermined distance each side of bends, reducers, or in-line valves. The distance is calculated as shown in AWWA (2004). The thrust design formula minus the cohesion and passive pressure terms is one of the better empirical formulas. If concrete is used, bend thrusts may be uniformly distributed from the curved barrel into the earth material.

Installation of a true expansion joint (in contrast to a bolted sleeve-type coupling) embedded in earth backfill is unlikely because of the reduced exposure to temperature variations and also because of the impracticality of allowing a pipe to slide longitudinally in the earth backfill.

These concepts are illustrated in Figs. 8-2 through 8-4. The points of application for pressure vectors are shown, along with the gravity and frictional loads acting on an anchor.

8.2.2 Loads Acting on Anchor Blocks

8.2.2.1 Loading Combinations Loads that act on anchor blocks are combined in a manner similar to the loading combination system assumed for penstock design (see Section 3.2). For each anchor design, the combinations listed below begin with the most constant and probable loadings. These combinations are then incremented by loads of lower probability of occurrence.

There are four groups of loading combinations for anchor block design that allow an efficient design system (Table 8-2). Group A is the most constant group. The combination of group A with group B is the basic design load for any anchor design. Additional combinations are obtained by separately adding loads from group C and D to the group A loads. Using the basic anchor design, the factor of safety under each combination of the second group of loads must be greater than the minimum specified in the design criteria. Otherwise, the anchor block must be revised to meet the criteria. Symbols (*DL*1, *DL*2, etc.) are taken from the nomenclature used in Section 3.2.

8.2.2.2 Components of the Loading Groups

1. Gravity loads
   These loads include the weights of the pipe, pipe contents, anchor block, and materials above the anchor block. If earth loads are used to assist anchor stability, the effects of erosion or future excavations must be considered. Live loading on an anchor is generally insignificant.

2. Internal pressure loads
   Axial pressure vectors are proportional to the array of static pressures described in Section 3.2. The basic design condition includes pump power failure in a pumping system or generator load rejection at a generating plant because these may occur as a normal operating condition.

   Intermediate and emergency transient pressures represent situations in which a load is applied to an anchor for a short duration but is cyclic during the surging time. This cycling may be significant for loading on soil as contrasted to a rock foundation. Hydrostatic tests during construction may cause major loading changes that do not occur during normal service. Another important distinction is that the penstock cross-sectional area must be based on outside shell diameter when calculating the axial load force at expansion joints or sleeve couplings because the pressure effectively acts on the ends on the shell. Reducing or expanding sections and line valves also develop substantial axial pressure vectors.

3. Momentum loading
   Momentum loading comes from a vector load in the direction of the upstream pipe toward the bend point applied at the point of intersection of the bend. The vector load can be calculated for each bend point on segmental fitting and can be applied accordingly, instead of a single vector load as applied to a mitered bend or curved bend.

4. Frictional loads
   These loads result from longitudinal penstock movements caused by temperature changes after original installation. Expansion joints and bolted sleeve-type couplings induce longitudinal forces by friction between sliding materials. Similar forces result from friction in the bearings under ring girders or saddle supports. All of these forces are almost constant and are unaffected by the degree of temperature changes.

5. Seismic loading
   Seismic loads are important because these forces could cause a disastrous anchorage failure. During an earthquake, the foundation is accelerated in a vibratory manner and penstock structures respond according to their inertia characteristics. An anchor block is rigid and generally responds in the same period of vibration as the foundation material. A penstock barrel is more flexible, and its seismic response applies loads to the anchor by beam action. See Section 1.6 for details on estimating seismic loadings.

6. Miscellaneous loadings

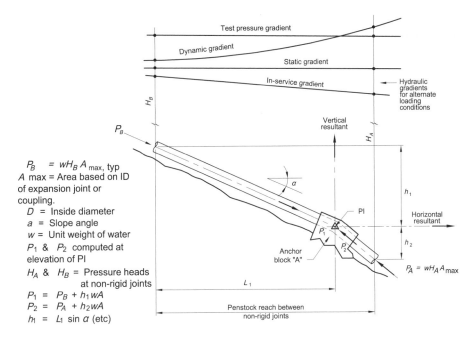

P_B = $wH_B A_{max}$, typ
A max = Area based on ID of expansion joint or coupling.
D = Inside diameter
a = Slope angle
w = Unit weight of water
P_1 & P_2 computed at elevation of PI
H_A & H_B = Pressure heads at non-rigid joints
P_1 = $P_B + h_1 wA$
P_2 = $P_A + h_2 wA$
h_1 = $L_1 \sin a$ (etc)

(A) Penstock Profile - Detail of Pressure Loads

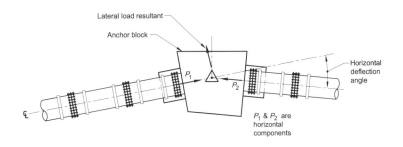

(B) Plan Showing Lateral Resultant

Fig. 8-2. Anchor block pressure forces

Other loadings include

a. Wind loads that act on the penstock and are ultimately passed into some anchorage;

b. Loadings caused by ice formation at sites subject to low temperatures, especially dead-load effects of icing should water leak out of the penstock; engineers also should consider the effects of ice formation at saddles or around other sliding supports;

c. Erection loads and effects of hydrotesting; these loads and effects require consideration and/or analysis at each anchor;

d. Buoyancy forces on buried anchor blocks;

e. Anchor loads during filling or draining; these loads are usually low; and

f. Thermal loading, which applies only to penstock barrels rigidly confined between

anchors. This condition must be carefully evaluated because large thermal strains may occur in the steel shell during filling or draining of the penstock because of the difference in temperature between water and steel.

8.2.3 Anchor Block Design

8.2.3.1 Design Approach   Anchorages for an aboveground penstock must accept, without significant movement or deflection, the combined effect of the three-dimensional combined loading applied to each bend. For a truly rigid anchorage, all the diverse loads must be effectively transferred through the concrete anchor block to the foundation. Ultimately, foundation resistance comes from the bearing and shearing strength of the earth material under the anchor block. The foundation material dictates the shape, size, and

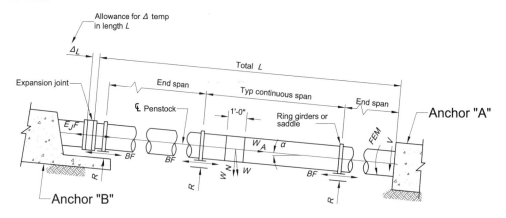

(A) Penstock with Expansion Joints

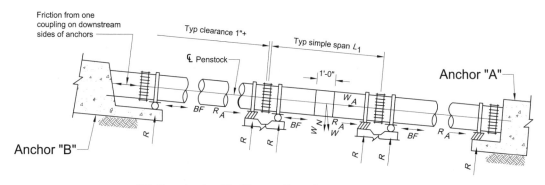

(B) Penstock with Sleeve Couplings

Notation:

(All loads and reactions shown are axial or normal to ⊄ of penstock.)

W = Total gravity weight of water and steel in 1 foot of penstock
W_N = Component of w normal to ⊄ of penstock (equals $W \cos \alpha$, etc)
W_A = Axial component of steel W only
        (Water component is absorbed into P_1 & P_2 at PI)
R = Load at each support
R_A = ΣW_A for one span
BF = Bearing friction
$E_J F$ = Expansion joint friction
FEM & V = Fixed loads on anchor

Notes:

1. BF and E, F do not depend on Δ temp.

2. Coupling friction assumed same as expansion joint. Apply in most adverse direction.

Fig. 8-3. Penstock loads affecting anchors

configuration of the anchor block. Therefore, the foundation material needs to be competent enough to resist the imposed reactions. Poor foundation materials, such as fractured rock and clays, may require remedial measures or alternate foundation designs.

For efficient anchor block design, load systems, including the dead load of the concrete, may be systematically analyzed by dividing the loads into three-dimensional vec-

tor components. Analysis is simplified if loads and gravity forces acting at an anchor are considered with one component in the vertical direction and the other two in fixed horizontal directions, such as the prime compass bearings. This procedure is well suited to spreadsheet analysis and tabulations by computer. If loads on an anchor are combined without including the weight of the concrete anchor, the vertical resultant gives an estimate of the minimal required dead

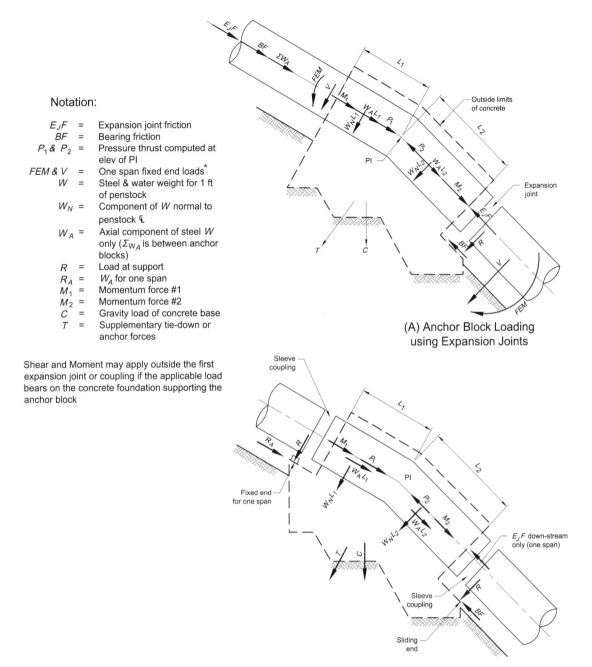

Notation:

$E_J F$	=	Expansion joint friction
BF	=	Bearing friction
P_1 & P_2	=	Pressure thrust computed at elev of PI
FEM & V	=	One span fixed end loads[*]
W	=	Steel & water weight for 1 ft of penstock
W_N	=	Component of W normal to penstock ℄
W_A	=	Axial component of steel W only (Σ_{W_A} is between anchor blocks)
R	=	Load at support
R_A	=	W_A for one span
M_1	=	Momentum force #1
M_2	=	Momentum force #2
C	=	Gravity load of concrete base
T	=	Supplementary tie-down or anchor forces

Shear and Moment may apply outside the first expansion joint or coupling if the applicable load bears on the concrete foundation supporting the anchor block

(A) Anchor Block Loading using Expansion Joints

(B) Anchor Block Loading using Sleeve Couplings

Fig. 8-4. Anchor block loading

load of the anchor to balance uplift forces. Similarly, the horizontal resultant estimates the minimal shearing forces applied to the foundation material. However, a major force remains because of the seismic loads on the mass of the anchor.

The magnitudes of specific loads applied to an anchorage show significant variations in accuracy. Even such parameters as the internal pressure value may be an approximation.

Conservative values must be used for the loads controlling the design to ensure adequate anchorage design.

8.2.3.1 Anchor Block Configuration Reinforced concrete is an efficient medium for transferring the anchorage load to the earth foundation. A common design configuration is complete encasement of the bend in concrete, but this is not absolute. An effective anchor design may use tie-down straps or rings, anchored in turn to a lower mass

Table 8-2. Anchor Block Loading Combinations

Type of load	Comments
Group A: Constant loading, unique at each anchor	
Gravity loads (DL1) (DL2)	Dead load components of steel, concrete, and water; Also fixed end beam moments and reactions from an adjacent, continuous penstock barrel
Friction loads (EJL) (SFL)	Sliding friction loads induced by temperature changes in the barrel; these originate at bearings and expansion joints or sleeve couplings
Other constraints (DL3) (DL4)	Earth backfill loading (backfill may be a constant load) Groundwater pressure or uplift (buoyancy)
Thermal loads (TL1)	Only if penstock is rigid between anchors
Static pressure loads (P_{N1})	Axial pressure vectors determined by maximum static water pressure in the penstock at each bend point of intersection
Group B: Variable loads, added to group A	
Transient pressure (P_{N2})	Positive transient pressure determined by hydraulic operating criteria
Momentum load	Caused by the velocity of water
Group C: Special variable loads, added singly to groups A and B without the transient pressures, and as applicable	
Seismic loads (EQ1)	Design basis criteria earthquake for the penstock location
Wind loads (LL1)	Normally only on an empty penstock
Snow loads (LL2)	Only at special locations (includes icing and transient pressure)
Construction loads (PT1)	From hydrotesting and any similar static loading
Group D: Loads from extreme pressure or seismic conditions, added to groups A and B without the normal transient pressure	
(P_{EM1})	Water hammer from gate closure in $2l/a$ s where l is length of transient segment (ft) and a is the celerity of the pipe material (ft/s)
(P_{EX2})	Water hammer from most adverse gate closure

of concrete. This design has the advantage of reducing the mass weight of the concrete under seismic loading. Weight and dimensions of an anchor block should hold the resultant vector of all foundation loadings to a position that does not overstress the earth foundation materials. Rocking is not allowed, and tie-down tendons may help by pretensioning an anchor block onto a foundation. With tie-down tendons or anchorages, adequate corrosion protection measures must be implemented to ensure the long-term behavior. Drilled piles or caissons have been used to provide the shearing and bearing resistance at a weak foundation. These additional components are effective in increasing the lateral resistance of a foundation under seismic loading and also in lowering the mass center of anchor blocks to an elevation at which overturning effects are more manageable.

Seismic loading produces the greatest lateral loads at the base of the anchor, at the contact with the foundation. If the sliding friction resistance of the earth under the concrete block is inadequate for such a loading, then other devices, such as shear keys or anchor tendons, may be used. These supplemental devices, which increase the lateral resistance of soil foundations, often are needed because resisting lateral forces by bearing against soil must allow for large relative displacements needed to mobilize passive soil forces, a practice that is hardly acceptable for an anchor design. Lateral soil pressures used to counter sliding loads should not exceed the value of the at-rest pressure of the soil; only the value of the active lateral pressure of the soil should be used.

Whenever a penstock pipe is completely encased at an anchor, sufficient reinforcing steel must be used to resist concrete shrinkage as the concrete cures. The reinforcement should also be capable of resisting the radial expansion of the pipe when pressurized.

For concrete encasement design, the reinforcement for radial pressures must consider the full range of pressures anticipated in service. The addition of extra reinforcing to control cracking is advised. Hydrotesting also may create high-stress conditions, which could promote cracking; checks should determine if the radial strain of the shell could crack the encasement. A minimal action is wrapping of the final 6 in. of pipe at the encasement edge with resilient material to limit this edge cracking at the transition zone of the pipe and anchor concrete.

REFERENCES

ACI. (2008). "Building code requirements for structural concrete and commentary." *ACI 318-08,* Farmington Hills, MI.

AWWA. (2004). "Steel water pipe: A guide for design and installation," 4th Ed. *Manual M11,* Denver.

Bouchayer, R. (1962). "Stresses on anchor blocks," *Transactions of the American Society of Civil Engineers*, Paper no. 3390, 127, 546.

CHAPTER 9

Appurtenances, Bends, and Transitions

Source: Photograph courtesy of National Welding Corp.; reproduced with permission.

Appurtenances are defined as accessory objects or apparatus to a pipeline. Examples of typical appurtenances found in a pipeline are expansion joints, couplings, manholes, valves, meters, nozzle outlets, and blowoffs. Often, expansion joints and couplings also receive consideration as pipe joints and may be shown in other sections of this manual.

9.1 EXPANSION JOINTS

9.1.1 Mechanical Expansion Joints

Mechanical expansion joints frequently are installed in exposed penstocks between anchor blocks to accommodate longitudinal movements caused by thermal conditions. Using a conventional design, the joints can accommodate up to 10

in. of total movement. Mechanical expansion joints can be installed in vaults but cannot effectively be installed in buried penstocks. Fig. 9-1 shows a typical mechanical expansion joint that is fully extended. AWWA (2007c) should be consulted for the design of mechanical expansion joints.

The mechanical expansion joint consists of a body, slip pipe, packing chamber, and end rings, with proper bolting to compress the rubber gaskets and lubricating rings to withstand internal design pressure. If mechanical expansion joints are installed in series without anchor blocks, limit rods are required to restrict the movement of each expansion joint. Expansion joints can be manufactured for any diameter pipe. If required, flanged ends can be supplied to assist with the installation of the expansion joints into the penstock. Special mounting rings also can be implemented into the design to

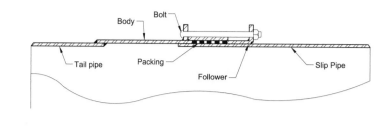

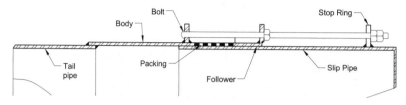

Fig. 9-1. Typical mechanical expansion joints

facilitate insertion into older penstocks that may have pipe ends of irregular dimensions.

These special mounting rings can be manufactured either to mate a special bolting flange or as solid rings welded to the ends of the mechanical expansion joint (to facilitate mating up to riveted or out-of-round older penstock sections).

The outer surface of the slip pipe that comes into contact with the packing chamber must be protected. Stainless-clad mild steel or stainless steel plate is used to prevent the joint from binding up. The material thickness for component parts of mechanical expansion joints is determined by the penstock operating pressure and diameter. The thickness of the slip pipe must be sufficient to resist loading forces introduced when packing is compressed. The body, slip pipe, and end ring components must be sized and tested where practical by cold expansion. When cold expansion is impractical and the diameter of the expansion joint is beyond the scope of available tooling, nondestructive examination (NDE) of the butt-joint welds is required. Because mechanical expansion joints do not effectively absorb any deflection or shear displacement, the adjacent pipe ends must be supported. Mechanical expansion joints must be installed according to the manufacturer's installation recommendations.

The longitudinal friction force between the slip pipe and body on a mechanical expansion joint is approximately 500 lb per linear foot of circumference. This force cannot be determined exactly because of the considerable number of variables, such as wall thickness, bolt loading, internal pressure, and imperfect nonmachined surfaces.

The sliding distance (d) of the slip pipe caused by temperature change is given by the following equation:

$$d = L \, \alpha \, T \tag{9-1}$$

where L = length of pipeline;

   α = coefficient of linear expansion ($6.5 \times 10^{-6}/°F$); and

   T = temperature change (°F).

9.1.2 Split-Sleeve Coupling Expansion Joints

Proprietary split-sleeve coupling expansion joints have been used on lower pressure (up to 150 psi), pipelines with thicknesses up to ⅜ in. and diameters up to 192 in. and in penstocks with expansion and contraction total movement of 4 in. or less. The advantage of using this type of expansion joint is that there is only one field joint to install, similar to using a coupling. The split-sleeve coupling expansion joint may be supplied in one or two pieces to facilitate the installation on existing or larger diameter penstocks.

The existing pipe ends on both sides of the joint require preparation for the use of split-sleeve coupling expansion joints. One pipe end is clad with 12-gauge stainless steel to provide for a slip surface for the polytetrafluoroethylene shoulder and O-ring gasket of the coupling to slide on. The other pipe end should be prepared with two round bar restraint rings welded to the pipe surface to fix the coupling expansion joint in place. Fig. 9-2(a) shows a typical split-sleeve coupling expansion joint.

Split-sleeve coupling expansion joints consist of a double arched center sleeve profile to contain two O-ring rubber gaskets. Each side of the coupling body has a shoulder attached. The fixed side uses a steel shoulder fully welded to the body, and the expansion side uses a polytetrafluoroethylene shoulder riveted to the body. Closure plates and fastening hardware close the coupling around the pipe ends. It is

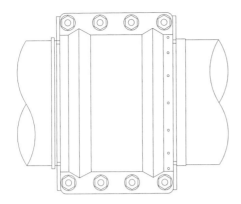

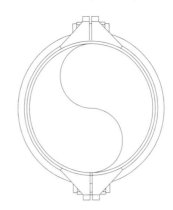

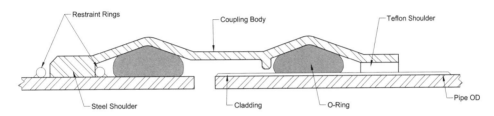

Fig. 9-2(a).   Split-sleeve expansion joint

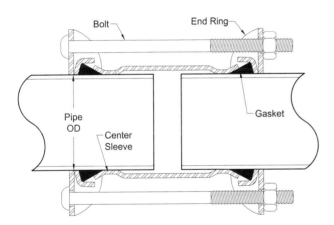

Fig. 9-2(b).   Bolted sleeve-type coupling

best to consult the coupling expansion joint manufacturer for detailed installation instructions.

Use Eq. 9-1 to calculate the total movement between fixed points, and only use one split-sleeve coupling expansion joint between any two fixed points.

9.2   SLEEVE COUPLINGS

9.2.1   Bolted Sleeve-Type Couplings

The bolted sleeve-type coupling (BSTC) consists of a center sleeve, two end rings, two wedge-shaped elastomeric sealing rings, and bolting hardware to compress the end rings against the gaskets to form a leak proof seal between the penstock shell and center sleeve [Fig. 9-2(b)]. The center sleeve may be of the same material and thickness as the penstock.

AWWA (2011a) should be consulted for the design, performance and manufacture of bolted sleeve-type couplings.

Fig. 9-3 shows various types of bolted sleeve-type couplings. Fig. 9-3(a) depicts the standard BSTC that is commonly used to connect similar size pipe ends. Fig. 9-3(b) is a transition coupling for minor pipe size differences and contains one normal-size wedge-shape gasket and one larger wedge-shape gasket. Fig. 9-3(c) depicts a reducing coupling where the center sleeve is fabricated to fit the two different-sized pipe ends. Reducing couplings are subject to a longi-

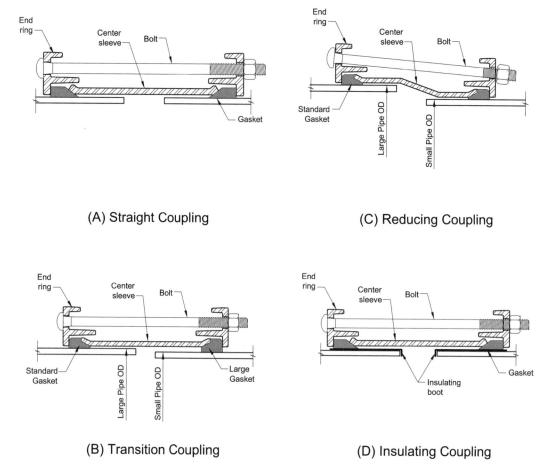

Fig. 9-3.   Types of bolted sleeve-type couplings

tudinal hydrostatic force (as they act similar to reducers), which causes them to slide unless they are restrained by external means. Fig. 9-3(d) is used on lower pressure (up to 150 psi) penstocks to isolate two sections of the penstock from each other.

BSTCs may be buried in earth; however, a hydrotest of the completed penstock must be performed before backfilling so that any leakage can be corrected by tightening in accordance with the manufacturer's installation instructions.

Where BSTCs are buried, proper corrosion protection must be considered to protect the coupling components and the gasket sealing area on the pipe ends.

When designing penstocks where BSTCs are required, consult AWWA (2011a) for the recommended gap between pipe ends, allowable laying deflections and other parameters.

Bolted sleeve-type couplings allow approximately 3/8 in. of longitudinal displacement of the pipe ends per joint and allow deflection as shown in Fig. 9-4 within the diameter and sleeve-length limits given in AWWA (2011a). Laying deflection is indicated by φ. The deflection angle must be equally split between both pipe ends (i.e., φ/2 on each end).

Fig. 9-5 illustrates deflections with couplings. Bolted sleeve-type couplings can accommodate some differential settlement only when two or more couplings are installed in series between properly supported pipe sections [Fig. 9-5(c)].

It is common to have the case of a penstock entering an underground structure, as shown in Fig. 9-5(c). Often the penstock may settle more than the structure and joint flexibility outside of the structure is needed. To prevent excessive shear on a single coupling outside of a structure, it is common to use two couplings. The following is a design example that shows how to design with BSTCs to provide this flexibility.

Design an 84-in. outside diameter (D_y) penstock for 2 in. of expected settlement outside of a structure. Use the performance limits for bolted sleeve-type couplings given in AWWA (2011a). Therefore, in accordance with Table 3 of AWWA (2011a), the maximum allowable deflection (φ) is 1 degree, but the maximum amount of in-service movement is 3/8 in.

1. Pipe gap at 1 degree for 84-in. pipe is: $D_y \tan \phi = 84$ (0.01745) = 1.47 in. However, this amount exceeds the limit for in-service movement of ⅜ in.
2. Find allowable joint deflection for the maximum in-service movement limit of ⅜ in.: $\tan^{-1}(\phi) = 0.375/84$, therefore, φ = 0.26 degrees.

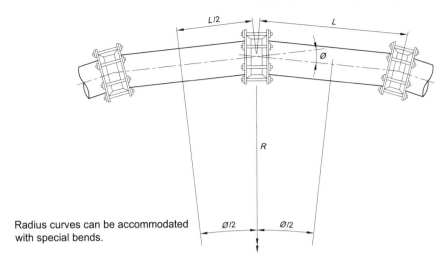

Radius curves can be accommodated with special bends.

Fig. 9-4. Bolted sleeve-type coupling laying deflection

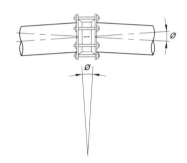

(a) Pipe Deflection at a single Coupling

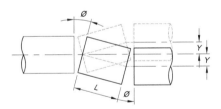

(b) Vertical/Lateral Deflection with two Couplings

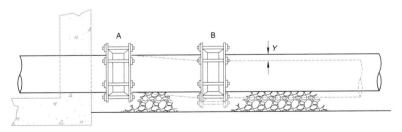

Ground Settlement Displacement 'Y' can be accommodated using two couplings 'A' and 'B'

(c) Typical Structure Connection with two Couplings

Fig. 9-5. Deflections with couplings

3. With the potential settlement, S, equal to 2 in., find the length of pipe, L, required between the two couplings: $L = S/\sin \phi = 2/(0.00454) = 440.7$ in., or approximately 37 ft of pipe.

9.2.1.1 Bolting Formulas and Design   Selection of the proper bolted sleeve-type coupling depends on the line pressure, temperature, and other service conditions. The pressure or design rating of the pipeline is one of the most important considerations in selecting the proper coupling. The pressure rating of a coupling is determined by the sleeve thickness, the minimum yield strength of the material used in its manufacture, and the gasket pressure developed by the tension on the bolts.

Eq. (4-1) is used to calculate the body thickness of the BSTCs. Gasket pressure is determined by an empirical formula and may be assumed to be at least 1.5 times the line pressure. The basic formula is

$$G_p = \frac{N_b \times 5,000}{G_a} \qquad (9\text{-}2)$$

where G_p = gasket pressure (psi);

N_b = number of bolts; and
G_a = gasket area (in.^{2}).

Five thousand lb is the average tension developed by a 5/8-in. unified coarse thread (UNC) bolt tightened to 70 ft-lb of torque. The 5/8-in. bolt is the standard size used by most manufacturers. Where bolts other than the 5/8-in. standard bolt are required, obtain the proper stress values from the BSTC manufacturer. The gasket area is the area at the heel of the gasket for the pipe size involved.

To assist in the design of sleeve thickness and bolt size and number for a particular internal design pressure, see the design examples for mechanical expansion joints and BSTC in Chapter 18.

A minimum number of bolts is necessary to provide uniform gasket pressure. The minimum number of bolts is determined by the equation

$$N_{b(\min)} = \left(D_y\right)(0.4384) \qquad (9\text{-}3)$$

where $N_{b(\min)}$ = minimum number of bolts and

D_y = pipe outside diameter (in.).

If BSTCs are used on steep slopes, the engineer should consider installing pipe stops in the middle of the center sleeve. Pipe stops prevent the coupling from creeping downstream as the penstock sections move with thermal changes. One disadvantage with pipe stops is that they make installation and removal of an adjacent section of penstock difficult because the complete coupling cannot be moved onto one of the pipe ends. In actual practice, pipe stops rarely are necessary.

For steep-slope installations, the penstock pipe sections should be installed from the bottom of the slope upward. This method prevents the pipe sections from creeping downhill as they are installed. As the pipe is installed upward, the gap between the pipe ends can be held constant by the use of temporary wood, rubber, or steel wedges. Where practical, BSTCs should be installed loosely on the corresponding pipe ends before transport to the job site. By doing this step, final installation is simplified, and any out-of-roundness of the pipe ends or coupling parts can be corrected under factory or shop conditions. See Chapter 13 for other installation requirements.

9.2.1.2 Segmented Bolted Sleeve-Type Couplings
Where diameter, transportation, and erection considerations do not permit one-piece BSTCs, the engineer should consider the use of segmented BSTCs (Fig. 9-6). Segmented BSTCs may be manufactured for any diameter of penstock. Sleeve thickness is determined by the hoop stress formula shown in Chapter 4 (Eq. 4-1). The bolting formula (Eq. 9-3) gives the number of bolts required to properly compress the elastomeric sealing rings to provide a leakproof design.

Center sleeves and end rings for segmented BSTCs are manufactured in three equal 120-degree segments. These segments are match marked with ends beveled for welding in the field. Struts or ties must be provided to hold the segments to the designed arc during transportation to the job site.

At the erection site, the segments are positioned for welding into a single sleeve. The segmented BSTC manufacturer must furnish the information for the proper welding procedure, including any fit-up drawings required to assist the installer. Welds on the field-welded, completed single-piece center sleeves or follower rings must undergo nondestructive examination (NDE).

After the completion of all NDE of the welds, the sleeve is positioned over the penstock ends. The procedures for installing BSTC gaskets must be followed. The three-piece followers (or bolting segments serving as followers and bolting segments) are then positioned against the gaskets. Complete the installation by following the instructions furnished by the BSTC manufacturer.

9.2.2 Bolted, Split-Sleeve Couplings

The following sections discuss the use of split-sleeve couplings as alternatives to BSTCs, both for flexible joints or restrained joints. The double arch cross section yields greater section modulus, providing for greater joint stiffness when compared to an equal thickness flat section profile. For installation guidance or instructions regarding this style of couplings, contact the split-sleeve coupling manufacturer.

9.2.2.1 Split-Sleeve Flexible (Nonrestrained) Couplings   Split-sleeve couplings can be designed to allow for pipe out-of-roundness, joint deflection, and minor differences in pipe outside diameter. A typical one-piece split-sleeve flexible coupling is shown in Fig. 9-7. The material

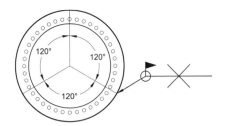

Typical Segmented Piece

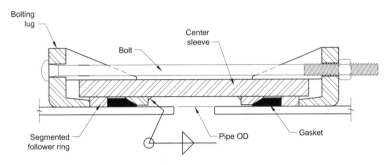

Fig. 9-6. Segmented bolted sleeve-type coupling

quality and performance characteristics of split-sleeve flexible couplings should conform to the requirements of AWWA (2011b). Eq. (4-1) is used to calculate the body thickness of the split sleeve coupling.

The closure plates that contain the fastening hardware can be oriented in any plane to ease the installation of split-sleeve flexible couplings. Split-sleeve flexible couplings have similar uses as BSTCs and can be used when laying penstocks on curves, as shown in Fig. 9-8.

Other uses of split-sleeve couplings include instances of dynamic movement as when the O-ring rubber sealing gasket is able to slide along the surface of the pipe end, assuming that the pipe end is smooth, to allow for in-service movement capability. See AWWA (2011b) for limitations on

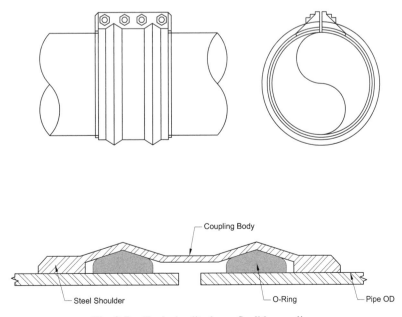

Fig. 9-7. Typical split-sleeve flexible coupling

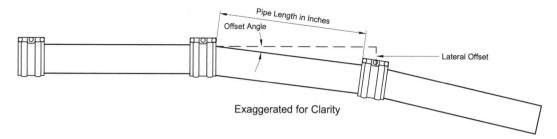

Fig. 9-8. Split-sleeve couplings for joint angular deflection

movement. For large-diameter pipe, split-sleeve couplings are available in two or more segments to help ease shipping, handling, and installation. Similar to a BSTC, when thrust exists across the joint, the pipe ends must be restrained from moving out of the coupling by some external means. Common methods of restraint are harnesses or fixed pipe supports designed to take the force of the pressure thrust (AWWA 2004). As an additional alternative, split-sleeve couplings may be designed to resist the thrust as shown in the next section.

9.2.2.2 Split-Sleeve Flexible (Restrained) Couplings
Split-sleeve restrained couplings are designed to resist the full thrust across the pipe joint caused by internal pressure. The body thickness is typically designed at a stress level of 2/3 of that used in the design of the body thickness of a split-sleeve flexible nonrestrained coupling or a BSTC. The design of the weld of the round bar to the pipe end is based

on a flare–bevel weld attachment in accordance with AWS (2010). This weld must be designed for the full bulkhead pressure thrust condition. A typical two-piece split-sleeve restrained coupling is shown in Fig. 9-9.

Similar to split-sleeve flexible couplings, the closure plates that contain the fastening hardware may be oriented in any plane to ease the installation of split-sleeve restrained couplings. Installation and service conditions must achieve uniform bearing against the round bar restraint rings.

9.3   MANHOLES AND SMALLER OPENINGS

Manholes provide access to the penstock shell interior by personnel and equipment. During construction, manholes allow access by the installer; after installation, they permit maintenance inspections, relining, or removal of rocks and sediment.

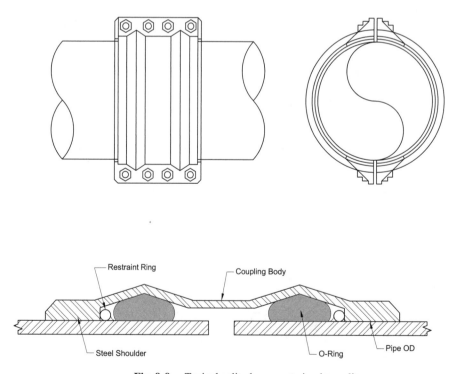

Fig. 9-9. Typical split-sleeve restrained coupling

Manholes allow inside air circulation. Circulation is vital for safe working conditions in conformance with regulations of the Occupational Safety and Health Administration (OSHA). Manholes placed at local crests on the pipe profile facilitate the use of hoisting equipment needed for painters' scaffolding and other maintenance operations. For adequate penstock maintenance, manholes must be located adjacent to line valves and should be near air-release and/or vacuum valves. Fig. 9-10 shows recommended manhole positions.

The dimensions of openings or manholes should be compatible with future maintenance needs; minimum dimensions of components of ladders, carts, hoses, hoists, or scaffolding; and to the expected volume of sandblasting grit or other debris removed through the opening. Small openings that accommodate the optimum lengths of welders' cables or other hoses are common outlets. Reductions in clearance caused by cement-mortar lining thickness also must be considered. However, OSHA requirements might supersede any of these requirements.

The typical diameter for a manhole opening is 24 in. (Fig. 9-11). Manholes or other openings must be installed 90 degrees to the main penstock shell and must have the shortest possible shaft length.

A penstock shell must be reinforced at most openings by collar or saddlelike plates welded around the opening. The controlling design procedure for this reinforcement is described in Paragraphs UG 36 and 37 of ASME (2010). Fig. 9-12 shows a typical but generalized basic design according to ASME (2010). The maximum size of opening that may be reinforced in this manner depends on the ratio of the diameter of the opening to the diameter of the main barrel, as shown in Table 9-1.

Openings with diameters larger than those given in Table 9-1 require special reinforcing of the penstock. Opening ratios up to $0.6D$ may be accommodated under Appendix 1-7 of ASME (2010), but over this limit, special designs described in Chapter 7 of this manual must be used.

Smaller penstocks also are affected by the criteria of Table 9-1. For example, the smallest diameter for practical personnel access through a penstock manhole is approximately 18 in. Table 9-1 shows that 36-in. pipe is the smallest diameter penstock that can accept an 18-in. typical manhole. Access to a penstock barrel less than 36 in. in diameter is best provided by installation of short, removable spool sections between two sleeve couplings.

ASME (2010) requires collar reinforcement for other openings smaller than manholes. An exception usually is made for openings smaller than 2 3/8 in. where the controlling factor passes from diameter to shell thickness.

Fairing plates prevent disruption of high-velocity flow in a penstock if disruption is objectionable. A design that holds the curved plate in position and yet allows rapid plate removal without major damage to protective lining material is required.

Steel pipe flanges and gaskets must conform to the requirements given in Section 9.8.

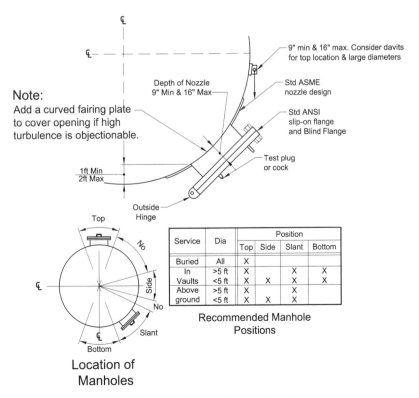

Service	Dia	Position			
		Top	Side	Slant	Bottom
Buried	All	X			
In Vaults	>5 ft	X		X	X
	<5 ft	X	X	X	X
Above ground	>5 ft	X		X	
	<5 ft	X	X	X	

Recommended Manhole Positions

Location of Manholes

Fig. 9-10.   Manhole locations and positions

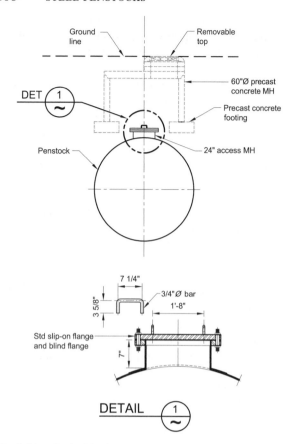

Fig. 9-11. Typical 24-in. access manhole for buried penstocks

9.4   STANDPIPES AND AIR VALVES

To ensure proper operation and protection, standpipes and air valves are required on penstocks for the following purposes:

1. They must admit air at the proper rate to accommodate the closure sequence of valves or gates and prevent collapse of the penstock; this design criterion is ordinarily used for sizing the diameter of standpipes and the large orifice of air valves. These standpipes and valves must be located wherever the hydraulic gradient may fall low enough to cause collapsing pressure in the penstock.

2. They must admit air in the event of an uncontrolled drainage for release of water from the penstock; this procedure involves uncertainties that must be evaluated for each penstock, but is not normally used as a criterion for sizing.

3. They must release air during watering-up of the penstock; this release requires placement of standpipes and air valves at high points on the profile or at other locations where air may collect during filling. Filling rates must be set based on the size of the standpipe or air valve.

4. They must release small amounts of accumulated air that come out of solution during normal operation; this

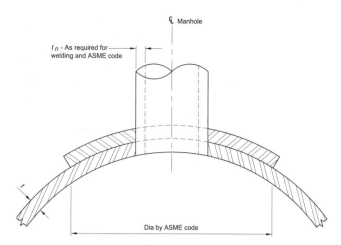

Fig. 9-12. Manhole reinforcement (limits of saddle-type reinforcement)

procedure requires the release of air accumulated at high points and at abrupt changes in pipe slope.

Adequate drainage and erosion protection must be provided at standpipe and air valve installations to accommodate accidental spillage during filling or normal operation.

9.4.1   Standpipes

Each standpipe provides a nonmechanical vent that allows the most rapid air flow into the penstock and a path for air release. A typical standpipe attachment to a penstock is shown in Fig. 9-13.

Standpipes that vent above ground must be protected from freezing if water stands in the pipe. Standpipes may be provided with bubblers or heaters or may be placed in a heated valve house for protection from snow and ice. Standpipes must be provided with hoods, appropriate grillwork, or screens to keep out debris. A tall standpipe is subject to large bending moments from wind and seismic forces

Table 9-1.   Manhole Maximum Allowable Opening Diameter

Inside diameter (D) of main penstock (in.)	Maximum diameter of opening allowed for basic design manhole (in.)
To 40	1/2 D
41–60	20
61–120	1/3 D
Over 120	40

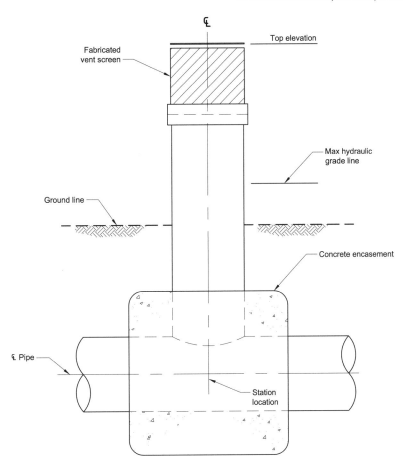

Fig. 9-13.   Typical standpipe

and must be reviewed for resonance by the methods used for steel stacks. Normal design for seismic forces considers the water at normal operating level. Allowable stresses with water at maximum level are given in Chapter 3.

9.4.2 Air Valves

Air valves provide a mechanical vent that allows smaller air flows in and out of a penstock. The three basic types of air valves are (1) air release, (2) air–vacuum, and (3) combination air-release–vacuum. Air-release valves have either large orifices designed for discharging air during filling or small orifices designed for releasing accumulated air during normal operation. The location of air valves is important to their operation, and they must be placed properly to provide the desired service.

Air–vacuum and combination air-release–vacuum valves must be sized to allow the release of air without pressurizing the penstock. The filling rate during watering-up must be considered. Manufacturers of air valves should be consulted for more specific recommendations on sizing.

As with standpipes, air valves must be protected from freezing. They are best installed below the frost line in valve houses or underground vaults that are properly ventilated and drained. Ventilation must be designed to prevent the vault or enclosure from becoming pressurized during the discharge of air and to allow air inflow during negative pressure situations.

The sizing of air valves depends on the operation of the penstock and the location, function, and individual characteristics of the air valve. Valve manufacturers should be consulted. Additional information can be found in AWWA (2001), AWWA (2007d), Parmakian (1950), and Lescovich (1972).

Air-valve body and trim materials and coatings must be selected to provide proper corrosion resistance to the water transported in the penstock and the environmental conditions in which the air valves are installed.

A minimum of two air valves must be used at any location to increase reliability. An isolating valve must be placed below the air valve to permit inspection and maintenance.

A typical air valve installation for a buried penstock is shown in Fig. 9-14.

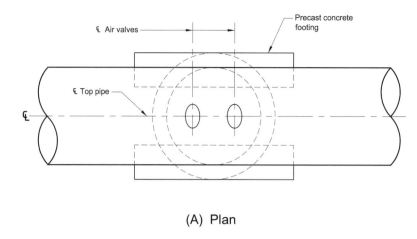

(A) Plan

(B) Elevation

Fig. 9-14. Typical air valve installation

9.5   ANCHOR BOLTS AND FASTENERS

A variety of anchor bolt sizes and lengths are used on penstocks. The largest types are tie-down tension anchors used to transfer bend forces downward into anchor block concrete. Smaller anchor bolts and fasteners provide tie-down forces that secure the components of penstock supports and bearings to individual concrete footings or smaller anchor blocks.

9.5.1   Tie-Down Anchors

Tie-down anchors are discussed generally in Section 8.2, which describes a penstock pipe partially encased in concrete at an anchorage. At these locations, restraint is always necessary without regard to bend direction. At a downward-to-upward bend, the net thrust is directed into the ground and the tie-down anchors prevent lateral movement or slippage. For an upward-to-downward bend, net thrust is upward and must be opposed by the tie-down bolts extending from a concrete anchor block,

Tie-down tension anchors and related members are attached around the penstock shell, as shown in Fig. 9-15. Bolt diameters are likely to exceed 1½ in., and the embedded length of such tension members must extend into the concrete a distance at least 40 times the diameter of the anchor bolt. Nuts and washers or shear plates are placed on the lower or embedded end bolt and surrounded with reinforcing steel.

Threaded connections are superior to welded ones for field assembly of the tie-down anchors, especially if reinforcing bars or rods of high-strength material are used for anchor bars.

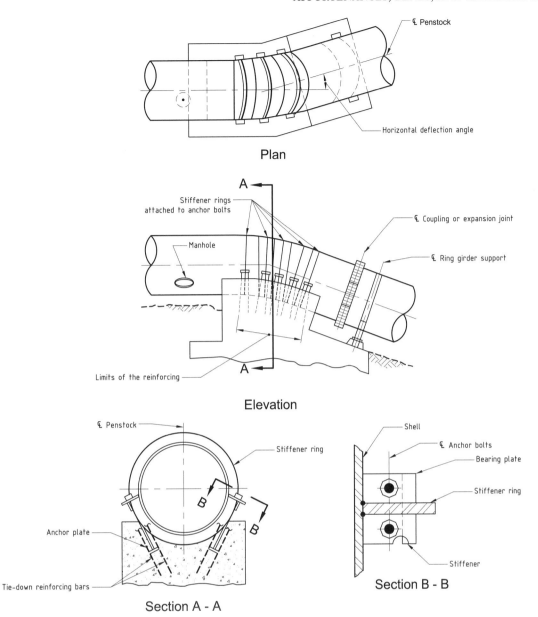

Fig. 9-15.   Tie-down anchors

Good construction practice requires that the completed tie-down assembly be attached to the penstock bend in its final position as fresh concrete is cast under the penstock. This step effectively secures unified action between the tension ties and the anchor block concrete, especially if the tie-down anchors are shipped in several pieces.

The use of penstock bend anchors as described implies that large forces will be transferred from the penstock shell to the concrete. To prevent buckling of the shell under this exterior loading, stiffeners are commonly used around the pipe at the points of tie-down attachment. For lighter loading, only wide flat straps may be needed.

9.5.2 Anchor Bolts

Anchor bolts are used at supports for penstock rocker bearings, base plates for valves, and bearings under ring girder bearings and saddles. Usually anchor bolts are less than 1 1/2 in. in diameter. Except when loaded under seismic forces that cause overturning, anchor bolts are most likely loaded by shear forces rather than tension. Shear load on these bolts may be reduced if lugs or shear plates are welded under base plates to better interact with the concrete of the base structure.

Since shear loading is most likely on the anchor bolts, the specifications for bolt material may vary. For most loading conditions, mild steel bolts conforming to ASTM (2007) specifications are adequate. When shear load on a bolt exceeds design limits for bolts in shear (see AISC 2010), a stronger steel bolt material should be considered.

If accuracy of bolt or fastener position is necessary to match shop fabrication, the usual practice is to install the anchor bolts inside pipe sleeves that extend at least 30 in. into the structural concrete. Only the bottom ends of the bolts are initially attached to the plates under the ends of the pipe sleeves (Fig. 9-16). The bolts may be adjusted somewhat in the field if the installed bolt location differs from the shop fabrication location. The void around a bolt in a pipe sleeve is filled with concrete grout after erection work has passed that point.

It is preferable to use leveling nuts on all anchor bolts. Leveling nuts transfer erection dead loads onto the threaded portions of the bolts. Each bolt in a sleeve acts as a long column before grouting; often each bolt must be laterally supported by shims in the sleeve before grouting.

In addition, bolt or sleeve diameters are selected so that the projected diameters under the lateral loading bear on the concrete at less than 500 psi. Furthermore, pullout forces

on the sleeves may require additional reinforcement and embedment cover on the adjacent outside faces of the concrete anchor block.

For lightly loaded anchor bolts less than 1 1/4 in. in diameter, it may be practical to set the penstock component in place and drill the anchor holes through the base plates, assuming that space is available for drilling. Expansion anchors or epoxy-grouted anchorages may be used as required for adverse vibration conditions.

9.6   ANCHOR RINGS, THRUST COLLARS, AND JOINT HARNESSES

9.6.1   Scope

Anchor rings, thrust collars, and joint harness lugs are fabricated from steel plates that are attached to the penstock circumference by continuous fillet welds. Special conditions may warrant the use of groove welds. Anchor rings and thrust collars provide resistance to thrust forces and are typically used for penstocks embedded in concrete. Joint harnesses

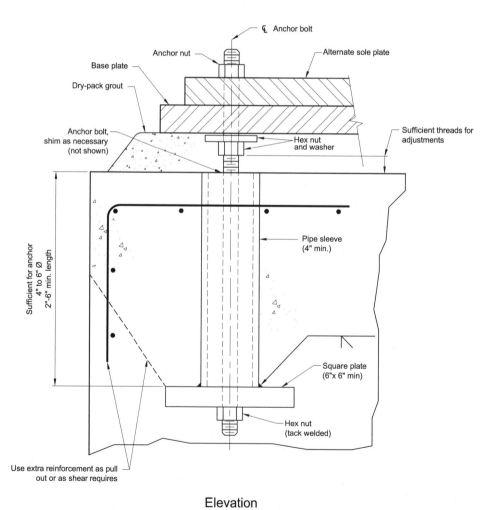

Fig. 9-16.   Typical anchor bolt in pipe sleeve

provide a means of restraining penstocks from movement caused by thrust forces at joints that are not capable of providing restraint.

9.6.2 Anchor Rings and Thrust Collars

Anchor rings and thrust collars that are embedded in concrete must be designed to resist thrust loads in accordance with established engineering principles (Fig. 9-17). The average bearing stress of the ring against concrete encasement must not exceed 0.6 times the specified 28-day compressive strength of the concrete. Anchor rings and thrust collars must be proportioned to transfer thrust loads from the penstock to the encasement primarily through shear. Anchor rings and thrust collars must be dimensioned and detailed according to Tables 9-2, 9-3, and 9-4, respectively. Consideration must be given to minimizing penstock wall thickness and welding requirements. It may be necessary to thicken the penstock shell or add wrapper plate reinforcement to reduce stresses in the penstock wall at anchor rings and thrust collars. Also, the thickened penstock shell or wrapper plate, if required, must extend beyond the limit of concrete encasement to limit secondary bending stress. Punching shear resistance of the concrete encasement must also be investigated.

The table values assume that anchor rings are fabricated from ASTM (2008), steel, the penstock is fabricated from ASTM (2010), Grade 70 steel, and E70XX welding electrodes are used. A 28-day concrete compressive strength of 4,000 psi has been assumed.

The anchor ring dimensions given in Tables 9-2, 9-3, and 9-4 (for penstocks 36 in. through 144 in. in diameter and pressures of 150, 250, and 350 psi) were developed using the following procedure:

1. The ring transfers loads to the concrete encasement with a contact pressure that varies linearly from zero at the tip of the ring to a maximum pressure at the ring–penstock connection resulting in a triangular pressure diagram with an average allowable ring contact pressure of 0.6 times the specified 28-day compressive strength of the concrete.

2. The ring resists a full "dead end" thrust force as the result of pressure (thrust = pressure times area of penstock cross section).

3. Thrust restraint from forces other than internal pressure (e.g., thermal) has not been included.

4. Maximum bending stress in the ring is limited to 75% of yield or approximately 27 ksi for a ring fabricated from ASTM (2008), steel.

5. Maximum principal and effective stresses in the penstock shell at the shell–ring junction, according to the Huber–Hencky–von Mises theory, are limited to the lesser of tensile strength/2.4 or yield strength/1.5; this amount is approximately 20 ksi for a penstock shell fabricated from ASTM (2010), Grade 70 steel.

6. Since thrust rings are embedded in concrete, the penstock is fully restrained from longitudinal and circumferential growth caused by internal pressure; thus, stresses in the penstock wall are affected by Poisson's ratio (v), which is assumed to be 0.3.

7. The depth of anchor ring embedment in the concrete encasement must be sufficient to resist punching shear forces when the thrust load is transferred from the ring to the encasement. The concrete thickness tabulated in Tables 9-2, 9-3, and 9-4 is based upon a thrust ring located in the center of the encasement such that the ring can act in either direction to resist thrust.

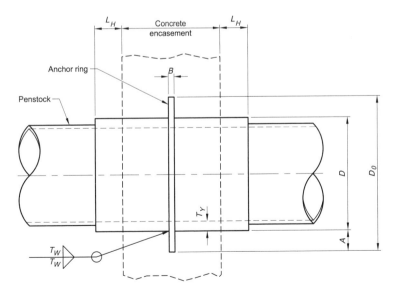

Fig. 9-17. Anchor ring and thrust collar detail

Table 9-2.   Anchor Ring Dimensions for Working Pressure of 150 psi

Diameter, D (in.)	Width of ring, A (in.)	Thickness of ring, B (in.)	Minimum embedment penstock thick., T_y (in.)	Extension of shell beyond encasement, L_R (in.)	Double fillet size, T_w (in.)	Minimum encasement thickness (ring in center) (in.)	Allow. thrust force (kip)
36	1.0	0.375	0.500	7	0.1875	25	153
42	1.0	0.375	0.500	8	0.1875	28	208
48	1.0	0.375	0.625	9	0.2500	31	271
54	1.0	0.500	0.625	10	0.2500	34	344
60	1.0	0.500	0.625	10	0.2500	37	424
66	1.5	0.625	0.750	12	0.2500	39	513
72	1.5	0.625	0.875	13	0.3125	42	611
78	1.5	0.625	0.875	14	0.3125	45	717
84	1.5	0.625	0.875	14	0.3125	48	831
90	1.5	0.625	1.000	16	0.3125	51	954
96	1.5	0.750	1.000	16	0.3125	53	1,086
102	2.0	0.875	1.125	18	0.3125	56	1,226
108	2.0	0.875	1.125	18	0.3125	59	1,374
114	2.0	0.875	1.250	20	0.3125	62	1,531
120	2.0	0.875	1.250	20	0.3125	65	1,696
126	2.0	0.875	1.250	21	0.3125	68	1,870
132	2.5	1.000	1.375	22	0.3125	70	2,053
138	2.5	1.000	1.500	24	0.3125	73	2,244
144	2.5	1.000	1.500	24	0.3250	76	2,443

The following sections show equations for determining the allowable stresses and other parameters for anchor rings and thrust collars.

9.6.2.1 Average Bearing Stress   The average bearing stress (σ_a) of the anchor ring on the surrounding concrete encasement is limited to allowable concrete bearing and is given by

$$\sigma_a = \frac{PD^2}{D_o^2 - D^2} \le 0.6 f_c' \qquad (9\text{-}4)$$

where σ_a = average bearing stress of the anchor ring on surrounding concrete (psi);

P = internal pressure under normal conditions (psi);
D = penstock outside cylinder diameter (in.);
D_o = ring outside diameter (in.); and
f_c' = concrete encasement specified 28-day compressive strength (psi).

9.6.2.2 Bending Stress   The ring, which resists thrust forces from the penstock, applies a bending moment to the penstock wall. If a triangular bearing pressure distribution, varying from zero pressure at the ring tip to maximum pressure at the ring–penstock wall connection is assumed, then the unit moment (M_o) at the base of the ring of width A is

$$M_o = \left(\frac{PD^2}{4}\right)\left(\frac{1}{D}\right)\left(\frac{A}{3}\right) = \frac{PDA}{12} \qquad (9\text{-}5)$$

where M_o = unit bending moment in ring at ring–penstock wall connection (in.-lb/in.);

P = normal design pressure (psi);
A = anchor ring width (in.); and
D = penstock outside cylinder diameter (in.).

The bending stress (σ_r) in the ring, assuming a ring of rectangular cross section, is given by

$$\sigma_r = \frac{6M_o}{B^2} \qquad (9\text{-}6)$$

Table 9-3.   Anchor Ring Dimensions for Working Pressure of 250 psi

Diameter, D (in.)	Width of ring, A (in.)	Thickness of ring B (in.)	Minimum embedment penstock thick., T_y (in.)	Extension of shell beyond encasement, L_R (in.)	Double fillet size, T_w (in.)	Minimum encasement thickness (ring in center) (in.)	Allow. thrust force (kip)
36	1.0	0.500	0.625	8	0.2500	34	254
42	1.5	0.625	0.875	10	0.3125	38	346
48	1.5	0.625	0.875	11	0.3125	42	452
54	1.5	0.625	1.000	12	0.3125	46	573
60	2.0	0.750	1.125	14	0.3125	50	707
66	2.0	0.875	1.125	14	0.3125	55	855
72	2.0	0.875	1.250	16	0.3125	59	1,018
78	2.0	0.875	1.375	17	0.3125	63	1,195
84	2.5	1.000	1.500	18	0.3160	67	1,385
90	2.5	1.125	1.500	19	0.3157	72	1,590
96	2.5	1.125	1.625	21	0.3367	76	1,810
102	3.0	1.250	1.750	22	0.3789	80	2,043
108	3.0	1.250	1.875	23	0.4012	84	2,290
114	3.0	1.375	1.875	24	0.3999	89	2,552
120	3.5	1.500	2.125	26	0.4419	93	2,827
126	3.5	1.500	2.125	27	0.4640	97	3,117
132	3.5	1.500	2.250	28	0.4861	101	3,421
138	4.0	1.625	2.375	30	0.5293	105	3,739
144	4.0	1.750	2.500	31	0.5269	110	4,072

where σ_r = bending stress in the ring–penstock wall connection (psi);

M_o = unit bending moment in ring at ring–penstock wall connection (in.-lb/in.); and

B = anchor ring thickness (in.).

The sum of the moments at the ring–penstock wall connection must equal zero; thus, the internal moments in the penstock wall, on either side of the ring, are one-half of the ring moment in Eq. (9-5), or

$$M_1 = \frac{M_o}{2} = \frac{PDA}{24} \qquad (9\text{-}7)$$

where M_1 = unit bending moment in penstock wall on each side of ring (in.-lb/in.) and

M_o = unit bending moment in ring at ring–penstock wall connection (in.-lb/in.).

9.6.2.3   *Longitudinal Stress*   The longitudinal stress (σ_1) in the penstock wall, which is completely restrained

from movement by the surrounding concrete encasement, is given by

$$\sigma_1 = \frac{PD}{4T_y} + \gamma\frac{PD}{2T_y} + \frac{6M_1}{T_y^2} \leq \sigma_n \qquad (9\text{-}8)$$

9.6.2.4   *Circumferential Stress*   The circumferential stress (σ_2) in the penstock wall, which is completely restrained from movement by the surrounding concrete encasement, is given by

$$\sigma_2 = \frac{PD}{2T_y} + \gamma\left(\frac{PD}{4T_y} + \frac{6M_1}{T_y^2}\right) \leq \sigma_n \qquad (9\text{-}9)$$

where σ_1 = longitudinal stress in the penstock wall (psi);

σ_2 = circumferential stress in the penstock wall (psi);
P = normal design pressure (psi);
D = penstock outside cylinder diameter (in.);
T_y = penstock thickness (in.);

Table 9-4.   Anchor Ring Dimensions for Working Pressure of 350 psi

Diameter, D (in.)	Width of ring, A (in.)	Thickness of ring, B (in.)	Minimum embedment penstock thick., T_y (in.)	Extension of shell beyond encasement, L_R (in.)	Double fillet size, T_w (in.)	Minimum encasement thickness (ring in center) (in.)	Allow. thrust force (kip)
36	1.5	0.625	0.875	9	0.3125	41	356
42	1.5	0.625	1.000	11	0.3125	47	485
48	2.0	0.875	1.250	13	0.3125	52	633
54	2.0	0.875	1.250	14	0.3125	57	802
60	2.5	1.000	1.500	16	0.3160	63	990
66	2.5	1.125	1.625	17	0.3241	68	1,197
72	3.0	1.250	1.750	18	0.3744	73	1,425
78	3.0	1.250	1.875	20	0.4056	79	1,672
84	3.0	1.250	2.000	21	0.4368	85	1,940
90	3.5	1.500	2.125	23	0.4640	90	2,227
96	3.5	1.500	2.250	24	0.4949	96	2,533
102	4.0	1.625	2.375	26	0.5477	101	2,860
108	4.0	1.750	2.500	27	0.5532	106	3,206
114	4.5	1.875	2.750	29	0.6055	112	3,572
120	4.5	1.875	2.750	30	0.6374	117	3,958
126	4.5	1.875	2.875	31	0.6692	123	4,364
132	5.0	2.125	3.125	33	0.6951	128	4,790
138	5.0	2.125	3.250	35	0.7267	134	5,235
144	5.5	2.250	3.375	36	0.7804	139	5,700

ν = Poisson's ratio (0.3 for steel);

M_1 = unit bending moment in penstock wall on each side of ring (in.-lb/in.); and

σ_n = allowable bending stress for normal conditions (psi).

9.6.2.5   _Equivalent Stress_   The combined Huber–Hencky–von Mises equivalent stress (σ_e) is given by

$$\sigma_e + (\sigma_1^2 + \sigma_2^2 - \sigma_1\sigma_2)^{1/2} \le \sigma_n \qquad (9\text{-}10)$$

where σ_1 is assumed to be in compression (–) and σ_2 is assumed to be in tension (+).

9.6.2.6   _Secondary Bending Stress_   The secondary bending stress (σ_b) in the penstock shell at the penstock–concrete encasement interface is given by

$$\sigma_b = 1.82\left(\frac{PR}{T_y}\right) \qquad (9\text{-}11)$$

where σ_b = secondary shell bending stress (psi);

P = normal design pressure (psi);

R = penstock radius (in.); and

T_y = penstock thickness (in.).

9.6.2.7   _Extension of Penstock Shell Beyond Concrete Encasement_   The length that the thickened penstock shell must extend beyond the concrete encasement limits is given by

$$L_R = 2.33\sqrt{RT_y} \qquad (9\text{-}12)$$

where L_R = length that thickened penstock shell must extend beyond concrete encasement (in.);

R = penstock radius (in.); and

T_y = penstock thickness (in.).

9.6.2.8   _Fillet Weld Size_   The size (T_w) of the double welds that attach the ring to the penstock shell, assuming use of E70XX electrodes, is given by

$$T_w = \frac{f_r}{\left[(0.3)(70,000)\left(\frac{\sqrt{2}}{2}\right)\right]} = \frac{f_r}{14,849} \qquad (9\text{-}13)$$

The following equations define f_r, f_b, and f_v:

Resultant shear

$$f_r = \sqrt{f_b^2 + f_v^2} \qquad (9\text{-}14)$$

Shear from bending

$$f_b = \frac{M_o}{\left(B + \dfrac{T_w}{2}\right)} \qquad (9\text{-}15)$$

Direct shear

$$f_v = \frac{PD}{8} \qquad (9\text{-}16)$$

where T_w = fillet weld size (in.);

f_r = resultant shear force in fillet weld (lb/in.);
f_b = unit shear force in fillet weld resisting anchor ring bending (lb/in.);
f_v = unit shear force in fillet weld resisting direct shear from anchor ring (lb/in.);
T_y = penstock thickness (in.);
M_o = unit bending moment in ring at ring–penstock wall connection (in.-lb./in.);
B = anchor ring thickness (in.);
P = normal design pressure (psi); and
D = penstock outside diameter (in.).

9.6.2.9 Minimum Encasement Thickness (Ring in Center) Punching shear strength of encasement, in kips, is

$$V_c = 2\sqrt{f'_c} \times d \times b_o \qquad (9\text{-}17)$$

Perimeter of failure surface at mid depth, in inches, is

$$b_o = \pi\left(D_o + d\right) \qquad (9\text{-}18)$$

Effective depth for shear (ring in circumference), in inches, is

$$d = \frac{h}{2} - 4 \qquad (9\text{-}19)$$

where D_o = outside diameter of ring (in.);

h = width of penstock encasement (in.); and
f'_c = concrete encasement specified 28-day compressive strength (ksi).

9.6.3 Joint Harnesses

Joint harnesses are used to transfer thrust loads across couplings or joints that are not capable of resisting thrust loads

along the penstock. The design of joint harnesses must comply with requirements given in AWWA (2004), for penstock sizes up through 96-in. diameter. Figs. 9-18 through 9-20 show typical joint harness configurations and details.

For penstocks larger than 96 in. diameter, the following procedure for thrust restraint using continuous front and back rings welded to the penstock wall is based on Blodgett (1963):

1. Calculate the section properties of the front and back rings. The back ring requires section properties to be calculated at two locations:
   a. Where thrust rod passes through the ring (net section) and
   b. At locations between thrust rods (gross section).

A portion of the penstock shell acts together with the rings. The effective width of the pipe shell is given by the following:

$$W_e = T_r + 1.56\sqrt{RT} \qquad (9\text{-}20)$$

where T_r = width of ring (in.);

R = radius of penstock (in.); and
T = penstock wall thickness (in.).

2. Calculate the force in each thrust rod as follows:

$$F_a = \frac{\pi D^2 P}{4N'} \qquad (9\text{-}21)$$

where D = penstock outside diameter (in.);

P = pressure (psi); and
N' = number of thrust rods.

3. The thrust rods create a bending moment because they are offset from the penstock shell. The bending moments are resisted by a force couple where radial outward forces, located where thrust rods pass through the ring, apply loading to the back ring and where radial inward forces, located at gusset plates, load the front ring. The radial force from the thrust rods is as follows:

$$F_a = \frac{e}{A} \qquad (9\text{-}22)$$

where F_a = thrust rod force (lb);

e = eccentricity (in.); and
A = distance between rings (in.).

4. The radial forces cause both direct stress (F_{ca}) and bending stress (F_{cb}) in the ring sections as follows:

$$F_{ca} = K_1 \frac{W}{A_e} \qquad (9\text{-}23)$$

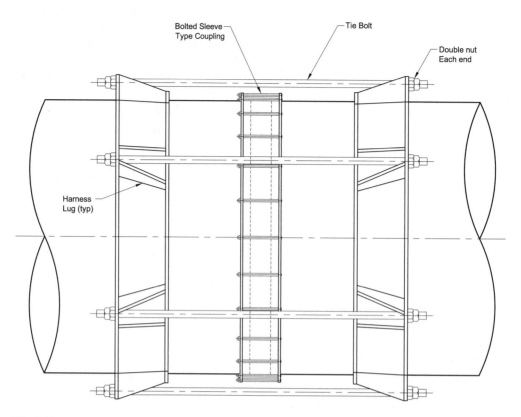

Fig. 9-18. Joint harness for single-sleeve coupling installation

$$F_{cb} = K_2 \frac{W}{S_c} \qquad (9\text{-}24)$$

where K_1 = direct shear coefficient;

W = radial load (lb);
A_e = ring cross section area (in.²);

K_2 = bending stress coefficient; and
S_e = ring section modulus (in.³).

The ring coefficients are calculated from equations given by Roark and Young (1989), formulas for circular rings, Table 17, case 7 (ring under any number of equally spaced radial forces). The coefficients are developed so that the back ring

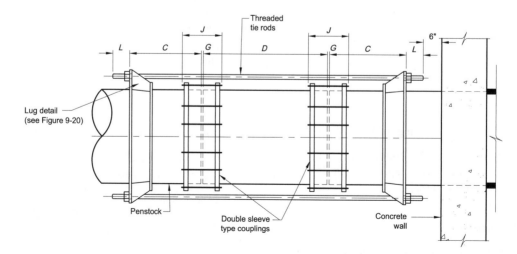

Fig. 9-19. Joint harness for double-sleeve coupling installation

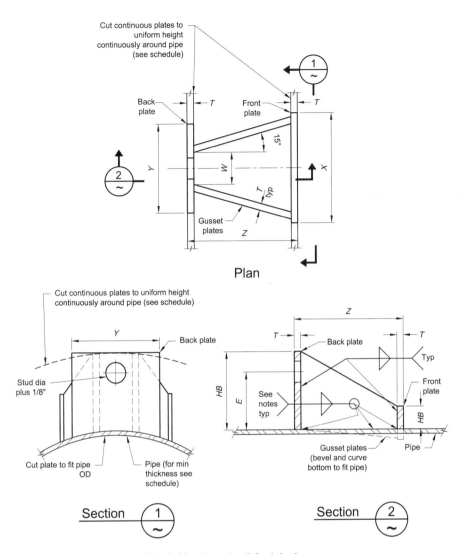

Fig. 9-20.   Lug detail for joint harness

stress can be calculated on ring cross sections both at the thrust rods and between the thrust rods. The front ring forces and ring coefficients are calculated with consideration for the 15-degree spread of gusset plates. Stress calculations for the back ring require that the net ring section properties must be used at the thrust rods and the gross ring section properties must be used between thrust rods.

5. Calculate the circumferential stress in each ring section caused by internal pressure acting on the effective shell width as follows:

$$F_{cp} = \frac{PDW_e}{2A_e} \qquad (9\text{-}25)$$

where P = pressure (psi);

   D = penstock outside diameter (in.);

   A_e = ring cross-sectional area (in.^{2}); and

   W_e = effective width of pipe shell (in.).

6. Add up the appropriate stresses to determine maximum total ring stresses.
7. Compare the calculated stress levels with allowable stress for assembly components and revise thrust restraint system as necessary.

9.7   PENSTOCK ROLL-OUT AND LIFT-OUT SECTIONS

For larger penstocks, especially those connecting into tunnels, full-diameter access to the interior is desirable through either a rollout or lift-out penstock section. This method is a good alternative to the severe dimensional limitations of manholes or mandoors. Removable sections provide space for wheeled and powered equipment to enter into the waterway for maintenance and repairs.

For the opening, a short spool piece of penstock is provided and connected into the main conduit with sleeve-type couplings. Sleeve-type couplings are much more flexible than flanges and less costly. The spool always must be supported by ring girders or some equivalent method, and the coupling middle rings must be furnished without pipe stops. A manhole provides short-time access. Some designs include rails on which to roll a heavy section or a simple lift-out piece with crane hooks.

Typical designs for roll-out sections are shown in Figs. 9-21 and 9-22.

9.8 FLANGED CONNECTIONS

9.8.1 Scope

This section covers steel pipe flanges for use on penstocks with reference to the following published flange standards:

1. AWWA C207, AWWA (2007a);
2. ASME B16.5, ASME (2009);
3. ASME B16.47, ASME (2011); and
4. MSS SP-44, MSS (2010).

The primary differences among the standards involve size and pressure rating limitations, flange type, and flange facing. The standard used for a particular project should be selected based on penstock diameter and design pressure, and by mating requirements with existing valves and fittings.

The following subsections provide a summary of basic flange considerations. In all cases, flanges must conform to the specifications of the applicable flange standard.

9.8.2 Pressure Ratings

For flanged connections outside the pressure class limits of the referenced standards, flanges should be designed according to Appendix 2 or Appendix Y of ASME (2010).

9.8.2.1 AWWA C207, AWWA (2007a) AWWA (2007a) applies to flanges and blind flanges ranging in size from 4 in. through 144 in. with pressure ratings up to 275 psi, and from 4 in. through 48 in. with a pressure rating of 300 psi, at atmospheric temperature. Pressure ratings are for customary conditions and temperatures in waterworks service. The pressure rating is based on the design of the maximum operating pressure plus the anticipated surge and water hammer pressure. Test pressures for flanges conforming to this standard must not exceed 125 percent of the ratings. Pressure ratings identified in AWWA (2007a), are given in Table 9-5.

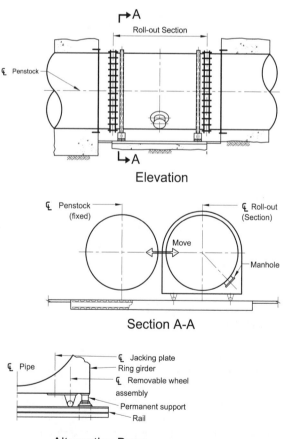

Elevation

Section A-A

Alternative Base

Fig. 9-21. Penstock roll-out section

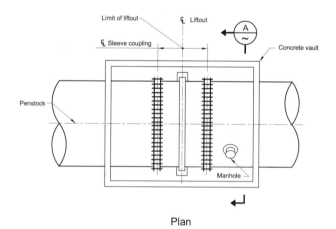

Plan

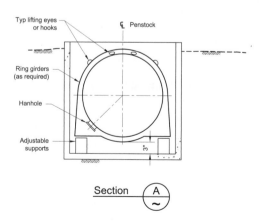

Section

Fig. 9-22. Penstock lift-out section

Table 9-5.   Pressure Ratings from AWWA (2007a)

Class	Nominal pipe size (in.)	Pressure rating (psi)
B	4–144	86
D	4–12	175
D	14–144	150
E	4–144	275
F	4–48	300

Table 9-7.   Pressure Ratings from ASME (2006)

Class	Nominal pipe size (in.)	Pressure rating (psi)
75	26–60	140
150	26–60	285
300	26–60	740
400	26–60	990
600	26–60	1,480
900	26–48	2,220

9.8.2.2   ASME B16.5, ASME (2009)   ASME (2009) is generally used for high-pressure, high-temperature mechanical piping; however, it also may be used for small-diameter penstocks. It applies to flanges and blind flanges in sizes up to 24 in. in rating classes 150, 300, 400, 600, 900, 1500, and 2500. The allowable pressure for each rating class, defined as the maximum allowable nonshock working gauge pressure, depends on the alloys used and the temperature of the penstock shell, which is generally the same as that of the fluid. Pressure ratings identified in ASME (2009) for carbon steel (Material Group 1.1) with a temperature between −20°F and 100°F are given in Table 9-6.

9.8.2.3   ASME B16.47, ASME (2006)   ASME (2006) applies to flanges ranging in size from 26 in. through 60 in. with pressure ratings up to 1,480 psi, and from 26 in. through 48 in. with a pressure rating of 2,220 psi. Pressure ratings are the maximum allowable nonshock working gauge pressure, at the specified temperature. ASME (2006) pressure ratings for temperatures less than 100°F are given in Table 9-7.

9.8.2.4   MSS SP-44, MSS (2010)   MSS (2010) applies to flanges and blind flanges ranging in size from 12 in. to 60 in. with pressure ratings up to 1,480 psi, and from 12 in. to 48 in. with a pressure rating of 2,220 psi, at atmospheric temperature conditions. Atmospheric temperature, for purposes of the standard practice, is defined to range from −20°F to 250°F. Pressure ratings are the maximum allowable nonshock working gauge pressure. MSS (2010) pressure ratings are given in Table 9-8.

9.8.3   Flange Types

Depending on the standard applied, flanges may be ring-type slip-on flanges, or welding neck type, as shown in Fig. 9-23. Slip-on flanges are attached to the penstock or appurtenance by two fillet welds. Welding neck flanges are butt-welded to the penstock.

Flanges conforming to AWWA (2007a) are ring-type (welding neck flanges may be used for pressure Class E, at the purchaser's option). Flanges conforming to ASME (2009) are slip-on or welding neck type. Flanges conforming to ASME (2006) or MSS (2010) are welding neck type flanges.

9.8.4   Facing

Flanges conforming to AWWA (2007a) are flat-faced, including hub flanges used for pressure Class E service. Flanges conforming to ASME (2009), ASME (2006), and MSS (2010) are of the raised-face type for all pressure classes. A 0.06-in. raised face is regularly furnished for pressure classes 300 and lower, whereas a 0.25-in. raised face is regularly furnished for pressure classes 400 and higher. Workers should use caution when connecting a raised-face flange to a flat-faced flange because this connection typically causes the raised-faced flange to crack or otherwise distort, resulting in leakage. When this situation cannot be avoided, the raised face should be machined off and a full-face gasket used, subject to the provisions of the flange standard.

Table 9-6.   Pressure Ratings from ASME (2009)

Class	Pressure rating (psi)
150	285
300	740
400	990
600	1,480
900	2,220

Table 9-8.   Pressure Ratings from MSS (2010)

Class	Nominal pipe size (in.)	Pressure rating (psi)
150	12–60	285
300	12–60	740
400	12–60	990
600	12–60	1,480
900	12–48	2,220

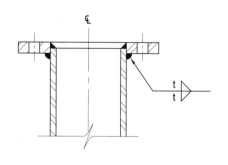

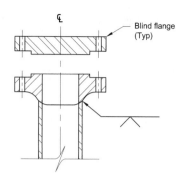

Slip on Ring Flange   Welding Neck Flange & Blind Flange

Fig. 9-23. Typical flange types

For proper gasket seating, the flatness of the machined facing surface of the flange can become critical for large, high-pressure flanges. The tolerances given in Table 9-9, relative to a plane located perpendicular to the penstock axis, are recommended.

9.8.5 Gaskets

Gasket materials should be selected in accordance with the provisions of the flange standard. Gasket design is based on expected bolt loading and service conditions and, where applicable, the need to provide an insulated joint. Rubber or compressed fiber ring-type gaskets generally are used for cold-water penstock service.

9.8.6 Drilling

Bolt holes are drilled to straddle the field top centerline, except where special mating conditions exist. When different standards are used for companion flanges, the engineer must verify that the drilling patterns are compatible. When companion flanges are required for flanged valves and fittings, the engineer may consider having the valve or fitting manufacturer supply the companion flange.

Table 9-9. Flatness Tolerance

Penstock diameter (in.)	Flatness tolerance (in.)
Less than 24	± 0.005
24–48	± 0.010
50–144	± 0.015
Greater than 144	± 0.020

9.9 BLOWOFFS

Blowoffs are provided at the low end of the penstock section to dewater the line. Sometimes, there are intermediate sections of penstock with sags that require additional blowoffs. A blowoff also must be provided just outside a tunnel portal to drain any tunnel leakage flow that may be present during penstock maintenance. In this case, the blowoff should be conservatively sized to account for unknown tunnel leakage. The exact location of blowoffs frequently is influenced by opportunities to dispose of the water. Where a penstock crosses a stream or drainage structure, usually there is a low point in the line; however, if the penstock runs under the stream or drain, it obviously cannot be drained completely into the channel. In such a situation, it is preferable to locate the blowoff connection at the lowest point that drains by gravity and, if necessary, provide a means for pumping out the section below the blowoff. Under normal circumstances, most of the penstock is drained through the plant. However, the blowoff components should be designed for full-head conditions in the event that draining through the plant does not occur. Where possible, the blowoff pipeline should be sloped to allow complete drainage.

9.9.1 Blowoff Valves

Blowoff valves, or blowoffs, should be provided with both a shutoff valve and a guard valve. If the penstock is above ground, the valves should be attached directly to the outlet of the penstock. For aboveground penstocks, freeze protection may be required for the blowoff valves. A pipe attached to the valves is used to route the discharge to a safe location. The discharge pipe usually requires installation of an elbow at the blowoff valve assembly, which must be blocked securely to avoid stresses at the attachment to the penstock.

Usually the blowoff is below ground. Because the operating nuts of the valve must be accessible from the surface, the valves cannot be located beneath the penstock but may be set with the stems vertical and just beyond the side of the penstock. If desired, the blowoff valves may be located within a manhole or valve vault to be more readily accessible for maintenance. The size of the valves and discharge pipe is dependent on the length of penstock to be dewatered and the time required to drain the line. Other considerations are the size of the outlet relative to the penstock size and the need to provide reinforcement.

9.9.2   Outlets

Typically outlets are welded to the penstock with reinforcing collars. Generally this work is done in the shop during penstock fabrication. Shop lining and coating of the outlets is satisfactory and more economical than work done in the field. Outlets should be checked to determine if reinforcement is required; however, for outlets larger than about 1/3 the diameter of the line, reinforcing should be given special consideration, even for small penstocks. Outlets that are encased in concrete should be wrapped in compressible material to prevent bending stresses.

The end of the outlet should be prepared to receive the valve or fitting to be attached. This may call for a flange, a plain end for a flexible coupling joint, a grooved spigot end for a bell-and-spigot joint, or a threaded end.

9.9.3   Blowoff Details

A typical blowoff detail for a buried penstock is shown in Fig. 9-24. In this example, two blowoff valves have been provided. The gate valve functions as a guard valve and is intended for watertight shutoff. The guard valve remains wide open during draining of the penstock. The ball valve also may be operated wide open, but could be used for flow control if necessary. The double-valve system is used because abrasion resulting from washing sediment out of the penstock may cause the ball valve seals to become worn, making tight shutoff impossible. If necessary, an orifice plate could be installed between the valve assembly and the outlet structure to further limit flow rate. If the penstock is installed at considerable depth, the blowoff may discharge into a manhole (Fig. 9-24) or to a riser pipe, and then into the drainpipe. In this case, the manhole must be pumped out to fully dewater the penstock. Where topography is favorable, the manhole may be omitted, and the blowoff piping may be routed to drain by gravity to a natural drainage way.

A concrete headwall structure typically is provided at the blowoff pipeline termination to dissipate the discharge water energy and prevent localized erosion within the channel accepting the discharge. Numerous headwall designs are possible. Two typical headwall configurations are shown in Fig. 9-25.

9.10   MAKEUP AND CLOSURE SECTIONS

9.10.1   Field Trim

Makeup or field-trim sections may be required in a penstock to adjust for the normal accuracy of surveys and pipe fabrication when a critical point in the pipe-laying sequence is reached, such as a major elbow or a bifurcation. To resolve this problem, the shop pipe is furnished in a length sufficiently long to allow for field trim.

9.10.2   Closure Sections

Closure sections are required when pipe must be laid between two fixed points or when pipe is laid from more than one starting point, as often occurs on long lines. In this case, a butt-strap closure, as shown in Fig. 9-26, may be used for pressures up to 400 psi.

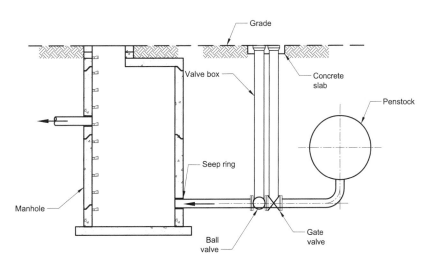

Fig. 9-24. Typical blowoff

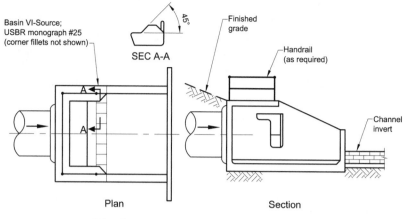

(A) USBR Impact-Type Energy Dissipator

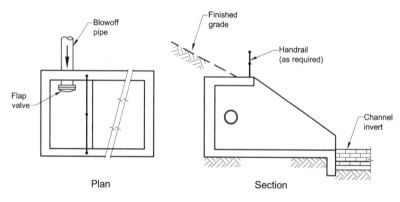

(B) Typical Headwall

Fig. 9-25. Typical blowoff headwall

For higher pressures a butt-weld closure section must be dropped into the line. Because closures usually involve adjustment for some degree of angular and axial misalignment between the two ends of the existing pipe, an outside backup bar may be used to provide a butt-weld joint that can accommodate this problem (Fig. 9-27).

When making closure welds, temperature stresses must be considered. The closure should be made after backfill has been brought up to the closure section and at a pipe tempera-

ture as close to the normal operating temperature as possible. This procedure usually occurs at the coolest part of the day.

9.11 CLOSURE HEADS

Closure heads commonly are used to block off the end of a pipeline, either as a temporary measure during hydrotesting and maintenance or as a permanent closure.

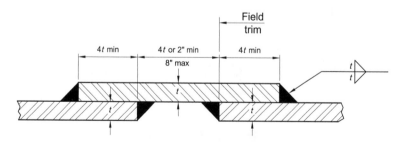

Fig. 9-26. Butt-strap closure

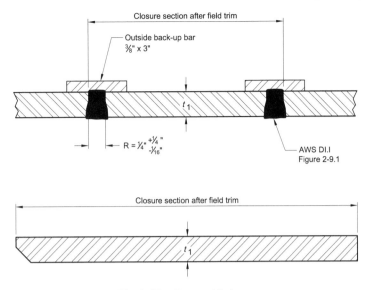

Fig. 9-27. Butt-weld closure

Closure heads may have the following construction:

1. Flanged-type connections;
2. Flat-disk closures; or
3. Elliptical, torispherical, and hemispherical dished closures.

Flanged-type closures commonly used for penstocks are slip-on type flanges that are welded to the penstock and a blind flange. Flanges can be of several classes or fabrication methods, including

1. Steel ring, hubless flanges made from rolled plate, billet, or curved flat bars;
2. Cast steel, using a welding hub; or
3. Forged steel, using a hub or a welding neck.

Blind flange-type closure heads can be used effectively with flanged pipe connections. This type of closure head can be economical for pipes up to 48 in. in diameter, with a nominal design pressure up to 150 psi.

Section 9.8 covers particular flange data, design criteria, and applications. Reinforced (i.e., stiffened) blind flange-type closures can be designed for pipe diameters greater than 48 in. and pressures in excess of 150 psi. This section is limited to the use, design, and application of dished closures and special, large, designed flanges.

In considering the selection between the blind flange type or the dished closure, the engineer must consider (1) functional use, (2) pressure considerations, and (3) size limitations.

Dished head closures usually are made of the same material as the penstock. However, different but compatible material should be considered to ensure the availability of these specialty items. The principal advantage of dished heads over flat-plate blind-flange closures is the large reduc-

tion of discontinuity between the shape of the head and the penstock cylinder. This advantage results in a substantial reduction of discontinuity stresses at or near the junction of the two members.

Shape and stress–strain measurements indicate that the best-suited shape for a dished head on a cylindrical penstock under internal pressure is an ellipse having a 2:1 major-to-minor axis ratio. Strain measurements on such thin-walled elliptical heads show that radial stresses are of minor importance. The principal stresses in the head act at right angles to each other and are identified as the hoop stress and the meridional stress. At the junction of the dished head with the cylinder, these stresses become the hoop (i.e., circumferential) stress of the shell and the longitudinal stress of the shell, respectively.

Elliptical heads having a 2:1 major-to-minor axis ratio are recommended for use in penstock design. For these types of heads, the required thickness *(t)* of the dished head at its thinnest point is given by the equation

$$t = \frac{PD}{2SE} \qquad (9\text{-}26)$$

where t = minimum required head thickness after fabrication and exclusive of any corrosion allowance (in.);

P = design pressure (psi);
D = inside diameter of the head at the skirt or inside length of major axis, measured perpendicular to the longitudinal axis (in.);
S = maximum allowable stress intensity value compatible with the design pressure (psi); and
E = lowest weld joint reduction factor of any joint in the dished head or at the head-to-shell joint (percentage in decimal form).

Tests and stress measurements show that a 2:1 ratio elliptical head made from the same material as the cylindrical shell and of the same thickness as the shell at the point of juncture will be as strong as, if not stronger than, the shell.

Ellipsoidal formed heads, subject to internal pressure, must meet the requirements of Fig. UW-13.1 of ASME (2010) as to fabrication and skirt length.

A typical ellipsoidal head installation is shown in Fig. 9-28.

9.12   BENDS (ELBOWS)

Changes in direction of flow are accomplished with curved pipe sections commonly called bends. Bends up to 24 in. in diameter may be smooth, wrought, or steel fittings, as specified in Section 2.4 or fabricated from mitered sections of pipe. Bends greater than 24 in. in diameter commonly are fabricated from mitered sections of pipe. The mitered pipe sections must be joined by complete joint penetration, single- or double-welded butt joints.

9.12.1   Fabrication

The radius of bends must be equal to or greater than one pipe diameter but need not be greater than three pipe diameters. Special situations, such as a bend immediately upstream of a turbine or a free discharge valve, may warrant a larger radius bend. Bends may be fabricated by mitering segments of a cone to produce a reducing bend or by mitering straight-pipe segments to produce a constant diameter bend. An analytical stress investigation, as recommended in Section 9.12.3 should be undertaken. The maximum deflection angle of the mitered segment must not exceed 22.5 degrees. Other methods that consider discontinuity stresses may be used.

9.12.2   Compound Bends

Compound bends are required where it is desired to change flow direction in both plan view and profile. Trigonometric calculations are necessary to determine the true angle of the bend and end rotations (USBR 1986).

9.12.3   Stress Analysis

Bends with a radius less than 2.5 pipe diameters must be designed with consideration given to concentration of hoop tension stresses along the inside edge of the bend. The following formula, given in AWWA (2004), is used to check stresses, the derivation of which is detailed in Sundberg (2009).

$$t = \frac{PD}{Sf}\left(\frac{D}{3}\tan\frac{\theta}{2} + \frac{S}{2}\right) \tag{9-27}$$

where t = required elbow wall thickness (in.);

   P = design pressure (psi);
   f = allowable tensile stress at design pressure (psi);
   D = outside diameter of elbow (in.);
   S = segment length along inside of elbow (in.); and
   θ = segment deflection angle (degree).

From basic geometry

$$S = 2\left(R - \frac{D}{2}\right)\tan\frac{\theta}{2}$$

where R = radius of elbow (in.).

Substitution of this equality into Eq. (9-27) yields the following simplified version of Eq. (9-27):

$$t = \frac{PD}{f}\left[1 + \frac{D}{3R - 1.5D}\right] \tag{9-28}$$

Stresses calculated according to this formula must not exceed the limits specified in Chapter 3 for P_m. To satisfy the formula, short radius bends may require thicker plate than adjacent sections of straight pipe. When required, increased plate thickness must, as a minimum, extend a distance past the tangent point of the bend given by

$$L = 0.78\sqrt{rt} \tag{9-29}$$

where L = distance past tangent point of the bend (in.);

   r = penstock radius (in.); and
   t = penstock thickness (in.).

9.13   GEOMETRIC TRANSITIONS

9.13.1   Diameter Changes

Changes in diameter are usually accomplished by the use of right conical shells placed in the straight tangent portion of a line or combined with a mitered bend, as shown in Fig. 9-29. Cone angles, diameter ratios, and miter angles shown in these sketches are recommended but not mandatory. At

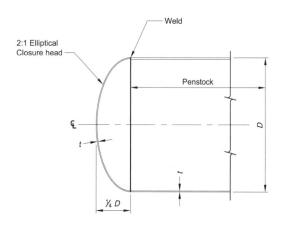

Fig. 9-28.   Elliptical closure head

angle changes in the profile of the penstock (e.g., cone–cylinder or miter-to-miter) greater than those recommended, special discontinuity analysis is necessary. One method for this analysis is given in the ASME (2010), Appendix 1-5. Stiffener rings sometimes are required at these junctions. Geometric layout of mitered conical sections and bends is given in AWWA (2007).

Note: Reducer lengths less than those in Fig. 9-29, however, should be reviewed for adequacy in accordance with ASME (2010), Section UG-36.

Fig. 9-30 illustrates a change in diameter affected by a contoured transition of parabolic form. Such transitions are often used at inlets to the penstock and at the inlet–outlet of surge tanks.

9.13.2 Cross-Section Shape Changes

Fig. 9-31 illustrates a square-to-circular transition. The transition is made up of flat triangular sides and oblique conical quarter sections at the four corners. Such shapes cannot resist the pressure of water forces with membrane-type stresses only and, therefore, must be stiffened. Moment-resisting frames can be used for stiffening. Each frame must be continuous around the perimeter and spaced to limit

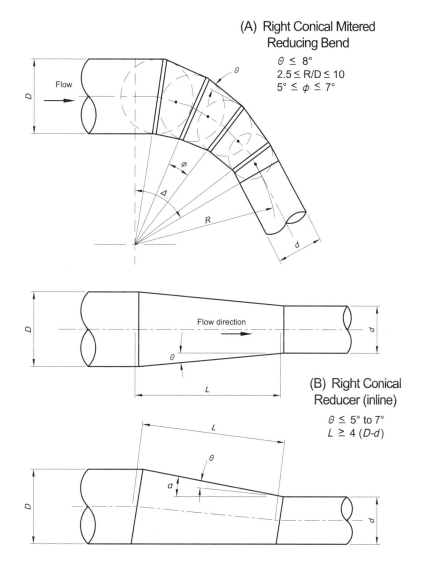

(A) Right Conical Mitered
Reducing Bend

$\theta \leq 8°$
$2.5 \leq R/D \leq 10$
$5° \leq \phi \leq 7°$

(B) Right Conical
Reducer (inline)

$\theta \leq 5°$ to $7°$
$L \geq 4\,(D\text{-}d)$

(C) Right Conical Reducer (offset)
Aligned invert for effective drainage

$\theta \leq 7°$
$L \geq 8(D\text{-}d)$

Fig. 9-29. Typical geometric transitions

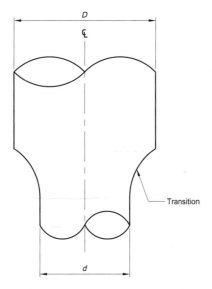

Fig. 9-30. Contoured change in diameter

bending stresses in the flat-plate liner between stiffeners. If the transition is embedded in structural reinforced concrete or mass concrete, anchor studs welded to the exterior side of the transition can be used in place of stiffener frames.

Other geometric forms can be used for geometric transitions, provided the form can contain pressure with essentially membrane forces only. Otherwise, the form must be stiffened. Appendix 13 of ASME (2010) contains rules for the design of noncircular cross-sectional conduits, both stiffened and unstiffened.

REFERENCES

AISC. (2010). "Specification for structural steel buildings." *AISC 360-10,* Chicago.

ASME. (2009). "Pipe flanges and flanged fittings: NPS 1/2 through NPS 24 metric/inch standard." *ASME B16.5-2009,* New York.

ASME. (2010). "Boiler and pressure vessel code, section VIII, division 1: Rules for construction of pressure vessels." New York.

ASME. (2011). "Large diameter steel flanges: NPS 26 through NPS 60 metric/inch standard." *ASME B16.47-2011,* New York.

ASTM. (2007). "Standard specification for anchor bolts, steel, 36, 55, and 105-ksi yield strength." *ASTM F1554-07ae1,* West Conshohocken, PA.

ASTM. (2008). "Standard specification for carbon structural steel." *ASTM A36/A36M-08,* West Conshohocken, PA.

ASTM. (2010). "Standard specification for pressure vessel plates, carbon steel, for moderate- and lower-temperature service." *ASTM A516/A516M-10,* West Conshohocken, PA.

AWS. (2010). "Structural welding code—Steel." *ANSI/AWS D1.1/D1.1M:2010,* Miami.

AWWA. (2001). "Air-release, air/vacuum, and combination air valves." *AWWA M51,* Denver.

AWWA. (2004). "Steel water pipe: A guide for design and installation," 4th Ed., *AWWA M11,* Denver.

AWWA. (2007a). "Steel pipe flanges for waterworks service, sizes 4 in. through 144 in. (100 mm through 3,600 mm)." *ANSI/AWWA C207-07,* Denver.

AWWA. (2007b). "Dimensions for fabricated steel water pipe fittings." *ANSI/AWWA C208-07,* Denver.

AWWA. (2007c). "Fabricated steel mechanical slip-type expansion joints." *ANSI/AWWA C221,* Denver.

AWWA (2007d). "Air-release, air/vacuum, and combination air valves for waterworks service." *ANSI/AWWA C512-07,* Denver.

AWWA. (2011a). "Bolted sleeve-type couplings for plain-end pipe." *ANSI/AWWA C219-11,* Denver.

AWWA. (2011b). "Bolted, split-sleeve restrained and non-restrained couplings for plain-end pipe." *ANSI/AWWA 227-11,* Denver.

Blodgett, O. W. (1963). *Design of weldments,* James F. Lincoln Arc Welding Foundation, Cleveland, Ohio.

Lescovich, J. E. (1972). "Locating and sizing air-release valves." *J. Am. Water Works Assn.* 64(7), 579–590.

MSS. (2010). "Steel pipeline flanges." *MSS SP-44-2010,* Manufacturers Standardization Society of the Valve and Fittings Industry, Vienna, Virginia.

Parmakian, J. (1950). "Air-inlet valves for steel pipe lines." *Transactions American Society of Civil Engineers,* 115, 438–443.

Roark, R. J., and Young, W. C. (1989). *Roark's formulas for stress and strain,* 6th Ed., McGraw-Hill, New York.

Sundberg, C. (2009). "Do the math—Spiral welded steel pipe is a good bet." *Pipelines 2009: Infrastructure's Hidden Assets: Proc., Pipelines 2009 Conf.,* ASCE, Reston, VA.

USBR. (1986). "Welded steel penstocks." Engineering monograph no. 3, Denver.

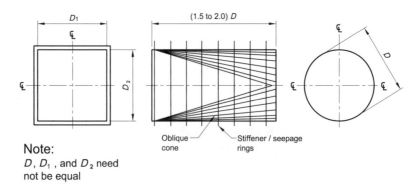

Note:
D, D_1, and D_2 need not be equal

Fig. 9-31. Square-to-circular transition

CHAPTER 10

Corrosion Prevention and Control

Source: Photograph courtesy of American Cast Iron Pipe Co.; reproduced with permission from American SpiralWeld Pipe LLC.

The prevention and control of metallic penstock corrosion are of paramount importance in providing reliable service and long life. There is an inordinate number of failures of steel penstocks caused by the lack of proper corrosion protection design, coating and lining applications, and maintenance of corrosion prevention systems. This chapter discusses aqueous, atmospheric, and underground corrosion mechanisms, tolerance for corrosion, applicable corrosion prevention and control measures, and penstock failure caused by corrosion. Discussion is limited to metallic corrosion of steel penstocks.

Increasing penstock wall thickness to allow sacrificial metal loss is an unreliable method of corrosion control and is not cost-effective. There is no assurance that corrosion will attack the wall uniformly, and in most cases the form of corrosion is localized pitting, which can result in premature penstock failure. The implementation of high-quality linings and coatings that are well maintained is much more reliable and economical. The use of sacrificial metal is now considered obsolete.

10.1 CORROSIVE ENVIRONMENTS

The corrosive environments to which a penstock is exposed include water, atmosphere, and soil.

10.1.1 Aqueous Corrosion

Metallic corrosion in aqueous solutions is a natural electrochemical process involving both the flow of electrons between metallic components and chemical reactions at the anodic (corroding) and cathodic (noncorroding) surfaces.

The driving voltage for corrosion current flow may be derived from galvanic (two-metal) coupling, or can be caused by local differences on the surface of a single metal or alloy. These differences or concentration effects include both metallurgical and electrolytic (water) influences. Included among the metallurgical causes are differences in metallic composition, heat treatment (including welding), and mechanical stress. Variations in the electrolyte can result from oxygen concentration, pH differences, salt gradients, and temperature heterogeneities.

The formation of potential differences results in anodic and cathodic surfaces. Current flows within the formed electrochemical circuit, and metal is lost. The current flows electronically from the cathode (positively charged) to the anode (negatively charged) and ionically (electrolytically) from the anode surfaces to the cathode surfaces through the electrolyte (any substance that conveys ions).

At the anodic sites, metal is oxidized to a positively charged metallic ion (cation) that enters the solution (corrodes). The electrons liberated at the anode are conducted metallically to the cathode surface. At the cathode, the electrons reduce oxygen to form hydroxyl ions and hydrogen ions to form hydrogen gas. The hydroxyl ions and hydrogen gas blanket the cathode surfaces and, in this manner, form an insulating film (cathodic polarization).

Relatively high concentrations of hydroxides at the cathode surfaces increase the local alkalinity so that conditions favorable for precipitation of calcium carbonate exist. The higher pH at the cathode is conducive to passivation of underlying surfaces, and corrosion products form a high-resistance barrier at the anode surfaces. These reactions increase the circuit resistance and correspondingly decrease corrosion activity. Also, lower water temperatures contribute to slower corrosion rates. Therefore, in static water, the original oxygen supply is consumed and corrosion rates decelerate. For this reason, metals and alloys exposed in deep water, even in normally highly aggressive seawater, corrode slowly. At the water surface and in the tidal and splash zones where oxygen is readily available, aggressive attack is experienced.

Under dynamic conditions, cathodic films may be swept away and oxygen may be replenished (depolarization), allowing the cathodic reaction and corrosion to continue. The dynamics of water motion and wetting and drying have similar depolarizing effects.

Thus, the activity at the cathode surfaces controls corrosion in neutral and near-neutral waters. Cathodic activity is limited by oxygen availability. Corrosion products on the anode surfaces increase the corrosion reaction by further restricting oxygen to underlying surfaces. Although oxygen is required for corrosion in neutral waters, contrary to common understanding, corrosion occurs where oxygen concentration is minimal.

A thin oxide layer can be formed on steel in powerful oxidizing environments, such as concentrated nitric and sulfuric acids. A similar passive film forms on iron and steel concurrently exposed to oxygen in highly alkaline conditions. The properties of the resulting oxide film, rather than those of the parent iron or steel, govern corrosion behavior.

Any phenomenon that disrupts or removes the products of polarization or passivation increases corrosion activity. If the disruption occurs uniformly and totally on all surfaces, the metal is uniformly lost (general corrosion) and, if the loss rate were known, sacrificial thickness could be incorporated during design to provide the necessary service life. Unfortunately, in practice, the protective films are disrupted in localized areas only, resulting in pitting corrosion.

Flowing water can remove polarization products and replenish oxygen, and certain anions, such as chlorides and sulfides, can penetrate and disrupt passive films. Chlorides in water penetrate and disrupt passive films. In this manner, galvanic cells are established, the intact passive film is the cathode, and the parent material exposed at disruptions in the oxide layer serve as the anode. As corrosion progresses, the environment at the anodic sites becomes depleted of oxygen and acidic as both oxygen and hydroxyl ions are consumed by reaction with metallic ions to form metal oxide and hydroxide corrosion products. The ensuing oxygen and pH concentration cells further accelerate corrosion.

Other potential but less frequently encountered forms of aqueous attack are dealloying microbiological corrosion and stress corrosion. Dealloying is a phenomenon common to cast iron. One of the alloying elements (carbon) is more noble than the other alloying metal (iron). Thus, intergranular galvanic cells are established that corrode the more active metal, iron.

Microbiological corrosion is the deterioration of a metal or alloy as a result of the metabolic activity of microorganisms. These organisms accelerate corrosion by promoting concentration cells, creating corrosive conditions as a result of their life cycle or decomposition and behaving as cathodic and anodic depolarizers. The two common corrosion-related forms are the iron bacteria and the sulfate-reducing bacteria.

The iron bacterium is aerobic (requires oxygen for its metabolic processes). It assimilates ferrous iron in solution and converts it to ferric iron, which precipitates in voluminous deposits (tubercles) on any surface, metallic or nonmetallic. If the deposits are nonuniform over the surface, they promote oxygen concentration cells.

The sulfate-reducing bacterium is anaerobic (thrives in oxygen-deficient environments). It requires sulfates in solution, which it reduces to sulfides using nascent hydrogen in the process. Because hydrogen is a cathodic polarizing agent, the sulfate-reducing bacterium, much like oxygen, is a cathodic depolarizer. Because hydrogen sulfide is generated, pH concentration cells also can result.

Stress corrosion cracking and hydrogen embrittlement are two forms of stress corrosion.

10.1.1.1 <u>*Stress Corrosion Cracking*</u> Stress corrosion cracking is the result of the synergistic effects of applied or residual tensile stress and corrosion. Corrosion pitting creates a weakened cross section. Cracking initiates at the base of the pit, where stresses are concentrated, and continues to grow into the cross section. Crack growth is self-sustaining and self-accelerating because it results in progressively increasing stress concentrations. The cause of crack initiation and growth is not clearly understood. Stress concentrations at the roots of pits may result in local strains sufficient to disrupt any passive or protective films formed previously by corrosion. This change establishes a new cell in which the pit surfaces behave cathodically to the freshly exposed metal anode. Repetition of this cycle results in progressively increasing stress until the end of the material's strength is reached and the material fails suddenly. The rapid strain rate occurring when the material strength is exceeded provides a brittle fracture. In any case, stress corrosion cracking is an anodic phenomenon, and cathodic currents are successfully used for its prevention and control.

10.1.1.2 <u>*Hydrogen Embrittlement*</u> Hydrogen embrittlement is a phenomenon caused by diffusion of atomic hydrogen into the metal. The absorbed hydrogen atoms, upon entry into the metal, may combine into hydrogen gas at dislocations in the metal microstructure. This phenomenon results in internally straining the metal and, if sufficient hydrogen is absorbed, brittle fracture of the internally stressed metal. Because hydrogen ion reduction is required, hydrogen embrittlement occurs at cathodes and is more likely in acidic electrolytes.

Both stress corrosion cracking and hydrogen embrittlement can result in failure of components at applied stress levels much lower than metal strength. Both phenomena result in sudden failure with no macroscopically visible warning signals. Susceptibility increases with increases in strength. Because phenomena result in adding strains to materials that may already be highly strained, the high-strength and ultrahigh-strength materials are primarily affected.

Paradoxically, although cathodic protection is applied to control stress corrosion cracking, it can generate hydrogen, thereby increasing the potential for hydrogen embrittlement. Usually this is not a problem when cathodic protection is applied in accordance with recommended practice.

10.1.2 Atmospheric Corrosion

The effect of corrosion on metal surfaces exposed to the atmosphere is the most visually evident form of corrosion. Atmospheric corrosion is a form of aqueous corrosion. Were it not for moisture, such as from precipitation and condensation, atmospheric corrosion would not occur. With atmospheric corrosion, two accelerators of aqueous corrosion are common: an abundance of oxygen and cyclic wetting and drying activity.

Corrosion activity occurs only during the wetted periods. Thus, designs must include provisions to reduce the duration of the wet periods. Such provisions include designs that promote free drainage, by elimination of moisture and water traps, and ensure that the bottom of an atmospherically exposed penstock does not contact potentially moist soil. Whether the penstock is exposed to the atmosphere or to soil, any dielectric coating on the pipe must be carried through partial concrete encasements, such as supports, anchors, and thrust blocks. Failure to coat interfacial surfaces between the steel and concrete results in accelerated corrosion at the edge of the concrete, attributable to a pH concentration cell. Because corrosion activity tends to be highly localized (i.e., resulting in pitting), the use of a corrosion allowance in the form of sacrificial thickness additions is not applicable.

Corrosivity of the atmosphere is controlled largely by the contaminants within the air. Industrial and marine atmospheres are among the most corrosive. Industrial atmospheres contain such contaminants as sulfur oxides and relatively large concentrations of carbon dioxide. Both result in acidic films on metals. Sea air contains chlorides and other salts, which are powerful corrosion accelerators. Acidic and salt films can concentrate from cyclic wetting and drying activity. Rural and inland industrial uses usually are relatively contaminant free and are not generally considered unusually corrosive.

10.1.3 Corrosion by Soils

Soil is an essentially neutral, aqueous electrolyte; thus, corrosion of ferrous alloys in soil is a special case of aqueous corrosion. The general cause of corrosion in neutral soil is attributable to cathodic depolarization or depassivation by the activity of oxygen. In oxygen concentration-promoted corrosion, the combined effect of oxygen and moisture causes corrosion. The driving voltage for the corrosion cell is caused by differences in oxygen available to all surfaces. The conductivity of the soil controls both the intensity and extent of attack. Oxygen concentrations in soil are attributable to differences in aeration, salt concentrations and their effect on oxygen solubility, soil permeability, and groundwater flow.

Stray currents are the second most prevalent cause of underground pipeline corrosion. In stray current corrosion, current from a foreign source such as a cathodic protection system, electrified railway, or improperly grounded welder is distributed in the earth. Current is collected at some surfaces (cathodes) of the pipeline and discharged from other surfaces (anodes) as it returns to the originating source. Again, as with galvanic corrosion, metal corrodes at the anodic sites. The extent and intensity of stray current corrosion are related to the driving voltage of the foreign power supply, the resistance of its circuit, the geometrical relationship between the source of currents and the penstock, the axial resistance of the penstock, the dielectric properties and continuity of the penstock, and the soil conductivity.

In any corrosion reaction, the magnitude of the current controls the rate of metal loss. The magnitude of current is related to the driving potential and the circuit resistance, in accordance with Ohm's law. Although the voltage difference between anode and cathode is normally small (less than 0.5 V), the circuit resistance also may be low, depending on the anodic and cathodic surface areas, their proximity, and the electrolyte conductivity. Larger potential differences, larger anode and cathode surface areas, and higher conductivities lead to macrocell corrosion (macro distances between cathodic and anodic surfaces) and localized corrosion. Small potential differences, small anodic and cathodic sites, and low conductivities result in microcell corrosion (microscopic distances between anodes and cathodes) and more uniform corrosion. Oxygen concentration and stray currents can produce macrocell corrosion.

The rate of anodic pitting in macrocell corrosion is related directly to the anodic current density (i.e., current discharge per anodic surface area). Large cathodes and small anodes result in high anodic current density and rapid pitting. Small cathodes and large anodes result in low anodic current density and slow, more general attack.

10.2 VULNERABILITY OF PENSTOCK STEEL TO CORROSION

Exposed surfaces of steel otherwise coated with a bonded dielectric coating are vulnerable to corrosion. The activity is limited to exposed surfaces only. Because coating defects, such as pinholes and holidays, are small, microcell corrosion occurs. Also, the area available to behave cathodically is minimized. This situation provides a more favorable cathode-to-anode area relationship, low anodic current densities, and reduced corrosion intensity. The magnitude of stray current collected and subsequently discharged is also reduced, depending on coating efficiency relative to bare penstock or to penstock coated with conductive coatings, such as concrete and mortar.

10.3 CORROSION PREVENTION AND CONTROL MEASURES

Corrosion activity can be prevented through design and by protective linings and coatings to minimize corrosion and can be controlled with cathodic protection.

10.3.1 Design for Corrosion Prevention

Penstocks are similar to pipelines, with three notable exceptions. They may be on steep profiles, they may convey water at high velocities, and unless proper preventative measures are taken, they may be grounded through the plant grounding system. The penstock and its appurtenances must be designed to minimize corrosion.

10.3.1.1 High Water Velocities Because of the high water velocities encountered in penstocks, internal surfaces must be as smooth as practicable. Accomplishing smoothness may require the elimination of high weld beads and other protrusions that could potentially upset the flow. Aside from head-loss considerations, elimination of protrusions minimizes potential damage from abrasion caused by suspended sediment and cavitation erosion. Because of this consideration, the protective lining surfaces also must be smooth. Spray application results in smoother surfaces than brush or roller application. Because of the potential for turbine runner damage from loose fragments, penstocks normally are not lined with cement-mortar. However, on flatter profiles, cement-mortar lining may be applicable, provided that the velocity does not exceed 20 ft/s and that there are not large volumes of transported sands and gravels.

10.3.1.2 Electrical Grounding Electrical ground mats for hydroelectric power plants are commonly constructed either of copper or copper-clad steel. Coupling the steel penstock with copper or copper-clad steel results in an adverse galvanic corrosion cell for the steel, wherein the steel corrodes and the copper is protected. Therefore, wherever practicable, the penstock should be electrically isolated from the intake shutoff gate (ISG) or valve and the penstock shutoff valve (PSV) at its downstream terminus. A common source of electrical shorts to the plant grounding system is at a penstock's wall penetration. Typically, to facilitate construction, the penstock pipe is welded to the steel reinforcement before placement of the concrete. To eliminate this potential, the penetration can be cased, the penstock pushed through the casing, and the annulus sealed using casing-insulating seals.

10.3.1.3 Design to Minimize Corrosion There are a number of things that should be taken into account in the design that can change the effectiveness of coatings and linings. These factors include, but are not limited to

1. Edges: Sheared edges created during fabrication are often not rounded and consequently cause problems in the coating process. Coatings generally have a tendency to shrink and pull back from edges, leaving a thinner, less protective coating.

   Edges should be smoothed to remove any sharpness, and then stripe-coated at least once during the coating process, preferably with each coat, to provide additional coating thickness at the edge.

2. Corners: Exterior corners are like edges; the same corrosion effect on coatings occurs. Corners should be smoothed to remove any sharpness and stripe-coated like edges.

   The problem is twofold with interior corners. Coatings applied over interior corners shrink in much the same way as at edges, tending to form a bridge with poor contact to the substrate. Stripe-coating these areas helps eliminate these problems.

3. Rivets, bolts, nuts, and overlapping plates: Small crevices between overlapping plates can allow the accumulation of moisture, and because it may be impossible to clean and coat these areas, corrosion can occur.

An example of overlapping plates is bell-and-spigot joints. Often, because of design requirements, only the interior or exterior girth joint is welded, not both. Applying caulking at the unwelded joints before lining helps to prevent moisture from getting between the two overlapping surfaces (this rule is applicable for liquid coatings and applies to mortar or shrink-wrap systems). If both interior and exterior bell and spigot are welded, then follow the recommendations in "Welds" (to be discussed).

Rivets, bolt heads, and nuts are all difficult to coat, and care is needed to ensure that all surfaces are adequately cleaned and coated.

4. Welds: Generally, welds are rough, discontinuous areas on a plane surface, and they may have many sharp edges. Often, welds are not cleaned properly, leaving weld spatter, slag, and acid flux residues. If these imperfections are not removed, proper coating cannot be achieved, and corrosion may develop.

Weld areas should be cleaned, ground smooth (depending on the operating conditions), and stripe-coated before coating.

Welds are often tested for cracks with chemical testing solutions that contaminate the surface. These surfaces need solvent cleaning per SSPC-SP1 (SSPC 2004) to ensure adequate cleanliness.

5. Intermittent welds: Plates welded with intermittent welds are similar to overlapping plates, creating areas inaccessible for coating. Moisture is able to migrate between the plates, where corrosion can occur.

The best way to seal the area is with a continuous weld, which should then be cleaned, ground, and stripe-coated before coating. If continuous welding is not practical, then caulking the open edge should be done before stripe-coating.

6. Construction aids: Once construction aids have been removed, the aids' connection points should be repaired as necessary and inspected for surface defects by appropriate nondestructive examination (NDE), the surfaces should be properly prepared, and a proper coating should be applied to avoid a premature failure of the surrounding coating.

10.3.2 Protective Linings and Coatings

Corrosive environments to which penstocks are exposed dictate the use of high-performance industrial coatings and linings in conjunction with premium surface preparation, application, and cure.

10.3.2.1 General No protective coating provides perfectly continuous coverage of the substrate. Bonded dielectric coatings can be perforated at damaged areas and pinholes. Mortar and concrete coatings can be cracked and spalled. Thus, some ferrous surface is exposed regardless of coating type.

Bonded dielectric coatings protect the steel substrate from corrosion by forming an electrically insulating barrier between the steel substrate and the electrolyte. The effectiveness of bonded dielectric coatings is a function of the bond developed between the coating and the substrate, the impermeability of the coating, and its continuity and durability. Although the coating may be perfectly continuous at the application plant, defects occur during handling, shipping, and installation, and in service after installation. The defects usually are in the form of pinholes and damaged areas.

Surfaces beneath bonded dielectric coatings cannot serve as anodic or cathodic sites. Depending on the degree of continuity (efficiency) of the coating, the reactive surface area can be greatly reduced. The reduction of available surface to serve as cathodic sites greatly reduces the cathode-to-anode surface area ratio, resulting in a reduction in corrosion activity.

The bonded dielectric coatings specified for steel penstocks typically exhibit efficiencies of 99% and greater. That is, only 1% or less of the surface area is exposed to behave anodically or cathodically. This action in turn reduces the corrosion activity, as discussed previously.

Portland cement concrete and mortar protect steel by a different mechanism than the bonded dielectric coatings. Concrete embedment and mortar coatings change the environment to which the steel is exposed. The highly alkaline (12.5 pH) environment provided by the hydrated cementitious materials, in the presence of oxygen, oxidizes (passivates) the steel surface to form a relatively stable hydrous ferric oxide. The passive film ennobles the surface, resulting in potentials similar to materials such as copper and stainless steel or about 0.5 V or more positive (cathodic) than steel exposed to neutral pH solutions. Thus, under sound mortar or concrete, the steel normally is protected from galvanic corrosion.

However, mortar and concrete are permeable to water and, being hygroscopic, attract moisture. Permeability of the coating varies, depending on its porosity, from location to location. Wetting and drying, caused by rising and falling groundwater within the pipe trench, result in the accumulation of a greater concentration of salts in the coating than the water from which they were derived.

Similar to dielectric coatings, mortar and concrete coatings of perfect integrity are not achievable in practice. They are brittle (strain-intolerant) materials and develop only minimal bonding to steel. These characteristics lead to cracking and spalling so that all surfaces of the steel are not exposed to the passivating, high-pH environment. Cracks and spalls allow unobstructed ingress of oxygen and depassivating ions, such as chlorides. Complete consolidation of the coating so that steel surfaces contact the passivating environment is not likely.

A common, unavoidable pH concentration cell occurs at mortared, rubber-gasketed penstock joints. A high water–cement ratio grout is required to fill a diaper form around the joint. Excessive drying shrinkage is experienced, and the mortar cracks. This cracking can be minimized if water reducing agents, to increase workability, are used in lieu of a high water–cement ratio.

Cementitious materials also are subject to loss of alkalinity by carbonation. Being ionically conductive, the underlying steel is vulnerable to stray current corrosion; therefore, the common causes of corrosion of steel embedded in concrete or mortar are failure to achieve passivation of surfaces initially, loss of passivity at localized areas, and stray current corrosion.

Because cementitious coatings are conductive, underlying steel surfaces are available to behave either cathodically or anodically. Because loss of passivity occurs at localized areas only, the cathode-to-anode area relationship is large, and high anodic current densities and intense corrosion activity result. The corrosion products of steel are more voluminous than the alloy. This phenomenon causes cracking and spalling of the mortar and concrete quite similar to cyclic freezing and thawing. Corrosion can be accelerated on these newly exposed surfaces.

10.3.2.2 Protective Linings Protective linings are materials used in the interior of penstocks to protect the metal from corrosion and wear.

Linings may be applied in the shop after fabrication or in the field at the time of installation. It is always necessary to field repair the lining after installation because of construction or installation damage, such as trucking, placement, or welding. It is recommended that these repairs be performed using the same materials as those used for initial application, or one that is recommended by the lining manufacturer.

Some lining materials may be adversely affected by cavitation if they are subjected to high water velocities. The type of lining material selected must take into account many factors, including cavitation, abrasive materials, application conditions, and service conditions.

The interior of penstocks can be lined with various materials, some which are listed in Table 10-1, provided the material is manufactured and applied in accordance with the applicable standards.

10.3.2.3 Protective Coatings Coatings for the exterior surfaces of penstocks protect against corrosion and are provided to meet aesthetic requirements. The coatings may be applied completely in the shop at the time of fabrication or in the field at the time of installation. Most coatings should be applied in the shop where there are facilities that can control application conditions.

Commercial coating materials and manufacturers of specific coating materials should be researched rather than relying solely on one set of standards. Organizations that provide specifications or standards for coatings include the American Water Works Association (AWWA), the Society for Protective Coatings (SSPC), ASTM International, the U.S. Bureau of Reclamation, the U.S. Army Corps of Engineers, the U.S. Department of Defense, and the National Association of Corrosion Engineers (NACE).

The following are recommended coating materials classified according to their use as (1) exposed penstock coatings (atmospheric), (2) buried penstock coatings (soil exposure), and (3) coatings for encasement in concrete.

10.3.2.3.1 Atmospheric Coatings The exterior of penstocks exposed to weather can be coated with one of the materials listed in Table 10-2, manufactured and applied in accordance with the listed standard. Paint material, including thinner, for coating must be made by the same manufacturer.

10.3.2.3.2 Coatings For Soil Burial The exterior of buried penstocks may be coated with one of the materials listed in Table 10-3, manufactured and applied in accordance with the listed standards.

10.3.2.3.3 Encasement in Concrete The coating applied to the exposed portion of the penstock must extend 6 to 12 in. into the concrete embedment. The use of waterstops should be considered.

10.3.2.4 Material Selection The engineer must ensure that the coating and lining system being selected is compat-

Table 10-1.   Penstock Lining Material

System	Listed standard
Coal-tar protective coatings and linings for steel water pipelines—Enamel and tape—Hot applied	AWWA C203 (2008a)
Cement-mortar protective lining and coating for steel water pipe—4 in. (100 mm) and larger—Shop-applied	AWWA C205 (2007a)
Liquid-epoxy coating systems for the interior and exterior of steel water pipelines	AWWA C210 (2007b)
Fusion-bonded epoxy coating for the interior and exterior of steel water pipelines	AWWA C213 (2007c)
Polyurethane coatings for the interior and exterior of steel water pipe and fittings	AWWA C222 (2008c)
Cement-mortar lining of water pipelines in place—4 in. (100 mm) and larger	AWWA C602 (2011a)

Table 10-2.   Exposed Penstock Coatings (Atmospheric)

System	Listed standard
Coal-tar protective coatings and linings for steel water pipelines—Enamel and tape—Hot	AWWA C203 (2008a)
Cement-mortar protective lining and coating for steel water pipe—4 in. (100 mm) and larger—Shop applied	AWWA C205 (2007a)
Liquid-epoxy coating systems for the interior and exterior of steel water pipelines	AWWA C210 (2007b)
Fusion-bonded epoxy coating for the interior and exterior of steel water pipelines	AWWA C213 (2007c)
Liquid coating systems for the exterior of aboveground steel water pipelines and fittings	AWWA C218 (2008b)
Polyurethane coatings for the interior and exterior of steel water pipe and fittings	AWWA C222 (2008c)

Precautionary notes:

1. For many liquid coatings, the time limit between coats must fall within a specific window, which is based on temperature. When subsequent applications are too soon, solvent entrapment may occur. When subsequent applications are too long, poor adhesion may occur. If the application time has been exceeded, the coating must be preconditioned with a brush-off blast before applying additional coats.
2. Consideration must be given to the effects of light-colored penstock coatings on ice buildup in cold climates and of dark-colored coatings on penstock expansion and contraction in hot climates.
3. Many coatings chalk when exposed to sunlight unless top-coated with an ultraviolet-tolerant coating, such as an aliphatic polyurethane. Chalking generally affects appearance only and not performance.

ible with the expected service conditions, the location of application, and the environmental and safety conditions.

The following should be considered in selecting various coating and lining systems:

1. The service conditions the system will be subjected to, such as atmospheric conditions for coatings, water quality, and bed load for linings, are important.

2. Application temperature is important for the type of system selected, in that most solvent-borne systems require higher temperatures to cure. This is important with multicoat systems for cure time between successive coats.
3. Compared with field-applied coating types, shop-applied coatings and linings yield themselves to a

Table 10-3.   Buried Penstock Coatings

System	Listed standard
Coal-tar protective coatings and linings for steel water pipelines—Enamel and tape—Hot applied	AWWA C203 (2008a)
Cement-mortar protective lining and coating for steel water pipe—4 in. (100 mm) and larger—Shop applied	AWWA C205 (2007a)
Cold-applied tape coatings for the exterior of special sections, connections, and fittings for steel water pipelines	AWWA C209 (2006)
Liquid-epoxy coating systems for the interior and exterior of steel water pipelines	AWWA C210 (2007b)
Fusion-bonded epoxy coating for the interior and exterior of steel water pipelines	AWWA C213 (2007c)
Cold Applied Tape coating systems for the exterior of steel water pipelines	AWWA C214 (2007d)
Extruded polyolefin coatings for the exterior of steel water pipelines	AWWA C215 (2010)
Heat-shrinkable cross-linked polyolefin coatings for the exterior of special sections, connections, and fittings for steel water pipelines	AWWA C216 (2007e)
Liquid coating systems for the exterior of aboveground steel water pipelines and fittings	AWWA C218 (2008b)
Polyurethane coatings for the interior and exterior of steel water pipe and fittings	AWWA C222 (2008c)
Fused polyolefin coating systems for the exterior of steel water pipelines	AWWA C225 (2007f)
Fusion-bonded polyethylene coating for the exterior of steel water pipelines	AWWA C229 (2008d)

broader range of coating types because of environmental controls.

4. Material selection to reduce maintenance repair or complete removal or replacement, particularly for lining systems, is important for a number of reasons, including, but not limited to
   a. Workers cannot adequately control the environmental conditions.
   b. Interior work must be treated as work in a confined space, as defined by the Occupational Safety and Health Administration (OSHA), and thus ventilation and material explosive levels must be monitored.
   c. 100% solid materials have economic and safety benefits because of minimizing volatile solvents. Material cure time is shortened because cure is developed by a chemical reaction, as opposed to an evaporative process.

10.3.2.5 Surface Preparation   The penstock surface must be prepared in accordance with the standard referred to in the project specifications for the material selected and the manufacturer's instructions. In addition, the procedures given below must be followed:

1. Surfaces to be encased in concrete must be cleaned and coated before concreting.
2. During or after abrasive blasting, any accumulations of slag, dirt, blisters, weld spatter, metal laminations, or other irregularities must be removed by appropriate mechanical means. Rough edges and welds must be smoothed or rounded and then brush-coated.
3. Bolts, nuts, and other appurtenances to remain uncoated must be cleaned and masked before coating of the penstock.
4. Piezometer and flowmeter holes must be plugged before the coating application.
5. Metalwork to be welded in the field should be left bare. Any coating or lining should be held back 6 in. from the end of each section of pipe to be field-welded.
6. Metalwork must be coated after welding.

10.3.2.6 Coating Application and Cure   Coating materials must be applied in accordance with the standard referred to in the project specifications for the material selected and the manufacturer's instructions. This standard includes, but is not limited to, material mixing, application, dry film thickness, handling, and potlife (i.e., the amount of time the coating can be used before it starts to solidify), if applicable. Limitations on ambient and substrate temperatures, relative humidity, and dew point must be observed as stated in the referenced standard or in the recommendations of the manufacturer of the coating material. The contractor must be experienced in the application of the type of coating specified. In addition, the following general procedures must be followed (some procedures do not apply to all coatings):

1. Until the coating is dry or cured, the penstock must be free of airborne dust. Exterior coating must not be applied during winds in excess of 20 mi/h unless enclosures are provided. Surfaces that become contaminated between applications of coats must be recleaned before application of the next coat. Surfaces to be coated or lined must have temperatures 5°F above the dew point.
2. A coating system agreed upon by the owner and the contractor must be used on bolts, nuts, and other appurtenances.
3. The curing time between coats must be rigidly followed.
4. Items that are normally factory-coated may be furnished primed or primed and top-coated, provided that such coatings are compatible with subsequent coatings to be applied in the field.
5. A one-quart sample representing the coating or lining material must be furnished before shipment for each type, batch, lot, and color of liquid and mastic used when quantities exceed 20 gal. The constituents of multiple-component coatings must be furnished separately.
6. When multiple coats are applied, successive coats must be shaded sufficiently to denote color differences between coats. The final coat color must meet the project specifications.
7. Penstock sections must be identified to facilitate proper erection without marking the exterior of any exposed part of penstock coatings.

10.3.2.7 Inspection   The owner's qualified representative, such as a NACE certified coating inspector, must inspect the coating and lining work to ensure that standards and project specifications are met. Coating and lining material, test reports, surface preparation checks, and inspection procedures must conform with the standards and project specifications. Shop-applied coatings must be holiday tested, inspected, and repaired before shipment. Coating and surface preparation must be documented by inspection reports. Reports must list work done, materials used, dry film thickness, batch numbers, date, weather conditions, and any deviations from project specifications or standards.

Field coatings and linings must undergo the same inspection as shop coatings and linings. Reports of tests and inspections must be made available to the owner.

10.3.2.8 Storing Coating and Lining Material   Materials must be stored in compliance with local regulations and manufacturers' recommendations and must remain in the original sealed containers until used. Materials must be used within the time limit recommended by the manufacturers. The container of each material must indicate the manufacturer's name, stock number, hazardous substances, type of coating, specification number, color, date of manufacture, batch number, and instructions for reduction, where applicable. Such information must be plainly legible at the time the material is used.

ASTM WK 23937
POLYUREA - NO AWWA STANDARD
but easier to Apply - SofterMaterial
"TAKES UP MORE WATER"

Coatings and primers must be delivered in container sizes ordered by the contractor. Coatings must be thinned with the coating manufacturer's thinner when thinning is permitted.

10.3.2.9 Regulations Many federal, state, and local jurisdictions have established abrasive, paint, and coating regulations. Coatings and linings and their application procedures must comply with the regulations in effect at the sites where the materials are applied or used.

For all locations, the manufacturer of the material must furnish the applicator and owner a manufacturer's material safety data sheet (MSDS) describing any components that may be hazardous. The manufacturer must certify that the materials comply with pollution, safety, safety equipment, and lighting regulations in effect at the point of application. Abrasives and coatings must be disposed of in accordance with regulations.

10.3.2.10 Safety The utmost important aspect of coating and lining of penstocks is safety. To prevent injury or a fatality on a project, therefore

1. Coating and lining materials and applications must be compatible with the environment in which they are used.
2. The selection process for the application contractor must focus on the past and current safety record of the application contractor, including application of like materials, and not just lowest cost.
3. The work area needs to be in compliance with the appropriate OSHA requirements (e.g., confined space (permitted or not), tunnel safety orders, fall protection, and ventilation).

10.3.2.11 Manufacturer's Information Submittals Copies of the manufacturer's recommendations for mixing and applying coating materials must be furnished to the penstock owner before application. The manufacturer's certificate of compliance (indicating both compliance with the standard to which the material was made and that the sample is from the actual batch or lot to be furnished) must be furnished for each material before its use in the work. The certificate must include material identification, quantity, batch number, date of manufacture, certified test results, and specifications (if available) for the material being used.

10.3.2.12 Handling Coated penstock sections must be protected against damage during transporting and handling in accordance with AWWA (2005), AWWA (2011b), or for shop-coated sections, in accordance with the appropriate coating standard.

10.3.3 Cathodic Protection

Cathodic protection is an electrochemical corrosion prevention and control method. It is the only method with the potential to eliminate totally the initiation of corrosion activity or halt ongoing corrosion. To achieve cathodic protection, a favorable electrochemical circuit is established by installing an external electrode in the electrolyte and passing direct current (conventional) from that electrode through the electrolyte to the structure to be protected. This electrical current polarizes the potential of the cathodic surfaces (relatively positive) on the structure to that of the anodic (more negative) surfaces. When this polarization is accomplished, there is no current flow between the formerly anodic and cathodic surfaces, and corrosion is prevented or eliminated. This procedure represents a balanced or equilibrium condition. In normal practice, sufficient current is passed to the surfaces so that the formerly anodic surfaces receive current from the electrolyte and their potential shifts in the more negative direction. In any event, the cathode surfaces require the majority of the current.

Achieving cathodic protection requires that all surfaces of the structure collect sufficient current to overcome the naturally occurring corrosion potentials. This process requires that all surfaces of the structure be electrically continuous. Although welded pipe joints provide for electrical continuity, rubber-gasketed joints do not. Thus, the latter must be fitted with joint jumper bonds.

Cathodic protection anodes must be placed in the same electrolyte as the structure. Because anodes placed inside the penstock would be an obstruction to high-velocity flow, internal surfaces of penstocks usually are not cathodically protected.

Cathodic protection is categorized into two types, depending on whether the source of the required current is galvanic or impressed current.

10.3.3.1 Galvanic Anodes Galvanic anodes commonly used for the protection of steel are made of magnesium or zinc. When coupled to steel, a favorable corrosion cell is established in which the zinc or magnesium corrodes and the steel is protected. Corrosion is not eliminated but is transferred to the galvanic anode, which sacrifices itself to protect the structure and is consumed. Hence, galvanic anodes are often referred to as sacrificial anodes. Galvanic-cell-generated voltage differences are relatively small. When coupled to steel, zinc and magnesium develop cell voltages of about 0.5 and 1.0 V, respectively. Because of these relatively low driving voltages, the cell current is normally very small. Typical output per anode is on the order of 50 to 100 mA. The corrosion rates of magnesium and zinc are 17 and 26 lb/ampere-year, respectively. As galvanic anodes are consumed, they must be replaced. For example, a 17-lb magnesium anode delivering 100 mA of current theoretically lasts 10 years.

10.3.3.2 Impressed Current In contrast to the sacrificial anode, the impressed current method requires an external source for direct current (DC) power. Typically, power is provided by a rectifier, which converts alternating current (AC) power to a pulsating form of DC energy. However, with the impressed current method, the anode is not required to provide a favorable voltage relationship with steel. Thus, more noble materials (materials that corrode or are consumed slowly with the discharge of current) can be installed as anodes. Common impressed-current anodes include graphite and high-silicon cast iron, which corrode at rates less than 2 lb/ampere-year. More exotic materials, such as the mixed-metal oxides, are consumed at even lower rates. Because the current is provided by an external source, driving voltage, and consequently current output, are virtually unlimited.

10.3.3.3 Protection Criteria   A potential of at least -0.85 V relative to a copper–copper sulfate reference electrode (CSE) is the commonly applied minimum criterion for protection. NACE (2007) provides additional minimum protection criteria. To prevent potential damage, such as cathodic disbondment and hydrogen embrittlement from hydrogen effects, the maximum (absolute value) potential is -1.10 V.

10.3.3.4 System Design   Obtaining cathodic protection within the potential tolerances requires not only sufficient current but also proper distribution of the current. The external electrode (either the galvanic or impressed current anode) is referred to as a cathodic protection ground bed. The magnitude of the current, its distribution (number of ground beds required), and, consequently, the energy requirement are related to the coating and its specific leakage resistance. For example, cement-mortar coated steel requires significantly more current than steel with a bonded dielectric coating. This difference results in substantial differences in costs of the installed systems and their operating costs. For a specific application, one type of protection (galvanic anode or impressed current) may provide a distinct advantage over the other. Thus, selection of a compatible protective coating, type of system to be used (galvanic anode or impressed current), and design of the cathodic protection system, including provisions for electrical continuity and isolation, should be made by an experienced cathodic protection specialist. NACE certifies professionals capable of designing cathodic protection systems.

10.3.4   Corrosion Monitoring

Corrosion monitoring systems provide for electrical continuity and electronic access, thus facilitating the application and monitoring of cathodic protection. Bonds are installed across rubber-gasketed (including coupled) joints. Electronic access is obtained by installation of surface test stations with leads originating from the pipe metallic components. Typically, test stations are installed on either side of insulating fittings at the upstream and downstream termini of the penstock, at buried line crossings, and at intermediate points along the penstock, such that they are no more than 1,000 ft apart. Corrosion monitoring systems are designed so that corrosion activity on the penstock can be evaluated through data generated by surface testing. Where penstocks are encased in a dam, test leads should be installed at a place where a reading can be taken. If the lead wire must be a long

distance to get the leads to a point where they can be read, the wire should be appropriately sized to reduce resistance.

One commonly used test is the determination of pipe-to-soil potential. To conduct this test, the pipe lead is connected to the positive terminal of a high-impedance voltmeter. The negative terminal is connected to a CSE contacting the soil surface directly over the penstock. The soil is saturated at the point of electrode contact to reduce contact resistance, thereby increasing the accuracy of the measurement. The premise for this test is that as the penstock corrodes, sampling over time will result in progressively higher negative potential values.

Complications arise because the reference electrode is remotely located from the penstock, typically a minimum of 5 ft. Use of the remote electrode results in determination of an average potential of underlying metal, which may be composed of both anodic and cathodic surfaces. If the anode surface is large relative to the cathode, the measured potential is dominated by the anode potential and produces the desired effect. However, if the cathode is large relative to the anode, for example, mortar-coated or concrete-encased, the measured potential is skewed toward that of the cathode.

Thus, potential measurements can be grossly misleading. That is, an increase in corrosion activity may be accompanied by little or no change in observed potential. Thus, monitoring is not as applicable to the mortar-coated and concrete-encased penstock as to dielectric coated steel. Because of the foregoing potential problems in testing and interpretation of test results, monitoring also should be conducted by an experienced corrosion engineer.

10.4   FAILURE CAUSED BY CORROSION

Typically, corrosion results in wall perforations and leakage. Repair requirements vary from weld repair of a perforation to total line refurbishment or replacement. Added to the cost of repair or replacement is the direct cost of any accompanying flooding. Failure of penstocks could result in serious disruption of electrical service. Penstock failures pose a serious threat to safety and can have serious economic consequences. Failures can be minimized through proper application of sound engineering principles to prevent and control corrosion.

REFERENCES

AWWA. (2005). "Steel water pipe—6 in. (150 mm) and larger." *ANSI/AWWA C200-05*, Denver.

AWWA. (2006). "Cold-applied tape coatings for the exterior of special sections, connections, and fittings for steel water pipelines." *ANSI/AWWA C209-06*, Denver.

AWWA. (2007a). "Cement-mortar protective lining and coating for steel water pipe—4 in. (100 mm) and larger—Shop-applied." *AWWA C205-07*, Denver.

AWWA. (2007b). "Liquid-epoxy coating systems for the interior and exterior of steel water pipelines." *ANSI/AWWA C210-08*, Denver.

AWWA. (2007c). "Fusion-bonded epoxy coating for the interior and exterior of steel water pipelines." *ANSI/AWWA C213-07,* Denver.

AWWA. (2007d). "Tape coating systems for the exterior of steel water pipelines." *ANSI/AWWA C214-07,* Denver.

AWWA. (2007e). "Heat-shrinkable cross-linked polyolefin coatings for the exterior of special sections, connections, and fittings for steel water pipelines." *ANSI/AWWA C216-07,* Denver.

AWWA. (2007f). "Fused polyolefin coating systems for the exterior of steel water pipelines." *ANSI/AWWA C225-07,* Denver.

AWWA. (2008a). "Coal-tar protective coatings and linings for steel water pipelines—Enamel and tape—Hot-applied." *ANSI/AWWA C203-08,* Denver.

AWWA. (2008b). "Liquid coating systems for the exterior of aboveground steel water pipelines and fittings." *ANSI/AWWA C218-08,* Denver.

AWWA. (2008c). "Polyurethane coatings for the interior and exterior of steel water pipe and fittings." *AWWA C222-08,* Denver.

AWWA. (2008d). "Fusion-bonded polyethylene coating for the exterior of steel water pipelines." *ANSI/AWWA C229-08,* Denver.

AWWA. (2010). "Extruded polyolefin coatings for the exterior of steel water pipelines." *ANSI/AWWA C215-10,* Denver.

AWWA. (2011a). Cement-mortar lining of water pipelines in place—4 in. (100 mm) and larger." *ANSI/AWWA C602-11,* Denver.

AWWA. (2011b). "Installation of steel water pipe—4 in. (100 mm) and larger." *AWWA C604-11,* Denver.

NACE. (2007). "Control of external corrosion on underground or submerged metallic piping systems." *NACE SP0169-2007,* Houston.

SSPC. (2004). "Solvent cleaning." *SSPC-SP1,* Pittsburgh.

FCAW

GMAW

CHAPTER 11

Welding

Source: Photograph courtesy of National Welding Corp.; reproduced with permission.

Welding of steel penstocks is critical to the success of hydro-electric projects. Welding codes formalize the application of welding technology so that a predictable outcome is possible. Application of welding codes to penstock installation provides a measure of quality assurance by requiring qualified welders, welding procedure specifications, and welding inspection.

Welding is used extensively in both shop and field fabrication of penstocks, and applicable provisions of the welding codes must be used in design and fabrication of penstocks. This manual, code design, and inspection provisions, in addition to production welding requirements, form the basis of welded steel penstock fabrication.

Welds that connect penstock sections must be designed to resist forces imposed on them. The full strength of the adjoining steel plate or coil can be attained by using complete joint penetration (CJP) butt-joint welds. If lesser strength is acceptable, then the more economical fillet welds are often used. Weld strength values can be determined by applying code recommendations.

11.1 MATERIALS AND MATERIAL TOLERANCES

11.1.1 Material Selection and Welding

Material selection, inspection, shop quality assurance, and installation processes must be clearly defined either through the use of applicable codes and standards or by calling out the requirements directly in the project specifications. Many possible materials may be suitable for steel penstock projects, depending on the application. Mechanical properties of carbon and high-strength, low alloy (HSLA) steels of interest for penstocks are discussed in other sections. Welding materials (such as electrodes, fluxes, and shielding gases) used for penstocks must comply with the requirements of the ASME (2010a), ASME (2010b), and the applicable welding procedure specification (WPS). Field-welding requirements and welder qualifications must be consistent with those for the shop fabricating process.

Welding is generally a requirement with these penstock materials, and therefore the base metal should be provided, in accordance with Chapter 2, and should dictate two key property requirements: weldability and toughness.

11.1.2   Material Tolerances

For an acceptable design thickness of the shell plate without any upward adjustment, it is important to apply minimum and maximum acceptance tolerances for shop and field weldment alignments at weld joints, as required by ASME (2010a).

Alignment and positioning of parts before welding, weld preparation, the welding, and nondestructive examination (NDE) processes must reflect the design intent and requirements. A description of NDE processes is listed in Chapter 14.

11.2   SUBMITTALS

Procedures for welding and NDE must be submitted before the start of work. Copies of the following should be submitted: procedure qualification record (PQR), welding procedure specification (WPS), welder performance qualification (WPQ), welding sequences, repair procedures, and welding rod control procedures. Welding procedures must be qualified in accordance with ASME (2010b) or AWS (2010).

Documentation must include a weld map or table identifying each type of weld joint, the assigned WPS, the parts being joined, the material thickness, and any requirements for preheat and postweld heat treatment.

11.2.1   Procedure Qualification Record

The PQR documents the welding parameters and conditions that occurred during welding of the test coupon and the results of testing the coupon. The PQR provides documentation of the essential variables and other specific information identified for each process used during welding, plus the actual results of testing the weld coupon. In addition, when notch toughness testing is required for procedure qualification, the applicable supplementary essential variables for each process are recorded.

11.2.2   Welding Procedure Specification

The WPS is a written document that provides direction to the welder or welding operator for making production welds in accordance with code requirements. The WPS provides the welding parameter ranges within which the welder or welding operator is allowed to operate. The purpose for qualification of a WPS is to determine that the weldment proposed for construction is capable of providing the required properties for its intended application. A WPS establishes the intended properties of the weldment, not the skill of the welder or welding operator.

11.2.3   Welder Performance Qualification

The WPQ establishes the basic criteria to determine the welder's ability to deposit sound weld metal, according to a WPS. The purpose of the performance qualification test for the welding operator is to determine the welding operator's mechanical ability to operate the welding equipment.

11.2.4   Welding Codes and Standards

Welding of shells and assemblies classified as pressure parts must be in accordance with ASME (2010a), AWS (2010), one of the ASME B31 standards (ASME multiple years), or AWWA (2005), as further described in Chapter 12. The welding processes used in the construction of penstocks are restricted to those outlined in this chapter. No production welding can be undertaken until the welding procedures have been qualified or are prequalified by AWS (2010). Only welders and welding operators who have been qualified in accordance with ASME (2010b) or AWS (2010) can be used in production and installation.

See Chapter 16 for specific requirements and qualifications.

11.3   JOINT TYPES AND CONFIGURATIONS

The engineer must evaluate the joint type that meets the project design requirements with consideration to the associated cost impact. See Chapter 3 for design considerations, the allowable stress intensities, and the appropriate corresponding joint types.

11.3.1   Butt Joints

Butt joints are created when the two pipe ends are in the same plane and do not overlap. They are frequently used for penstocks, and, depending on the inspection method, they can provide the highest joint efficiency of all available joint types. Welded joints required to be examined by radiography and ultrasound must be CJP butt welds.

11.3.2   Butt Joint with Backing

Where clearance between the pipe exterior and obstructions, such as a tunnel wall, is too restrictive or the pipe diameter is too small to permit adequate working access for welders and NDE technicians, the use of a butt joint with steel backing, as shown in Fig. 11-1, is recommended. Butt joints with steel backing should not be radiographically inspected because of the restricted access and the inherent difficulty of interpreting the radiographic test results with this type of joint configuration. This joint type allows the installer the most latitude, although the joint design criteria should take this joint configuration into account. The steel backing thickness should be a minimum of 1/4 in., although a heavier thickness may be required to prevent melt-through.

11.3.3   Fillet Welded Lap Joints

The fillet welded lap joint is created when one pipe end overlaps the adjoining pipe end. Fillet welded lap joints are often used to join penstock sections in the field because they are generally more economical. Also, this joint type facilitates changes in pipe alignment. The double-welded lap joint and single-welded lap joint are assigned more conservative joint efficiencies of $E = 0.55$ and $E = 0.45$, respectively. It is recommended that fillet welded lap joints should not be radio-

graphically inspected because of the weld geometry. Also, Table UW-12 in ASME (2010a) limits fillet welded lap joints on pressure vessels to 3/8 in. maximum shell thickness for longitudinal seams and 5/8 in. maximum shell thickness for girth welds, although fillet welded lap joint girth seams have been successfully used on many penstocks with shell thicknesses up to 1 in. thickness. Refer to Chapter 3 for joint efficiencies allowed for fillet welded lap joints and limitations.

11.3.4   Single-Welded Lap Joints

Single-welded lap joints are frequently used as an economical method of construction when the penstock design allows the joint efficiency provided by this joint type. The assembly of a welded lap joint is much easier than the assembly of butt-welded type joints because of the overlapping members. Single fillet welded lap joints may be welded from the inside or outside.

11.3.5   Double-Welded Lap Joints

Double-welded lap joints allow the same economical methods of construction as the single-welded lap joint but include a weld on both the interior and exterior. This second weld improves the joint efficiency and facilitates performing an air test to ensure that the joint is leak tight.

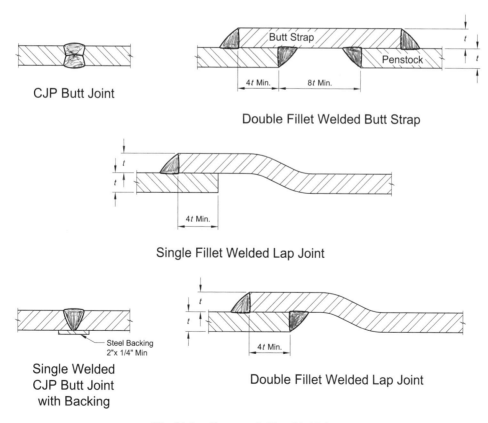

Fig. 11-1.   Common field-welded joints

11.3.6 Plug Welds

Plug welds are a method of improving joint efficiency for circumferential fillet welded lap type joints. They may be used in lap joints, in reinforcements around openings, and in nonpressure structural attachments. They should be properly spaced to carry their proportion of the load but must not be considered to take more than 30% of the total load to be transmitted. Fig. 11-2 shows examples of plug welds.

11.4 JOINT CONFIGURATIONS

Fig. 11-1 shows the common joint configurations. Refer to ASME (2010a) for joint configurations of nozzle and manway penetrations, attachments, and corner joints. Fig. 11-3 shows a selection of joint configurations.

11.5 WELDING PROCESSES, EDGE PREPARATION, PREHEAT, NOTCH-TOUGH WELDING, AND ASSEMBLY

11.5.1 Welding Processes

Welding of penstock sections is commonly performed using manual shielded metal arc welding (SMAW), semiautomated flux cored arc welding (FCAW), or submerged arc welding (SAW) processes. The FCAW process is subdivided into two variations: FCAW-S, or self-shielded flux cored arc welding, and FCAW-G or flux cored arc welding with gas shielding. Other arc welding processes, including gas metal arc welding (GMAW) and electroslag welding (ESW), have been used for penstocks.

11.5.2 Edge Preparation

Joint edge preparation is done by machining, thermal cutting, gouging (including electric arc cutting and gouging processes, plasma arc cutting, and gouging), chipping, or grinding. Oxygen gouging is not to be used on steels that are ordered quenched and tempered or normalized. Edges to be welded should be clean or cleaned by grinding and should form the joint angle required by the applicable welding procedure. Surfaces on which weld metal is to be deposited must be smooth, uniform, and free from fins, tears, cracks, and other discontinuities that would adversely affect the quality or strength of the weld. Surfaces to be welded and surfaces adjacent to a weld must be free from loose or thick scale, slag, rust, moisture, grease, and other foreign material that would prevent proper welding or produce objectionable fumes. Steel and weld metal may be thermally cut, provided a smooth and regular surface free from cracks and notches is secured, and provided that an accurate profile is secured. Cut surfaces and adjacent edges must be left free of slag.

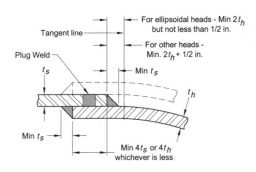

(a) Double Fillet Lap Weld with Plug Weld

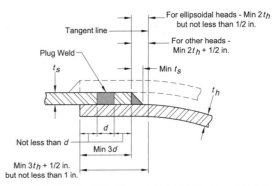

(b) Single Fillet Lap Weld With Plug Weld

Notes:
1. Plug weld holes should have a diameter not less than t + 1/4 in. and not more than $2t$ + 1/4 in. where t is the thickness in in. of the plate or attached part in which the hole is made.
2. Plug weld holes should be completely filled with weld metal when the thickness of the plate, or attached part, in which the weld is made is 5/16 in. or less; for thicker plates or attached parts the holes should be filled to a depth of at least half the plate thickness or 5/16 in. of the hole diameter, whichever is larger, but in no case less than 5/16 in.
3. Additional design information may be found in ASME BPVC Section VIII, Division 1, ASME (2010a).

Fig. 11-2. Plug welds
Source: Reprinted from ASME (2010a), by permission of the American Society of Mechanical Engineers. All rights reserved.

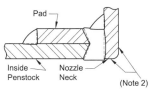

(A) Nozzle with Pad Reinforcing Section

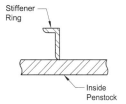

(B) Attachment

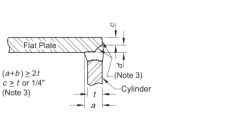

(C) CJP Corner

$(a+b) \geq 2t$
$c \geq t$ or 1/4"
(Note 3)

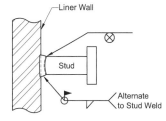

(D) Stud-Welded Attachment

Notes:

1. When lap welded penstock sections with differing wall thicknesses are joined the weld size is only required to be the thickness of the thinner member.
2. If justified by VT, examine surfaces further by MT before and after welding to detect presence of lamellar tearing.
3. If justified by VT, examine weld prep by MT before welding. Repair defects. After welding, again examine weld surface by MT.

Fig. 11-3. Configurations for nozzle and manway penetrations, attachments, and corner joints
Source: Reprinted from ASME (2010a), by permission of the American Society of Mechanical Engineers. All rights reserved.

Edges of plates prepared for welding should be visually examined for signs of mill-induced discontinuities, laminations, shearing cracks, slag, and other imperfections. Pieces showing signs of laminations should be repaired to the satisfaction of the engineer. Where laminations are found, ultrasonic examination (UT) must be used to ensure that the laminations do not extend beyond the allowances of the appropriate ASME or ASTM codes.

11.5.3 Preheat and Interpass Temperature

The minimum preheating temperature for welded joints is established by the controlling code. A metallurgical or welding engineer should be consulted about mandatory or recommended preheat requirements. The WPS for the material being welded specifies the minimum preheating temperature under the PQR requirements of ASME (2010b) or AWS (2010).

Successful welding results depend upon essential variables, which include preheat and interpass temperature maintenance to prevent weld cracking. Also, preheat and interpass temperature maintenance can be beneficial in reducing thermal gradients that contribute to distortion from welding. Preheat and interpass temperature values are controlled by the parent material selected, its carbon equivalent, plate thickness, and other variables. Guidance for preheat and interpass temperatures is provided in AWS (2010) or ASME (2010a). In general, higher strength and thicker base metals require higher preheat and interpass temperatures. As an alternate, Annex I of AWS (2010) provides optional methods for determining welding conditions to avoid cold cracking; several methods are presented to avoid cold cracking, including heat-affected zone (HAZ) hardness control and hydrogen control. Preheat and interpass temperatures must be maintained for a distance of at least 3 in. in all directions from the point of welding and should be checked just before initiating the arc for each pass. Temperature-indicating crayons, which change color when the correct temperature is attained, or infrared temperature detectors can be used to measure preheat and interpass temperatures.

11.5.4 Notch-Tough Welding

Notch-tough welding (i.e., welding that provides welds with good notch toughness) is recommended when there is concern for protection against brittle fracture. Notch-tough welding dictates additional essential variables that are intended to limit heat input during production welding to values no greater than heat inputs experienced during PQR. Welding codes recognize several methods for limiting heat inputs, including electrical characteristics and weld bead size.

11.5.5 Assembly and Thermal Effects

During assembly and welding, attention must be given to minimizing distortions caused by welding. It may become necessary to alternate or change weld locations during the welding process to control distortion. When welding circumferential joints of a large-diameter penstock, adequate stiffening must be provided to avoid weld cracking. In making welds under conditions of severe external shrinkage restraint, once the welding has started, the joint should not be allowed to cool below the minimum specified preheat until the joint has been completed or sufficient weld has been deposited to ensure freedom from cracking.

Decreases in member length caused by temperature effects may be aggravated by weld-related shrinkage, which can reach as high as 1/8 in. for each butt-welded joint.

11.6 FIELD-WELDED JOINTS

Common field-welded joints are typically CJP butt-joint or welded lap joint types, as indicated in Fig. 11-1. Although the butt joint offers greater strength when compared to a lap joint, the fit-up time associated with alignment of edges to be welded is much greater, resulting in greater installation cost, which often favors lap joints. Welding codes recognize that welder skill levels are greater for butt-joint welds than for fillet welds.

Double-welded butt joints, which result in the highest weld strength, require welder access from both sides of the joint. The first pass (root pass) of an open butt weld generally does not provide proper shielding from the atmosphere and frequently results in an unacceptable weld quality. However, having access to both sides of the joint allows the welder to gouge out the root pass from the opposite side and apply sound weld metal. Weld passes after the root weld are shielded and therefore do not have the same weld quality issues.

Single-welded butt joints are often used where welder access is limited to only one side, for example, in steel tunnel liners. This joint configuration is generally provided with a backing bar to improve the quality of the finished weld and assist in assembly.

Welded lap joints are often the joint of choice because they offer reasonable strength and economical means. The joints are easy to assemble, and welder skill level is less than that required for butt-joint welds. Minor changes in alignment can be accomplished by "pulling" the lap joints before welding.

Double-welded butt-strap joints are often used for field station adjustment. They are comparable in strength and economy to double-welded lap joints; however, plain pipe ends are used rather than sized bell-and-spigot joint ends. The butt straps are normally furnished in two 180-degree halves that are field-spliced using CJP butt-joint welds.

11.7 BIFURCATIONS

Fig. 11-4 shows a common joint configuration for bifurcations. Detail 1 of Fig. 11-4 shows an acceptable detail at the junction of a bifurcation shell (or skin) to the center reinforcing girder. Other details have been successfully used; however, the one shown is a good illustration of such a joint. Special precautions are necessary to avoid or minimize susceptibility to lamellar tearing. These special precautions include use of weld overlay and ultrasonic examination before and after welding; use of material conforming to low sulfur practice; manufacture of the material by forging (to ultrasonically tested quality); and the requirement that the material meet specific tensile properties in the three orthogonal directions (i.e., specific through-thickness properties of elongation, impact energy, and tensile strength). The joints between the bifurcation shell and center reinforcement are CJP welds and receive 100% UT, radiographic examination (RT), or magnetic particle examination (MT).

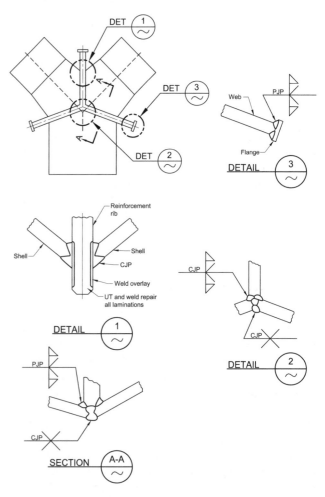

Fig. 11-4. Welded joints—bifurcations

11.8   TAPERS

When the transition is formed by adding additional weld metal beyond what would otherwise be the edge of the weld, such additional weld metal buildup must undergo examination by MT or liquid penetrant examination (PT) and must be included in the examination for the butt weld. The butt weld may be partly or entirely in, or adjacent to, the tapered section (Fig. 11-5).

11.9   ATTACHMENT PIPE WELDS

Circumferential joints in 24-in. nominal pipe size and smaller in diameter may be joined by complete joint penetration single-welded groove butt welds without backing strips. Single fillet welded lap joints are acceptable for circumferential joints in penstocks, provided the joint meets the design requirements.

11.10   GROUT CONNECTIONS

Grout connections are often covered with a steel plate and seal welded after they are no longer needed. An alternate method to seal this penetration would be to seal weld a weldable-grade grout plug directly to the steel penstock. In either case, the creation and usage of the grout connections frequently causes contamination of the area, which could affect the weld quality. Added precautions must be taken to clean cutting fluids, grout, coatings, and other contaminants from the grout connection area before welding to avoid unacceptable weld quality. Carbon steel or stainless steel plates have been used successfully for grout plug cover plates, although in the event that a stainless steel cover plate is used for this purpose, then a welding procedure must be prepared to accommodate the dissimilar metal weld.

11.11   POSTWELD HEAT TREATMENT

Weld procedures must conform to ASME (2010b) or AWS (2010) and must include any preparations required for postweld heat treatment in accordance with that same code.

Penstock sections and parts with this requirement must be given a postweld heat treatment at temperatures not less than those specified in Table UCS-56 of ASME (2010a) or in AWS (2010). These codes provide requirements on the need for postweld heat treatment, which is generally only required for certain steel types or plate thicknesses.

11.12   WELD INSPECTION AND ACCEPTANCE CRITERIA

Field-welding, inspection, and acceptance of penstocks must be in accordance with ASME (2010a), where qualification of WPSs and WPQs are required in accordance with ASME (2010b). Alternately, AWWA (2011) and AWS (2010) may be used. Refer to Chapter 14 for specific inspection requirements.

Welding inspection is necessary to confirm that penstock design requirements have been met. NDE of welding provides a means of examining welds without destroying them and is preferred for field inspection when compared to destructive testing of welds, which requires a testing laboratory and is often performed to qualify welding procedures. Common NDE methods include visual examination (VT), MT, PT, UT, and RT. VT is the most common NDE method used to verify penstock design conformance, and it must be performed by an AWS QC1 certified welding inspector (CWI) (AWS 2007), who is the focal point of welding inspection activities during penstock installation. VT should be performed as necessary before assembly, during assembly, during welding, and after welding to ensure that code requirements are met. Personnel who perform NDE, other than VT, are often certified to Level II of ASNT (2011).

REFERENCES

ASME. (multiple years). B31 Standards of pressure piping, New York.
ASME. (2010a). "Boiler and pressure vessel code, section VIII, division 1: Rules for construction of pressure vessels." New York.
ASME. (2010b). "Boiler and pressure vessel code, section IX: Welding and brazing qualifications." *PD-190,* New York.
ASNT. (2011). "Personnel qualification and certification in nondestructive testing." *Recommended Practice No. SNT-TC-1A-2011,* The American Society of Nondestructive Testing, Columbus, Ohio.
AWS. (2007). "Standard for AWS certification of welding inspectors." *AWS QC1,* Miami.
AWS. (2010). "Structural welding code—Steel." *ANSI/AWS D1.1/D1.1M,* Miami.
AWWA. (2005). "Steel water pipe—6 in. (150 mm) and larger." *ANSI/AWWA C200-05,* Denver.
AWWA. (2011). "Field welding of steel water pipe." *ANSI/AWWA C206-11,* Denver.

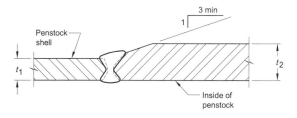

Fig. 11-5.   Taper or change in thickness transition (butt weld)

CHAPTER 12

Manufacture

Source: Photograph courtesy of Northwest Pipe Co.; reproduced with permission.

ᴌ-2ᴏᴏ

12.1 GENERAL

The following manufacturing standards and rules apply specifically to penstocks and penstock parts that are fabricated by welding and constructed of carbon steel, low-alloy steel, high-alloy steel, or heat-treated steel. Specific welding requirements are found in Chapter 11.

12.1.1 Fabrication Standard

Fabrication procedures and tolerances of penstocks must conform to the provisions of ASME (2010), one of the

ASME B31 standards (ASME multiple years), or AWWA (2005). Although these standards and codes are referenced for minimum requirements for fabrication procedures and tolerances, all other provisions of this manual must also be met.

12.1.2 Manufacturing Methods

12.1.2.1 Straight Seam Straight-seam penstocks may be shop- or field-fabricated. They are manufactured from plates that are edge-broken (crimped) and then rolled or

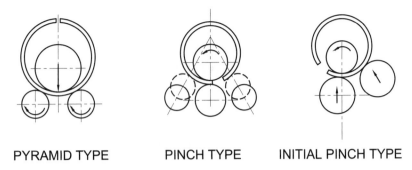

PYRAMID TYPE   PINCH TYPE   INITIAL PINCH TYPE

Fig. 12-1. Bending rolls

formed, resulting in a continuous and uniform curvature using a plate bending roll or U-ing and O-ing forming presses. Several types of bending rolls are available, including (1) pyramid bending rolls, (2) pinch bending rolls, and (3) initial pinch bending rolls (Fig. 12-1). For the pyramid and initial pinch bending rolls, the plate edges must be crimped or edge-broken to the proper radius before rolling. Fig. 12-2 shows U-ing and O-ing presses used for forming. Plates for shell sections must be formed to the required shape by a process that does not unduly impair the physical properties of the material.

Plates for penstock shells must be sized to minimize the number of longitudinal and circumferential seams in the completed section. Penstock courses must be assembled with the longitudinal seams staggered a minimum of 15 degrees.

12.1.2.2 Spiral Seam   Spiral-seam penstocks are manufactured in the shop using steel in coiled form. Currently, wall thicknesses through 1 in. are available. The diameter of the penstock is determined by the angle of the coil feed to the axis of the penstock in relation to the width of the coil. A schematic of a typical spiral mill is shown in Fig. 12-3. In this process, the coil is unrolled, the coil edges are prepared as required, and the coil is helically formed to a true circular shape and welded in a continuous operation. The penstock is cut to length as required by an automatic cut-off device.

Pipe ends for the required field joints may need expansion or sizing to meet the tolerances given in Section 12.3.3.

12.2   QUALIFICATION OF MANUFACTURER

The manufacturer must demonstrate that it has the shop facilities and the fabrication experience with specialty work of similar size and type within the past five years, as well as an internal quality control system similar to ISO 9000 (ISO 2008) or SPFA Pipe Quality Audit Certification Program.

12.3   MANUFACTURING SPECIFICATIONS

Fabrication of the penstock and penstock parts must be performed in accordance with the provisions of the ASME (2010), ASME B31 standards, (ASME multiple years) or AWWA (2005). However, hydrostatic testing is not mandatory and is addressed in Chapter 14. The level of inspection and testing required is determined by the design in Chapter 3 and the testing and inspection requirements in Chapter 14. The fabricator is not required to stamp any part of the penstock with the official ASME code symbol applicable to pressure vessels.

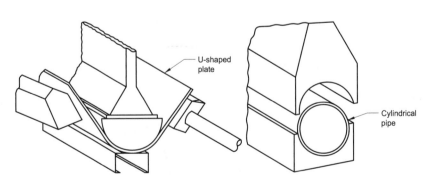

Fig. 12-2. Forming presses

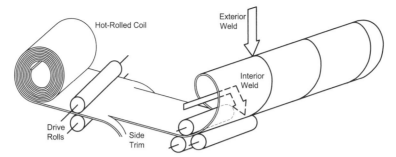

Fig. 12-3. Schematic diagram of process for making spiral-seam pipe

12.3.1   Material

Materials used for steel penstocks and penstock parts must be furnished in strict accordance with the proper standards and design requirements, and specifically in accordance with the requirements of Chapter 2.

12.3.1.1   Material for Pressure Parts   Material subject to stress caused by pressure must conform to one of the specifications given in Chapter 2.

12.3.1.2   Material for Nonpressure Parts   Material for nonpressure parts, such as supports, thrust rings, lugs, and clips, need not conform to the specifications for the material to which they are attached.

12.3.2   Field Joint Ends

Ends of supplied penstock sections must be of the type given in the project specifications.

12.3.2.1   Field Joint End Types   AWWA (2005) describes the following types of ends:

1. Plain-end pipe,
2. Beveled ends for butt welding in the field,
3. Ends fitted with flanges,
4. Ends fitted with butt straps for field-welding,
5. Ends for mechanical couplings,
6. Lap-joint pipe ends for field-welding, and
7. Bell-and-spigot ends with rubber gaskets.

12.3.2.2   Field Joint Ends Manufacturing Tolerances

1. Out-of-Roundness
   For pipe with $D/t < 120$, where D is the nominal pipe diameter and t is the pipe wall thickness, the difference between the major and minor outside penstock shell diameters at pipe ends must not exceed 1%.
2. Outside Diameter
   The outside diameter tolerance, based on circumferential measurement (πD), is $\pm 1/4$ in. for the following:
   a. Plain-end pipe;
   b. Beveled ends for butt welding in the field;
   c. Ends fitted with flanges;
   d. Ends fitted with butt straps for field-welding;

   e. Ends for mechanical couplings, which must have tolerances within the limits required by the coupling manufacturer; and
   f. Lap joints for field-welding (circumferential difference between inside of bell and outside of spigot must not exceed 0.40 in.).
3. Squareness of Ends for Butt Welding
   Ends of penstock sections must not vary by more than 1/8 in. at any point from a plane normal to the longitudinal axis and passing through the center of the section at the end.

12.3.3   Pipe Barrel Tolerances

12.3.3.1   Out-of-Roundness   For pipe with $D/t < 120$, the difference between the major and minor outside diameters must not exceed 1.5%.

12.3.3.2   Outside Diameter   The outside circumference of the pipe shall not vary more than $\pm 1\%$, but not to exceed 3/4 in. from the nominal outside circumference based on the diameter specified.

12.3.3.3   Alignment   The maximum deviation from a straight line, over a pipe length, shall be 0.2% of the pipe length, as measured with a taut wire or string.

12.3.3.4   Length   Lengths must be within $\pm 1/2$ in. of the specified length.

12.3.3.5   Minimum Wall Thickness   The wall thickness at any point in the penstock shall not be less than the thickness required by the project specifications, minus 0.01 in. Local discontinuities, such as roll marks, in the parent metal of welded pipe shall be considered defects when the depth of the discontinuity is greater than 10% of the nominal wall thickness.

REFERENCES

ASME. (multiple years). B31 Standards of pressure piping. New York.

ASME. (2010). "Boiler and pressure vessel code, section VIII, division 1: Rules for construction of pressure vessels." New York.

AWWA. (2005). "Steel water pipe—6 in. (150 mm) and larger." *ANSI/AWWA C200-05,* Denver.

ISO. (2008). "Quality management systems." *ISO 9000,* Geneva, Switzerland.

CHAPTER 13

Installation

Source: Photograph courtesy of National Welding Corp.; reproduced with permission.

No two penstock installations are truly identical. Subtle variations require unique approaches best suited to a particular installation. There are, however, general installation requirements and procedures common to all penstock systems. This chapter discusses those general requirements and procedures, leaving the consideration of unique requirements and solutions for particular projects to the engineer and installer.

13.1 HANDLING, SUPPORTS, AND TIE-DOWNS

13.1.1 Handling

Compared to the fabrication site, the harsher environment at the construction site requires greater care in the unloading, storing, and handling of penstock sections. This issue is particularly true for precoated penstock sections. The coating should be protected both during shipping and at the job site, or repairs may be necessary. Pipe with exterior paint must be holiday tested in accordance with NACE (2006).

To avoid damage to coatings, coated pipe must not be placed directly on the ground. It must be supported properly when in storage so as to spread the supporting load over a large area. If necessary, painted penstock sections may be rolled only along supports on unpainted areas or by protecting the pipe or cushioning the supports. When lifting penstock sections for unloading or installation, fabric or fabric-protected slings must be used for exterior painted pipe. If available, lifting lugs welded to stiffener rings or ring girder supports may be used. Every effort must be made to sling

153

the penstock sections for pickup at their balancing points. Penstock sections must not be dragged along the ground or bumped. While being fitted up, penstock sections must be supported in such a manner as to protect the coating and stabilize the pipe section.

13.1.2 Supports

Both temporary and permanent supports may be needed during penstock installation. Supports include temporary wood blocking, steel or concrete saddles, internal stiffeners or spiders, external lugs, and anchor ties.

Also, steps must be taken to prevent excessive deformation of the penstock sections, particularly after the lining and coating have been applied. Internal bracing may be used to prevent overstressing the shell and also to prevent excessive deflection and out-of-roundness. Padding supports and tie-down straps may be necessary to prevent damage to the lining and coating.

Internal stulling with wood or bracing with steel members may be necessary for buried penstock under backfill conditions. Table 13-1 lists pipe stulling criteria. The stulling criteria apply to all pipe, whether bare, painted, or cement-mortar lined. Wood stulls must be cut to a length equal to the pipe inside diameter minus 1/4 in. Cedar shingles can be used as wedges to provide a snug fit. Framing anchors are used to tie posts together (as radii of a circle). Based on common practice, wood stulls with felt or carpet padding are used for painted pipe with inside diameters up to 84 in. and without padding on cement-mortar lined pipe through 120

in. inside diameter. The pipe stulling should not be removed until the compacted backfill is placed to a depth that provides adequate lateral support for the pipe.

Table 13-2 gives steel bracing requirements. Steel stulls must be cut to a length equal to the pipe inside diameter minus 1/8 in. and inserted in sleeves at one end. Sleeves and the other end of the stull must be tack-welded to the pipe shell.

Pipe stulling and bracing configurations are illustrated in Fig. 13-1. Temporary bracing can be applied internally or externally to the shell.

Internal bracing, also called spider bracing, may consist either of a simple cross pattern attached directly to the steel interior by welding or a complex system of adjustable rods attached to a central ring with steel lugs welded at the interior of the liner. This bracing is used when the steel pipe conforms to specified tolerances for out-of-roundness and it is necessary to maintain these tolerances during shipment and handling.

The adjustable rod (tension spider) allows adjustment of the out-of-roundness tolerances in the shop or in the field. However, this procedure is difficult with thick-walled penstocks because of the rigidity of the pipe.

Bracing and attachments may be provided externally to the penstock in the form of stiffener rings. These rings may be required under design and, as such, perform a double function. External ring bracing can be used effectively for penstocks designed for saddle support, for buried construction, or for tunnel liners.

Bracing, when required to support the penstock sections for hydrostatic testing and concreting, must be minimized

Table 13-1. Pipe Stulling Criteria—Wood

	Pipe diameter/Stull size					
Diameter to thickness ratio (D/t)	D = <24 in. 2 in. × 6 in.	D = 24 in. to <30 in. 3 in. × 3 in.	D = 30 to <48 in. 3 in. × 3 in. for 30 in. and 4 in. × 4 in. for larger diameters	D = 48 in. to <60 in. 4 in. × 4 in.	D = 60 in. to <84 in. 4 in. × 4 in.	D = 84 in. to 120 in. Cement-mortar lined pipe only 4 in. × 4 in.
$D/t \leq 120$	No stulls	No stulls	No stulls	No stulls	No stulls	No stulls
$120 < D/t \leq 160$	Brace between bunks	2 stulls vertical	2 stulls crossed	2 stulls crossed	2 stulls, 3 legs	2 stulls, 3 legs
$160 < D/t \leq 200$	Brace between bunks	2 stulls vertical	2 stulls crossed	2 stulls, 3 legs	3 stulls, 3 legs	3 stulls, 3 legs
$200 < D/t \leq 230$	—	2 stulls vertical	2 stulls crossed	3 stulls, 3 legs	3 stulls, 3 legs	3 stulls, 3 legs
$230 < D/t \leq 288$	—	—	2 stulls, 3 legs	3 stulls, 3 legs	3 stulls, 3 legs	3 stulls, 3 legs

Notes: D = nominal pipe diameter; t = pipe wall thickness; stulls should be placed 15 to 20% of the total pipe length from each end, but no less than 4 ft in from end; and shipping bunks are to be located near stulls.

Table 13-2.   Pipe Bracing Criteria—Steel

Pipe inside diameter (in.)	Required r (in.)	Nominal brace diameter (in.)	Brace outside diameter (in.)	Brace minimum wall thickness (in.)
90	0.75	2.0	2.375	0.154
96	0.80	2.5	2.875	0.203
102	0.85	2.5	2.875	0.203
108	0.90	2.5	2.875	0.203
114	0.95	2.5	2.875	0.203
120	1.00	3.0	3.500	0.216
126	1.05	3.0	3.500	0.216
132	1.10	3.0	3.500	0.216
138	1.15	3.0	3.500	0.216
144	1.20	4.0	4.500	0.237

Notes: Based on an $L/r = 120$, L = length of the brace, r = radius of gyration = $(I/a)^{1/2}$, I = moment of inertia = $\pi(R^4 - R_1^4)/4$ where R = brace O.D. and R_1 = brace I.D.; a = area of brace wall = $\pi(R^2 - R_1^2)$. For D/t ratio = 120, no stulls are required where D = nominal pipe diameter and t = pipe wall thickness. For D/t ratio of 120 to 160, 2 stulls with 3 legs each are required. For D/t ratio of 160 to 288, 3 stulls with 3 legs each are required. Braces should be placed 15 to 20% of the total pipe length from each end, but no less than 4 ft in from end. Shipping bunks are to be located near braces.

to the greatest extent practical or designed in a manner that does not restrain the shell. Bracing resulting in potential restraint must be located at low-stress shell areas of complex structures, such as wyes.

When no longer needed, bracing and attachments within the shell must be completely removed, the areas of weldment must be ground flush with the shell plate material, and the surfaces must be inspected by nondestructive examination (NDE) and then coated as required.

During the installation of the penstock, consideration must be given to ensure that permanent supports, such as thrust blocks or thrust-resisting saddles, are installed in conjunction with the penstock to accept any gravitational or temperature-related loads. This situation is particularly true for exposed penstocks on steep terrain and penstocks using flexible field joints, such as couplings and expansion joints. If thrust considerations are not accomplished during installation when those conditions are present, joints are likely to open during construction and/or testing of the penstock.

13.1.3   Tie-Downs

Temporary ties may consist of simple lug pieces welded to the shell surface with provisions for fastening a steel cable or tie rod, or they may consist of complex ring girder type tie-downs, if needed to prevent shell movement during construction. Plate straps, anchor bolt footings, and lifting sling lugs, or other external attachments are used to accommodate working platforms for construction and for positioning and adjusting the sections into alignment with adjacent structures.

As with bracing, temporary ties are required primarily for shipment and to expedite construction. Ties that have served their intended use must be removed, and the weldment must be ground smooth with the shell surface, inspected by NDE, and then coated as required.

Permanent tie-downs consisting of strap anchors can be used to hold down an aboveground penstock to each of its supports. These straps should be wide enough and suitably padded to limit stress applied to the pipe. They are intended to limit pipe movement. Tie-downs may also be required for penstocks fully encased in concrete as a means of preventing flotation during concrete placement and curing.

Penstocks or tunnel liners to be encased in concrete must be designed for the buoyant and concreting loads resulting from concrete placement. In an installation through reinforced concrete, the reinforcing bars must not be used as temporary support for the penstock during installation. The penstock must not contact the reinforcing bars because of the risk of galvanic corrosion to the penstock.

13.2   THERMAL EFFECTS

During penstock installation, the effects of temperature must be considered. The effects of temperature variations are most severe for an aboveground penstock installation exposed to direct sunlight, less severe for a buried penstock during its

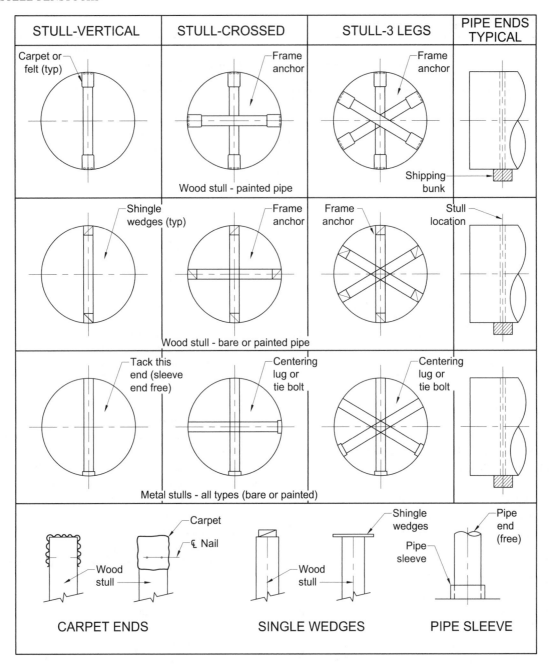

Fig. 13-1. Pipe stulling and bracing configurations

installation, and least severe for a penstock installed free in an open tunnel or embedded in concrete, i.e., a steel liner. The most serious thermal stresses occur because of the penstock temperature differential from the time of installation to the time of operation. Temperature-induced movement plagues every penstock builder because of its potential to cause both temporary problems and permanent damage to the penstock and its supporting structures. Ideally, penstock installation should be done at a temperature that is constant and close to the mean temperature of the site. Unfortunately, this procedure is rarely feasible.

Temperature changes affect alignment, straightness, and length of penstock sections. Direct sunlight on an exposed penstock results in the lighted side being hotter. Temperature differentials can be as high as 70°F, with the side exposed to direct sunlight reaching temperatures of 140°F or higher. These conditions cause penstock deflection and/or curling, resulting in large side-lurch forces on the supports. This problem is especially critical on long-span, ring-girder-supported installations. Even for cantilever-type installations, the end of the cantilever can fluctuate several inches off alignment with the variation in night and day temperatures.

Temperature side-lurch, expansion and contraction, and curling conditions are calculable for both the dewatered and watered conditions. The engineer must prepare an adjustment table for all three axes to be used by the installer during installation.

Side-lurch forces and uplift conditions may be greater during construction and the dewatered condition because there is no water to act as a heat sink. High side-lurch or uplift forces have been known to cause support displacement and damage. A dewatered penstock and its supports must be inspected for potential damage before refilling.

It is important to make frequent checks of penstock alignment and length during the installation. These checks are best done before sunrise (dawn) or after sunset (dusk), when the installation has reached a more uniform temperature over the entire shell surface. These checks then must be compared with the design assumption criteria and alignment for the mean temperature at the site and corrections made as needed. Should there be increases or decreases in length that exceed allowable tolerances, adjustments to the penstock length must be made either by cutting off any excess length or by adding a filler piece of pipe. Decreases in length caused by temperature effects may be aggravated by weld-related shrinkage, which can reach as high as 1/8 in. for each butt-welded joint.

Long, continuous runs of exposed penstocks are subject to greater temperature effects. Because concrete footings or piers usually are placed before the penstock erection, adjustments in the base plate and ring girder support and fastening design are necessary. Allowance should be made for misalignment in all directions at each footing or pier. The degree of allowance for directional movement varies with the installation. By providing this type of flexibility, the number of field adjustments for shrinkage, length increase, or misalignment can be reduced greatly on long penstock runs.

Expansion joints designed to allow change in length also help protect the installation against temperature effects. These expansion joints can be located so that the penstock length upstream and downstream of the joint undergoes an essentially equal thermal expansion and contraction or deflection and rotation condition. On steep slope installations, sliding of coupling joints can be prevented by attaching external restraints to the pipe. Actual expansion joint and coupling joint openings must be evaluated for the temperature at the time of installation relative to the temperature differential in the design and must be installed accordingly. The engineer should provide the installer with a table of installation temperature versus expansion joint settings to ensure that the joint is properly installed to allow for expansion and contraction as the engineer intended.

The temperature effects on an exposed penstock can be reduced by the application of a reflective type of paint or by insulation. Insulation may need to be considered in cold regions where there have been cases of freezing of the water within the penstock, even under flow conditions. Similarly,

warming water may have detrimental effects caused by temperature-related stress conditions at shell restraints.

The installer must position the pipe-support attachments to allow axial movement on the support in proportion to the pipe temperature at the time of installation to that considered the ultimate movement by the engineer. This information must be made available through communication between the engineer and the installer during installation.

During cold weather, lowered temperatures affect the penstock welding and protective coatings. For certain types of coatings, it is inadvisable to store unprotected pipe through the winter season before installation. Also, to prevent damage to the coating, special care is needed in handling or transporting coated pipe during cold weather. The coating manufacturer's recommendations must be followed.

Pipelines installed on slopes, particularly above ground, tend to slide down slope and, under certain conditions, to resist uphill expansion. Pipes at the bottom of a slope have been crushed under direct axial compression when, upon expansion caused by temperature change, they did not expand uphill as expected. Use of frequent anchoring on long slope intervals reduces the weight of pipe supported at each anchorage. Expansion joints must be properly located to reduce these thermal compressive stresses.

Steel liners embedded in concrete also experience the effects of temperature variations. Axial length changes caused by temperature changes ($15\,°F$ to $35\,°F$ temperature differentials) produce shearing loads on external fasteners and attachments. Also, temperature variations cause radial shell growth or shrinkage and can influence the stress conditions in the shell. Usually, axial and outward radial movements are not considered critical to the design. For axial movements, it is reasonable to assume that temperature changes affecting the steel shell will be gradual and, as such, the shell and surrounding concrete will react similarly, without the presence of relative differential movements. This assumption holds even with heavy thrust rings or seal rings at the shell exterior.

13.3 FIELD JOINTS

Field-welded joints are typically complete joint penetration (CJP) butt joint or welded lap joint types, as indicated in Fig. 13-2. Although the butt joint offers greater strength when compared to a lap joint, the fit-up time associated with alignment of edges to be welded is much greater, resulting in greater installation cost that often favors lap joints. Welding codes recognize that welder skill levels are greater for butt-joint welds than for fillet welds.

Double-welded butt joints that result in highest weld quality require welder access from both sides of the joint; this access permits the welder to gouge out the root pass from the initial weld and apply sound weld metal against the remaining weld metal that acts as backing for subsequent welding.

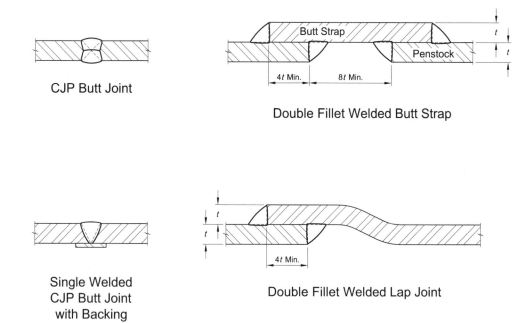

Fig. 13-2. Common field-welded joints

Single-welded butt joints are often used where welder access is limited to only one side, for example, in steel tunnel liners.

Welded lap joints are often the joint of choice because they offer reasonable strength yet economy since the joints are easy for installers to assemble and welder skill level is less than that required for butt-joint welds. Minor changes in alignment can be accomplished by pulling the lap joints.

Double-welded butt-strap joints are often used for field station adjustment. They are comparable in strength and economy to double-welded lap joints; however, plain pipe ends are used rather than sized bell-and-spigot joint preparation. The butt straps are normally furnished in two 180-degree halves that are field-spliced using CJP butt-joint welds.

Single-welded lap joints and butt straps can be used for less critical applications, where joint strength is not a limiting factor.

13.4 FIELD WELDING

For field-welding criteria, see Chapter 11.

13.5 CONCRETING

Concreting of structures for exposed penstock and buried pipeline, such as footings, piers, and anchor blocks, should follow the criteria, guidelines, and working practices established by the American Concrete Institute (ACI). For the loads usually encountered in these installations, concrete with a minimum of 3,500 psi, 28-day unconfined compressive strength, is usually satisfactory. Good material, admixtures, and mix practice are necessary to achieve this strength. Other desirable characteristics include workability, durability, and freeze–thaw and alkali-reactivity resistance.

Concreting of steel liners presents more difficult installation conditions compared to readily accessible structures. Concreting of steel liners requires

1. Preparation of structures for concrete placement,
2. Clearances for concrete placement around the liner,
3. Installation of slick lines for placement of concrete,
4. Supports and tie-downs to maintain shell alignment and to prevent flotation during concreting, and
5. Support, if required, to prevent buckling of the liner.

Preparation of the foundation and excavated tunnel rock surfaces for the placement of concrete depends on the purpose of the concrete. For exterior aboveground and buried penstocks, concrete provides structural support. For steel liner embedment, the concrete serves not only as a structural support for the filled penstock, but it also distributes the internal pressure to the surrounding rock. The concrete also maintains the physical stability of the steel liner against external pressure and thermal-related forces. Although the steel lining provides the desired hydraulic flow and water-sealing attributes and is the predominant load-carrying member against internal pressure, it is desirable to provide a mass of solid, continuous concrete in close contact with surrounding rock. The rock surface must be cleaned and timber or other degradable and nonessential support material must be removed. Possible voids should be filled by grouting.

Water seepage must be controlled to prevent damage to the fresh concrete before it sets. Sidewall flow guides and invert drains can be used for this purpose.

High water flows may require grouting, damming, and/or pumping to remove water before concrete placement. Drains and other water control features not required as part of the permanent operating system should be marked for location and later grouted. Excessively high inflows of water must be controlled by grouting the rock face as the tunnel is advanced.

Concrete with 1 1/2 in. maximum size aggregate may be used. The maximum size depends on the concrete pump used, the thickness of the concrete lining around the steel liner, and the existence of obstructions such as stiffener rings, external shell drain pipes, and thrust rings around which the concrete must flow. It is necessary to select concrete placement equipment that does not involve high-velocity discharge or cause concrete segregation.

The concrete mix for nonstructural backfill would typically have a minimum 2,000 psi, 28-day unconfined compressive strength. It is preferred that low heat of hydration cement be used to prevent overheating in areas susceptible to freeze–thaw deterioration. The use of additives such as pozzolan can be effective in reducing this heat. The inside of the steel liner should be spray-cooled continuously with water during setting. If there are vibration difficulties or any external stiffeners and appurtenances that may inhibit good distribution of the concrete, fly ash, superplasticizers, or air entrainment can be used to enhance concrete placement. Whenever possible, concrete placement should begin at the invert and at both lower quadrants and slowly work up the sides of the liner with the grouting progress monitored by observing grout flowing out of the grout holes. The crown must be vented to help prevent trapped air from forming voids. Pipes used for concrete placement must be well submerged in fresh concrete. This embedment may vary from 3 to 5 ft for thin concrete linings in tunnel boring machine (TBM) tunnels to 5 to 10 ft for the thicker linings of drill-and-blast tunnels. Sloping cold joints may be permissible. For a vertical joint, the sections being concreted may be bulkheaded at the downstream end. Bulkhead forms must be stripped 72 h after placement (sooner if field conditions allow). Green cutting of cold joints is preferred, depending on the installation.

Tie-down is important in steel liner installation, not only to prevent flotation but also to prevent the formation of hard spots on the exterior of the shell. Cable tie-downs and adjustable steel struts with some compressible wedging against the shell can be used. Wood wedging must be avoided because it could lead to corrosive action from chemicals within the wood. Pin-connected tie-downs that permit some outward radial movement at the shell also are acceptable. Hard-welded or braced hold-downs must be avoided.

13.6   GROUTING

Grouting requirements discussed in this manual apply primarily to steel tunnel liners embedded in concrete. Grouting applications for steel liners include (1) contact grouting of the voids between the concrete backfill and rock intersurface and (2) void or skin grouting of voids between the steel liner and surrounding concrete.

13.6.1   Contact Grouting

Contact grouting generally is performed in the upper 180 degrees of the tunnel and must be incorporated in the design and detailing of the steel liner. Grouting plugs are provided in the shell during the shop fabrication process. Customarily, contact grout plugs are located at the top shell surface, 15 and 60 degrees on either side of the vertical central axis of the shell. These plugs can be located either at intervals of 10 ft along the same plane that is perpendicular to the liner axis or can be staggered at intervals of 5 to 10 ft on either side of alternate transverse planes. Grouting plug details are shown in Section 6.6.3. The welded steel disk has proven to be an excellent detail and helps prevent the cracking associated with seal welding around threaded plugs.

Contact-grouting installation and implementation procedures during construction are the same as those imposed for the concrete-lined section of the tunnel. Grout holes are drilled through the concrete at a depth of about 1 to 2 ft into the surrounding rock. Drilling is done through the grout plug or through pre-placed grout sleeves.

The grout pipe connection is made to the threaded grout boss plate, and contact grouting is performed using a sand–cement grout mix. Grout pressures may range from 30 to 100 psi. Higher pressures have been used. The pressure should be checked to ensure that the steel liner does not buckle because of external local pressure.

13.6.2   Void or Skin Grouting

Void or skin grouting fills voids that develop between the steel liner and concrete backfill because of shrinkage caused by concrete curing and from temperature adjustments through the steel shell and surrounding media. Void grouting must be done at pressures not exceeding 30 psi.

By sounding the steel shell, the peripheral extent of voids can be established and marked on the shell interior. After a void area is located, 1/2 in. to 3/4 in. holes are drilled through the steel at the lower and upper void area extremities. A flowable, nonshrink, anticorrosive grout is pumped through the lower hole and allowed to vent through a riser pipe installed at the upper hole. After the grout has set, both holes are plugged with threaded plugs and capped with a welded steel plate, similar in detail to the grout plug plate.

Another method is to provide grout holes along the invert of the shell at a maximum of 10 ft on centers, or one each between stiffeners. Vent holes also are provided. Grouting between the steel liner and concrete is done first, followed by grouting between the concrete lining and rock (followed by high-pressure consolidation grouting of the rock, if required).

Care must be taken with skin grouting to ensure that the application pressure does not create a buckling or denting effect. Internal supports may be used to aid in eliminating this potential problem.

13.7   TRENCHING, BEDDING, AND BACKFILL

Trenching, laying, bedding, and backfilling methodology, approach, and techniques depend on factors such as character and purpose of the penstock; its size, operating pressure, and operating conditions; the location of the penstock relative to urban areas; the type of terrain to be traversed by the installation; the depth of the trench; and the character of the trench soil and backfill.

Trenching, bedding, and backfilling details for buried penstock designs are as numerous and different as for aboveground installations. A typical example is given in Fig. 13-3. Factors to consider in the trenching, bedding, and backfilling of buried penstock include

1. Surface and subsurface exploration;
2. Soil and rock classification;
3. Excavation and/or blasting techniques;
4. Effects of backfill on penstock coatings;
5. Material sources and disposal;

6. Refurbishing and/or reconstruction of disturbed grounds, facilities, and structures;
7. Safeguarding and bracing of excavations;
8. Closing of abandoned conduits;
9. Specification for and care of backfill material;
10. Backfill operations for pipe and structures; and
11. Rock refill and sand backfill criteria.

The engineering properties of the excavated soils and the backfill are important. The use of soil mechanics produces safer and more cost-effective results.

Trenching depth must be based on design requirements that have taken into consideration frost line depth, groundwater surface level, potential buried utilities, and any existing civil government guidelines and conditions. Trench width should provide for minimum safe working clearances, with consideration given to the need for the working space that may be required during tamping operations or for local installation of joints and/or appurtenances. It is important to grade the bottom of the trench as accurately as practicable. Backfilling is not permitted until impermissible, local foundation conditions have been treated.

The bottom of the trench must be free of stones and hard lumps and well graded. For large pipes, the trench bottom may be shaped for arc contact instead of line contact. If rigid blocks are used for pipe support (which is not recommended), they must be removed before backfill. Rigid blocks form hard spots that can damage pipe coatings. Also, they can create a galvanic cell and encourage corrosion. If blocking of pipe is necessary, bags filled with sand should be used; they can be broken at the appropriate time to pre-

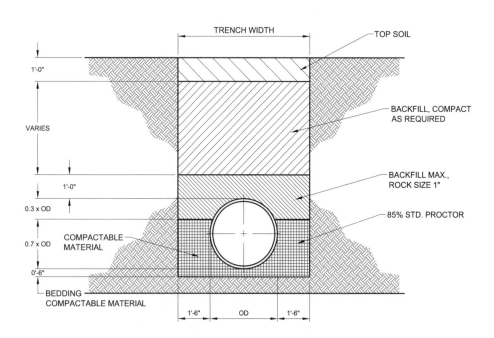

SUGGESTED BACKFILL FOR WELDED STEEL PIPE

Fig. 13-3.   Typical trench section

vent load concentration. Alternatively, a layer of pea gravel or sand 3 in. or thicker can be placed on the bottom.

Important in the design of buried penstocks is a good understanding of the earth loads on steel pipe and the principles of how vertical and lateral loads resulting from trench fill or from fill cover correlate with the flexibility and rigidity of the pipe. General backfill requirements include (1) use of a material that is free of boulders, rock, or other unsuitable substances and (2) the deposit and compaction of the material in the trench along both sides of the pipe to an elevation of not less than 70% of the pipe outside diameter. Material for backfill may be taken directly from the excavation unless the material is unsuitable or of insufficient quantity, in which case imported sand or stone is required within 12 in. of the pipe. Backfill must not be dropped directly on the pipe. The material must be placed along each side so that the surface elevation at the sides is even, with the differential on each side not exceeding 12 in. Backfill containing hard material larger than 1 in. in diameter must not be placed within 12 in. of the penstock. The backfill must be tamped in horizontal layers of thicknesses varying from 6 to 12 in. Layer thickness depends on the fill properties, degree of compaction required, and method of compaction. Hand or power tampers can be used to compact cohesive material to a density of not less than 90% of the maximum laboratory dry density, as determined by ASTM (2007). Free-draining, cohesionless material, such as sand or gravel, may be compacted by tampers, rollers, surface vibrators, or internal vibrators. In some instances, the use of materials such as soil cement or flowable fill can be an economical option to a conventional backfilling operation.

The importance of equally distributing fill material along each side of the penstock increases as the penstock becomes larger or more flexible. In well-draining native soil, jetting and vibrating sand bedding in larger lifts through the use of specially designed stingers may provide the compactive effort specified by the designer. This procedure must be properly evaluated and approved before application. For large-diameter, thin-wall pipe for low-pressure design, the use of stiffeners provides a well-designed and economic solution for buried pipe. Careful placement of backfill is required because the backfill provides the support between the penstock and the undisturbed earth. Primary attention in backfilling must be given to the haunches of the pipe

Backfill that is 12 in. above the top of the pipe may contain coarser material but should be free of organic material or other objectionable matter that could cause improper consolidation and potential settlement. Flooding may be used for this portion of the backfill, provided that precautions are taken against penstock flotation (caused by deliberate or accidental flooding of the trench).

REFERENCES

ASTM. (2007). "Standard test methods for laboratory compaction characteristics of soil using standard effort (12 400 ft-lbf/ft^3 (600 kN-m/m^3))." *ASTM D698-07e1*, West Conshohocken, PA.

NACE. (2006). "Discontinuity (holiday) testing of new protective coatings on conductive substrates." *NACE SP0188-2006*, Houston.

CHAPTER 14

Inspection and Testing

Source: Photographs courtesy of National Welding Corp.; reproduced with permission.

14.1 GENERAL

14.1.1 Owner/Engineer

Because the amount and type of nondestructive testing required affects the basic design, testing requirements must be determined early during the design stage. Testing and inspection alone do not ensure the integrity of the penstock. They provide only additional assurance that the completed penstock does in fact conform to the engineer's considerations.

14.1.2 Contractor

The supplier, fabricator, and installer are responsible for quality control programs that provide the degree of quality control/quality assurance required by the project specifications. Inspection and testing must be documented (see

Chapter 16). Inspection by the owner or the engineer does not relieve the fabricator or installer of the responsibility of conforming to the project specifications.

14.2 SHOP OR FIELD INSPECTION

Penstock components are subject to inspection at the place of fabrication and/or installation. The inspector must be allowed to view any or all of the operations and have access to report forms and test results, such as radiographs.

14.2.1 In-Process Inspection

Before the start of manufacture, the inspector must review the testing requirements of the specification and the applicable codes and standards for tolerances.

14.2.1.1 In-Process Inspections of Fabrication In-process inspections must include the following where applicable:

1. Review of the purchase order and specification document requirements for materials, fabrication, welding, examination, testing, marking, tagging, cleaning, painting, and preparation for shipment.
2. Review of the suppliers' quality assurance programs for conformance with the project specifications and discussions with the suppliers' personnel about the quality verification activities to be performed during the course of the assignment.
3. Review of the shop drawings and other vendor documents that require submittal before fabrication, including a determination that they have been approved by the engineer, are properly stamped, and released for fabrication.
4. Review of mill test reports for all materials used in the fabrication of the penstock sections to verify that the materials are in compliance with the specification and applicable ASTM standards for chemical composition, mechanical properties, and Charpy impact tests, if applicable.
5. Verification that the pressure and attachment material identification, heat number, and thickness correspond with the certified mill test reports provided for the project. In addition, the owner may request the certified mill test reports for other materials. Copies of certified mill test reports must be available for inclusion in the project file.
6. Verification that welding procedures and welders have been qualified and are approved in accordance with the requirements of the specification, drawings, and applicable codes and standards. This step may require an in-process review of the welding procedure specification (WPS) and supporting procedure qualification record (PQR) for conformance with the specification, drawings, and applicable codes and standards, particularly if there were design or personnel changes after the original submissions.

    Welders and welding operators must have current certifications and must be able to produce acceptably sound welds with the processes, materials, and welding procedures to be used in production.
7. Verification that welding electrodes, filler materials, and fluxes are identifiable, in compliance with the applicable WPSs, and properly stored and maintained in a clean and dry environment.
8. Check of all in-process welding for conformance to approved welding procedures, drawings, and applicable codes. This check includes all welding parameters, preheat, and interpass temperature requirements.
9. Where welding repairs are required, verification that the contractor has an approved repair and inspection procedure and follow-up testing plan to ensure satisfactory defect removal and welding of the affected area.
10. Check of the plate edges.
11. Inspection of shell courses for concentricity and roundness and verify that the fit-up gaps of longitudinal and circumferential joints are within the tolerances given by the specifications.
12. Inspection of the various parts of the penstock to ensure that all appurtenances have been provided and that the projections and orientations are in accordance with approved shop drawings.
13. The inspector's witnessing of nondestructive examination (NDE) of materials or welds (spot-check) on a random basis or as specified by the purchase order and review of the required radiographic film to ensure that there are no indications exceeding the limitations specified by the applicable codes and standards.
14. The inspector's verification that any required preheat or postweld heat treatment has been performed in accordance with the approved procedures. The inspector must obtain copies of time–temperature charts of each stress-relieving operation for inclusion in the project file.
15. For elbows, measurement of each chord to ensure that the units are fabricated as specified and have the desired radii.
16. For wye branches, verification of branch angle(s), projection, and orientation.
17. For couplings and mechanical expansion joints, review for compliance with the project specifications of the mill test reports provided for the couplings and gasket materials; documentation of NDE (if required). Couplings and gaskets must be checked for physical damage, and dimension checks must be made to verify compliance with the drawings. The longitudinal or spiral seam welds on both ends of the penstock sections must have been ground flush in accordance with the project specifications and drawing requirements.

14.2.1.2 Coating and Lining Inspection Requirements
When corrosion protection, internal lining, or external surface coating is required in the project specifications, a careful inspection for strict compliance with the project specifications, referenced standards, and Chapter 10 of this manual is mandatory. Details of inspection procedures for coating materials and their application are given in Chapter 10.

14.2.2 Final Shop Inspection

Before shipment, final shop inspections must be completed.

1. Ensure that required test reports, as-built drawings, and other documentation generated during the fabrication of the penstock are complete and available.

2. Perform final inspection on the penstock sections, and verify that all deficiencies have been corrected and that each section has been properly marked.

3. Verify the identification marks on each penstock section for conformance with specification and drawing requirements. Confirm that the field top or bottom of special sections have been marked and are clearly visible on both ends of each section.

4. Verify that each penstock section has been properly braced to prevent damage during transit and handling. Obtain assurance that the penstock sections will be correctly blocked for shipment and that coated penstock sections are loaded on padded bunks or saddles as required.

14.2.3   Final Field Inspection

The entire penstock, including coatings and linings, must be inspected for damage that may have occurred during shipment and erection. Damage must be repaired.

14.2.3.1   Damage   Damaged areas, such as scratches, gouges, grooves, or dents as determined by the inspector, must be corrected as specified in the project specification.

14.2.3.2   Field-Weld Inspection   Field welds must be inspected for conformance with the project specifications and to the applicable requirements of ASME (2010b) or AWWA (2011).

14.2.3.3   Field Inspection of Linings and Coatings   Coatings and linings, including those applied in the shop or field, must be examined for damage that may have occurred during shipment and erection according to the appropriate requirements of the specification, referenced standards, and Chapter 10 of this manual. Damage must be repaired.

Cathodic protection systems must be tested for electrical continuity. If an impressed current system is in place, the power source must be given appropriate performance tests. Documentation in the project file must detail operating procedures.

14.2.3.4   Bedding and Backfill   Bedding and backfill must be inspected for conformance to the project specifications during placement and after completion.

14.3   NONDESTRUCTIVE EXAMINATION

The type of NDE for butt-joint and fillet welds in pressure-retaining parts is established by the engineer in conjunction with the weld joint reduction factors described in Section 3.5.1.

The engineer must document the required nondestructive examinations for these weld joints in a fabrication and installation NDE data sheet. For weld joints that are not described in Section 3.5.1 or where supplemental NDE is required, minimum nondestructive examination requirements are listed in Table 14-1.

Nondestructive examination for acceptance of any material subject to hydrostatic pressure testing must be performed before the hydrostatic tests. Repaired defects must be retested by the same NDE methods used for the original tests. If the engineer determines that the specified NDE is not possible because of conditions encountered in the work, another method of inspection should be chosen by the engineer. The engineer may require additional tests and examinations.

14.3.1   Radiographic Examination

Radiographic examination (RT) must be in accordance with ASTM (2010) or Article 2 of ASME (2010a).

Table 14-1   Minimum Nondestructive Examination Requirements

Location	$f_y < 55$ ksi	55 ksi $< f_y < 75$ ksi	$f_y > 75$ ksi
Corner joints and fillet of pad to shell	100% MT[a]	Blend grind[b] and 100% MT	Blend grind and 100% wet MT
Corner joints	100% MT[a]	Blend grind and 100% MT	Blend grind and 100% wet MT
Fillet weld	10% spot MT	100% MT	Blend grind and 100% wet MT
Stud weld	Visual	10% bend test	Blend grind and 100% wet MT
Butt joints	Section 3.5.1	Blend grind and 100% MT	Blend grind and 100% wet MT
Fillet welded butt straps	Section 3.5.1	Blend grind and 100% MT	Blend grind and 100% wet MT
Backed-up butt joint	Section 3.5.1	Blend grind and 100% MT	Blend grind and 100% wet MT
Double-fillet lap joint	Section 3.5.1	Blend grind and 100% MT	Blend grind and 100% wet MT
Repairs and material/overlay	100% MT[a]	Blend grind and 100% MT	Blend grind and 100% wet MT

Note: f_y = Minimum specified yield strength.
[a]MT applies to all final weld surfaces only.
[b]Blend grind is defined as the elimination of surface irregularities and faring of edges of the weld.

Film must be marked with the date, owner's project number, contract number, piece or section number, and weld number. Procedures must include an identification system that ensures traceability between the radiographic film and the weld examined, as well as clear location of weld defects. Radiographs and interpretation reports must be submitted to the owner for permanent retention.

Acceptance standards for welded joints examined by radiography must be in accordance with Paragraph UW 51 or UW 52 of ASME (2010b).

Defects disclosed by radiography and determined unacceptable in accordance with the project specifications must be repaired and reradiographed.

14.3.2 Ultrasonic Examination

UT and acceptance standards for welds other than spiral welds must be in accordance with Paragraph UW 53 and Appendix 12 of ASME (2010b). For spiral-welded pipe used in penstocks, ultrasonic test procedures and acceptance standards must be in accordance with API (2009).

14.3.3 Magnetic Particle Examination

MT techniques are described in Article 7 of ASME (2010a) and in ASTM (2008). Acceptance criteria must conform to the requirements of Appendix 6 of ASME (2010b).

14.3.4 Liquid Penetrant Examination

Liquid penetrant examination (PT) procedures must conform to the requirements of ASME (2010a). Acceptance criteria must conform to the requirements of Appendix B of ASME (2010b).

14.3.5 Visual Examination

Visual examination (VT) procedures must conform to the requirements of ASME (2010a).

The following visual indications are unacceptable:

1. Cracks;
2. Undercut on surface greater than 1/32 in. deep;
3. Lack of fusion on the surface;
4. Incomplete penetration;
5. Convexity of fillet weld surface greater than 10% of the longer leg plus 0.03 in.;
6. Concavity in groove welds;
7. Concavity in fillet welds greater than 1/16 in.; and
8. Fillet weld size less than indicated or greater than 1.25 times the minimum indicated fillet leg length.

Disclosed defects that are determined unacceptable in accordance with the job specification must be repaired and reexamined.

14.3.6 Areas Requiring Special Consideration

1. Where required by the project specifications or the weld procedure, weld bevel preparation in wye branches and penstocks must undergo magnetic particle examination (MT) before welding. Laminations and other linear defects must be repaired in accordance with Paragraph 9.4 of ASTM (2011).

2. C-girders used for reinforcement of wye branches are subjected to through-thickness loading where the shell plates are welded to them. If the plate has laminations in the loaded area, lamellar tearing can occur. The likelihood of this defect occurring increases with plate thickness. Therefore, C-girder plates exceeding 1 in. in thickness must undergo ultrasonic examination (UT) in the area within 6 in. of the shell-to-C-girder weld. UT and evaluation must comply with ASTM (2007), with the additional requirement that the 6-in. band adjacent to the weld area must undergo 100% UT. Any defect that shows a loss of back reflection that cannot be contained within a 1-in.-diameter circle is unacceptable. Laminations must be repaired in accordance with Paragraph 9.4 of ASTM (2011). Shell-to-bar joints in wye branches must also be examined for laminations in a similar manner.

   Exposed edges of C-girders must undergo MT after postweld heat treatment.

3. After postweld heat treatment, if required, weld surfaces must undergo MT in accordance with Appendix 6 of ASME (2010b). If, in the opinion of the inspector, major repairs are required, stress relief again may be required after repairs.

14.4 HYDROSTATIC TESTING

14.4.1 General   *OPTIONAL - SIZE Limit*

Hydrostatic testing is not a mandatory requirement for penstocks. The need for hydrostatic testing is determined by the engineer and included in the project specifications. The following should be considered in determining the need for hydrostatic testing:

1. Site location, automatic shut-off systems, and head;
2. Risk to the public and damage to property in the event of a failure;
3. Structural complexity (complicated weldments, such as wye branches, may warrant testing to verify their fabrication). The engineer may consider 100% RT or UT, and MT in lieu of hydrostatic testing; and
4. The extent and type of NDE performed in the shop and field.

14.4.1.1 Component Testing   Pipe for mitered bends may be hydrostatically tested before making mitered cuts.

Girth welds in mitered bends that have not been hydrostatically tested must be tested for defects using 100% RT or UT.

Fittings and attachments, such as nozzles, ring girders, and anchor rings, may be fabricated into previously hydrostatically tested pipe provided they are tested using the appropriate NDE procedure after welding. Otherwise, fittings and attachments must be welded to the shell before hydrostatic testing.

14.4.1.2   Types of Hydrostatic Testing   If it is determined that hydrostatic testing is necessary, the location and time of testing must be established in the project specification. Hydrostatic testing can be performed either in the fabrication shop or in the field (after the penstock has been completely installed) or by a combination of both. The following must be considered in deciding whether to perform shop testing or field-testing.

1. Hydrostatic testing of straight pipe sections can be performed using the manufacturer's hydrostatic testing machine. This method is usually less expensive than using test heads.
2. Shop testing does not test the system in its final installation condition.
3. Shop hydrostatic testing is not mandatory for penstocks subjected to 100% radiographic or ultrasonic examination, or if there is to be field hydrostatic testing of the completed system.
4. Manufacturer's hydrostatic testing machines may not be able to test very large-diameter and high-pressure penstocks. In these cases, welded test heads can be used or hydrostatic testing can be performed in the field after installation.
5. In-place testing may be more expensive than shop testing, and if rupture occurs, the failed section must be repaired or removed and replaced.
6. Testing in the field on a steeply sloped system may overpressure the lower portions and underpressure the upper portions. When testing after installation, the penstock may have to be tested in sections to ensure an adequate test pressure at all points without overpressure at other points.

14.4.2   Shop Hydrostatic Testing

Hydrostatically shop test penstock sections to a pressure described below as required by the project specifications. Conform to the test procedures described in AWWA (2005), in addition to the requirements listed below. If hydrostatic testing machines are used, the end seals should not produce any significant inward pressure on the ends of the pipe section.

Slowly fill the pipe sections to be tested with water and slowly increase the pressure until the required test pressure is reached. Hold the test pressure for 5 min, release, then reapply the pressure and hold at the required test pressure while the pipe section is observed for leaks. Defects caus-

ing leakage must be repaired, and the pipe section must be tested again.

The hydrostatic tester must be equipped with a calibrated recording gauge to record the test pressure and the duration of time that pressure is applied to each pipe section. Provide records or charts as required by the project specifications. Complete shop hydrostatic testing before application of linings and coatings.

The shop test pressure must not exceed 1.5 times the working pressure, defined as the normal operating condition, P_{N2}, in Chapter 3, or a pressure that produces a stress 80% of the minimum specified yield strength of the steel, whichever is less, as determined by Eq. (14-1).

$$P = \frac{2St}{D} \qquad (14\text{-}1)$$

where P = hydrostatic test pressure;

$\quad S =$ 0.8 times the minimum specified yield strength of the steel;

$\quad t =$ wall thickness; and

$\quad D =$ inside diameter of the pipe.

14.4.3   Field Hydrostatic Testing

For buried penstocks, complete the hydrotest and/or watering-up before burying couplings. To facilitate construction or to avoid overpressuring, the penstock may have to be tested in sections, using bulkheads for isolation. For field tests, some loads imposed on the structure may not be the same as those during operating conditions. Check the installation before testing to determine that anchorages and bulkheads can withstand the test pressure. Install air valves or standpipes to prevent vacuum collapse in the event of failure of a portion of the penstock under test. During the filling of the penstock, follow watering-up procedures described in Chapter 15.

Hold the test pressure for a minimum of 2 h and long enough to ensure that the entire section under test can be inspected and any leakage detected and the leakage rate measured. The leakage rate is determined by metering the water required to maintain pressure within 5 psi of the test pressure. If the penstock being tested has cement mortar linings, consideration must be given to the absorption of water into the mortar during watering-up and also during pressurization. Depending on the nature of the installation, the time required to perform the hydrostatic test may vary considerably. However, the duration of testing must be sufficient to ensure the detection of any problems. This duration may range from 2 to 24 h. Upon reaching test pressure, readings from pressure gauges must be recorded at 10- to 30-min intervals, depending on the duration of the hydrotest. Continuous surveillance for leakage must be performed during the first 30 min, and the test section must be visually inspected for leakage at 1-h intervals thereafter. Perform the hydrostatic test during daylight hours when practical. If any portion of the test is performed during night hours, suitable lighting for inspection must be provided.

Hydrostatic tests are accepted on the basis of the leakage rates described in the project specifications. If a break or unacceptable leakage occurs during any of the testing operations, the test must be terminated and the engineer must be notified.

Defects disclosed during the hydrostatic test must be repaired, and the section must be retested.

Hydrostatically field-test all or a specified length of the penstock from a minimum pressure of 1.1 times the working pressure defined as P_{N2} in Chapter 3 to a maximum pressure of 1.5 times the working pressure. The field-test pressure must not exceed a pressure that produces a stress of 80% of the minimum specified yield strength of the steel, as determined by Eq. (14-1).

14.4.4 Hydrostatic Test Inspection Verification

For both shop and field tests, the inspector must observe the testing and check the procedures, verifying the following parameters:

1. The test pressure and duration of test are as specified for the penstock sections.
2. The calibrated pressure-measuring devices provide for the recording of pertinent data (i.e., pressure, time, and temperature, as specified).

3. The gauges or transducers used to monitor the pressure testing show evidence of current calibration traceable to the National Institute of Standards and Technology.
4. Seams, nozzles, and manholes are inspected during the test for leaks, and findings are recorded.
5. The penstock section is drained, dried, and preserved as specified after completion of testing

REFERENCES

API. (2009). "Specification for line pipe," 44th Ed. *ANSI/API Spec 5L,* Washington, DC.

ASME. (2010a). "Boiler and pressure vessel code, section V: Nondestructive examination." New York.

ASME. (2010b). "Boiler and pressure vessel code, section VIII, division 1: Rules for construction of pressure vessels." New York.

ASTM. (2007). "Standard specification for straight-beam ultrasonic examination of steel plates for pressure vessels." *ASTM A435/A435M-90(2007),* West Conshohocken, PA.

ASTM. (2008). "Standard guide for magnetic particle testing." *ASTM E709-08,* West Conshohocken, PA.

ASTM. (2010). "Standard guide for radiographic examination." *ASTM E94-04(2010),* West Conshohocken, PA.

ASTM. (2011). "Standard specification for general requirements for steel plates for pressure vessels." *ASTM A20/A20M-11,* West Conshohocken, PA.

AWWA. (2005). "Steel water pipe—6 in. (150 mm) and larger." *ANSI/AWWA C200-05,* Denver.

AWWA. (2011). "Field welding of steel water pipe." *ANSI/AWWA C206-11,* Denver.

CHAPTER 15

Start-Up

Source: Photograph courtesy of Richard Stutsman, RDS Consulting; reproduced with permission.

15.1 WATERING-UP

One of the most important elements of starting up a new facility with penstocks is the watering-up process. This chapter will discuss initial watering-ups and subsequent watering-ups, including the elements that go into each.

15.1.1 Initial Watering-Up

The initial watering-up of a new or a replacement section of a penstock generally is the most critical watering-up operation because it is the first time that the penstock components have been fully assembled and tested together under operating conditions. This initial watering-up must be performed under slow rates and controlled conditions.

15.1.1.1 Start-Up Procedures Before completion of the project, detailed project start-up procedures must be developed and approved. These procedures include, but are not limited to, powerhouse mechanical and electrical equipment testing (micro and macro), water conveyance system watering-up and dewatering, and operational testing. Before watering-up, the emergency closure system must be completely operational, and an emergency contingency plan must be prepared. The scope of the work depends on the scope of the project (e.g., new project or penstock replacement).

In addition, a detailed inspection procedure, the civil features monitoring program (CFMP), must be prepared and instituted. A CFMP is a document that outlines all of the critical civil features that should be inspected: before, during, and after watering-up or start-up, their frequency, their acceptance criteria, the responsible personnel, communication responsibility, and a sign-off (accountability) for all elements of the program. Its purpose is to ensure that during this process (watering-up and start-up), all components are performing within the original design criteria and/or parameters.

15.1.1.2 Before Watering-Up The following are examples of work to be completed before watering-up the system:

1. A baseline survey must be taken of monuments to be used for short- and long-term monitoring (e.g., anchor blocks, tunnel portals, and penstock supports).
2. Baseline data must be taken using equipment such as inclinometers, piezometers, and extensometers.
3. Emergency closure systems (ECSs) must be complete, dry tested, and fully operational. An ECS is a closure system, such as a butterfly valve, wheel gate, or some other system that is designed to close (manually or automatically) under flow conditions and to stop water flow downstream of its location.
4. The interior of the penstock system must be inspected in a walk-down to ensure that construction debris and tools have been removed from the interior and that all personnel have vacated the interior.
5. The exterior of the system must be inspected in a walk-down to ensure that manholes and valves have been closed and that air valves are open.
6. Clearances with the start-up engineer or powerhouse foreman must be signed off verifying that the system is ready for watering-up.
7. Penstocks may require special cleaning and disinfection procedures.
8. Valve indicators must show the valve positions correctly.
9. Air valve floats must work properly.

15.1.1.3 During Watering-Up These are examples of work to be completed or performed during the initial watering-up of the system:

1. A slow filling rate must be initiated to ensure full control of the system. An operator must be stationed at the ECS during the entire watering-up process in the event that a manual closure of the system is required. Should a problem develop that would require rapid dewatering, a slow watering-up might prevent a potential disaster. A suggested maximum range for this initial filling rate for inclined penstocks and shafts is 20 to 50 vertical ft/h, which usually allows adequate control for filling operations. For horizontal tunnel liners, or tunnels, a conservative fill rate less than 30 vertical ft/h should be used.
2. A system must be set up to provide communication among the inspection crew, the start-up engineer, and all operators.
3. The filling operation should be performed only during daylight hours, for two reasons. The first is to provide for the safety of the personnel involved in the inspection program. The second is to provide better visibility so that problem areas can be identified and acted upon quickly. Intermediate hold points must be considered to allow for a thorough inspection and evaluation. If practical, at least 4 to 6 h of daylight should be available after completion of filling. During the nighttime hours, intake head gates and ECS must be closed for safety reasons. The bypass fill gates may be closed or left open.
4. After the penstock has been filled completely, the head gates are usually closed and the system is allowed to soak for 24 h before any operational testing of the powerhouse equipment. This step is particularly important for penstock sections that include pervious sections, such as tunnels, concrete, and wood stave. This procedure allows for (1) porous elements to absorb water in a controlled and observed condition and to stabilize and (2) monitoring of water loss, caused either by absorption or leaking compontents since no new water is being introduced into the system. A periodic inspection must be made during this period to identify any problems. Additional personnel must be on site or on call in the event that problems develop. During the initial filling process, items such as sleeve couplings, valves, and other bolted connections may require retorquing to stop leaks. Torquing must be done according to the manufacturer's recommendations.
5. For buried penstocks with mechanical couplings, bell holes should be left open around them for inspection of leaks and retorquing of couplings.
6. For buried penstocks that have been fully pressurized, the intake shutoff gate (ISG) or penstock shutoff valve (PSV) should be closed and monitored for any pressure drops to verify that there are no leaks. This process may require accurate pressure monitor-

ing equipment located at the ISG or PSV to monitor pressure drop.

7. After completion of CFMP inspection procedures, the start-up engineer's start-up procedure must be signed off, indicating that the penstock is complete and ready for operational testing.

15.1.1.4 Recommended Inspection Checks The following are examples of inspection checks that should be made and documented if applicable during watering-up:

1. Read floatwell water surface elevations. (A floatwell is an open vertical well in which the static water surface of a penstock or tunnel can be measured).
2. Inspect appurtances, such as valves, manholes, sleeve coupling or expansion joints, and air–vacuum valves, for leaks.
3. Read inclinometer, piezometer, extensometer, and other monitoring systems.
4. Check for short-term settlement in foundations.
5. Check for cracking or distortion of the penstock support system.
6. Check for any leaks (exposed penstock and tunnel adit plugs or gates) or wet soil, any increase in groundwater, and any increase in weir or piezometer readings (for liner or buried penstock).
7. Compare movements or deflections to analysis.
8. Survey elevations.

15.1.2 Maximum Dewatering Rates

For inclined penstocks, the dewatering rate must not exceed 50% of the controlling air valve flow capacity or 100 vertical ft/h, whichever is less.

For tunnels and horizontal tunnel liners, the maximum dewatering rate must not exceed the range of 10 to 60 vertical ft/h, depending on their design and construction characteristics. Depending on the rock formation (bulk modulus) and the liner design, lower rates may need to be considered. If a differential pressure monitoring system is installed across the tunnel liner, these dewatering rates may be exceeded, provided that an adequate factor of safety against buckling is maintained.

15.1.3 Subsequent Watering-Up

Subsequent penstock watering-up generally is not as critical as the initial watering-up operation, since the system has already been tested and observed under controlled conditions. Where there have been long outages, with the penstock in a dewatered condition, watering-up also must be performed under slow rates and controlled conditions. This requirement is because of appurtenances that may have dried out, such as expansion joints and couplings, or shifted because of thermal changes. However, subsequent watering-up also must be performed under controlled conditions.

The following must be considered for subsequent watering-up operations:

1. The filling rate can be much higher than for the initial watering-up. A reasonable rate for a subsequent watering-up of the penstock is 100 vertical ft/h. Again, the maximum dewatering rate must not exceed 50% of the controlling air valve flow capacities.
2. The initial and subsequent watering-up reports must be reviewed for the occurrence of any unusual events or circumstances.
3. A walk-down of the exterior of the system must be conducted to ensure that manholes have been closed, air valves and vacuums are operational and open, and drain valves are closed.
4. After the system has been fully pressurized but before being returned to operation, the entire system must be walked down, and any abnormal items or conditions must be noted and corrected if necessary.
5. Clearance with the start-up engineer, powerhouse supervisor, or other designated person must be signed off to certify that the system is ready to return to operation.

15.2 OPERATIONAL TESTING

15.2.1 General

As discussed in Section 15.1.1, one phase of the start-up procedure involves operational testing. This testing consists of performing load-rejection and load-acceptance testing of the turbine, pump, or combined pump and turbine. Physical testing must be conducted for all penstock operating conditions (normal condition) that are likely to occur. Pressures transmitted to the penstock must be compared against the penstock design criteria. For example, if the penstock is designed for a normal operational pressure (load rejection) of 10% over static pressure, then the turbine and generator unit must be fully tested as described below to ensure that the pressures in the penstock do not exceed 10% over static pressure.

Operational testing would generally be done only for new installations or where substantial turbine and/or pump modifications have been made that would affect pressure rise or drop, or where branches into the penstock were made to add additional turbines and/or pumps. The purpose of these tests is to confirm that the pressures and reactions to the penstock are within the design limits for the system (e.g., pressure rise and pressure drop).

15.2.1.1 Head Loss If flow-measuring devices have been installed on or near the turbine or generator, penstock or system head loss can be measured directly; otherwise, conventional pressure gauges can be used. The results then can be compared against the original design criteria.

Flow-measuring devices, methods, and systems in common use include the following:

1. Acoustic flowmeters

Acoustic (or ultrasonic) flowmeters measure the water velocity by the interaction of an interrogating sound wave and the water. The most common type for penstock flows is based on the transit-time method. Transit-time flowmeters are described in AWWA (2010). This method is based on the principle that an acoustic pulse sent upstream travels slower than one sent downstream. By measuring the time difference between the upstream and downstream pulses, the average velocity of the water crossing the path of the pulse can be determined vectorially. There are several methods for using transit-time ultrasonic flowmeters. These methods include (1) permanently mounted wetted systems, in which the transducers are inserted into the pipe wall and come in contact with the water, (2) strap-on systems, in which the transducers are temporarily mounted on the outside of the pipe, and (3) path systems for the acoustic wave (single- and multiple-path along the pipe diameter, or along chordal paths). A schematic representation is shown in Fig. 15-1.

Under proper conditions, accuracies better than 1% can be obtained, particularly for multiple-path systems. Under the proper conditions, typical single-path strap-on systems have claimed accuracies of between 1% and 3% (without calibrating in place against another flow standard). However, larger variations have been observed, depending on the installation. These systems have limitation in that (1) they require a uniform flow profile for accurate measurements, typically at least 10 pipe diameters of undisturbed, straight pipe upstream of the meter; (2) the pipe inside diameter must be accurately determined; and (3) bubbles or other particles in the fluid can cause problems because of reflection or scattering of the signal.

2. Gibson method

This method, also called the pressure-time method, is based on Newton's second law of motion, in which the pressure forces acting on a mass of water are related to the change in momentum of the water as it decelerates. For penstock flow measurements, the difference in pressure between two cross sections is measured as a downstream valve or gate is closed. By integrating the pressure difference over the time period, and knowing the appropriate pipe geometry data, the steady-state flow rate before shut-off can be determined. Modern instrumentation and computer numerical methods can be used to simplify the data collection and analysis. Accuracies of about 1% are generally assumed for the method. Limitations include the need for properly located pressure taps, the need to

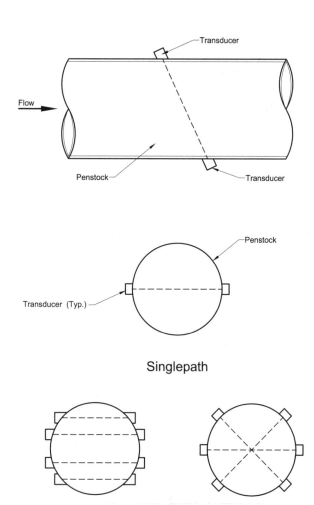

Fig. 15-1. Schematic representation of transit-time flowmeter

shut the flow down to get the measurement (not an on-line method), and the somewhat complex data reduction required. This test procedure is fairly expensive.

3. Current meters

The current meter measures a point velocity using a propeller-type meter whose rotational speed is proportional to the water velocity. A large number of current meters usually must be used to obtain a good overall volumetric flow rate. Under proper conditions, accuracies of 1% can be obtained. Limitations include the need to mount a large number of current meters inside the penstock (if not in an open canal situation) and the need for a uniform flow profile.

4. Salt velocity method

The salt velocity method works on the principles that salt in solution (brine) increases the conductivity of water and that the salt introduced at one point travels as a "slug" for some distance down the penstock. The passage of this slug of brine past electrodes at two locations downstream of injection can be recorded, and the time the slug takes to travel the known distance between electrodes can be determined. The distance and time of travel are used to calculate the average flow velocity, which when multiplied by the penstock cross-sectional area, results in the volumetric flow rate. Under proper conditions, accuracies of 1% can be obtained. Limitations include the need to mount electrodes at two locations inside the penstock and a brine injection system at another location, and the need for accurate dimensional measurements between electrode stations.

5. Dye dilution flow method

The dye dilution flow method measures the volumetric flow rate directly by using the principle of conservation of mass of a tracer dye injected into the flowing fluid. Rhodamine WT dye typically is used as the tracer. A known concentration of dye is injected at a steady rate at one location, and a fluorometer is used to determine the mixed concentration at a second location sufficiently downstream to ensure complete mixing. The ratio of concentrations at the two locations and the measured injection rate are used to calculate the water flow rate. Under proper conditions, accuracies between 1% and 2% can be obtained. Limitations include the need for a sufficient distance between the injection and sampling locations (at least 100 penstock diameters of undisturbed straight pipe is recommended) and the need for great care and accuracy in the making of dye standards for fluorometer calibrations.

6. Pitot tubes

The pitot tube measures point velocities by determining the difference between the total and static pressures at a point in the flow. A traverse must be done across the cross section, typically along two diameters, to obtain a sufficient number of velocity points to calculate a good average. Twenty points along each diameter typically are used. The flow rate is then calculated by multiplying the average velocity by the cross-sectional area. A U-tube manometer typically is used to measure the differential pressure produced by the pitot tube. Under proper conditions and with a calibrated pitot tube, accuracies of about 2% can be obtained. Limitations include the difficulties in traversing across a large pipe diameter (such as pitot tube support problems and vibrations), the need to obtain a good measurement of the penstock inside diameter, and the need for a good flow profile (at least 10 pipe diameters of undisturbed, straight pipe upstream of the pitot tube).

7. Winter–Kennedy method

The Winter–Kennedy method uses the differential pressure between two suitably located taps in the turbine spiral case to determine the flow rate. The flow rate is approximately proportional to the square root of the differential pressure. Winter–Kennedy flow measurements typically are used for index testing, where absolute flow measurements are not needed. These relative flow rates can be used to determine the shape of the turbine efficiency curve relative to some design point. For absolute flow determination, the constant of proportionality between flow and differential pressure, as well as the actual exponent for the differential pressure (it may not be an exact square root relationship) must be determined by calibration against another known flow method. The main limitation of this method is that accurate absolute flow measurements can be obtained only if a calibration is done using another known flow method.

8. Thermodynamic method

The thermodynamic method directly measures turbine efficiency using the principles of conservation of energy. Turbine losses are manifested by a rise in the temperature of the water passing through the turbine, by appropriate temperature and pressure measurements at the inlet and outlet of the turbine, and by using an energy balance analysis. Turbine efficiency can be calculated. The water flow rate can then be calculated from the turbine efficiency by using separate measurements of net head and generator power output and assuming that the generator efficiency is known. Accuracies of better than 0.5% are claimed for the turbine efficiency. Flow rate accuracy depends on the efficiency accuracy, as well as on the accuracy of the net head, generator power, and generator efficiency values. Limitations include the need to accurately determine temperature differences between the turbine outlet and an inlet test chamber, and the relatively complex instrumentation system required.

9. Weir method

The weir method uses a V-notch, trapezoidal, or rectangular sharp-crested plate interposed into a free surface flow. The measurement of the water head above the weir crest (measured upstream of the weir) can be used to calculate the flow rate. The weir generally is placed in the tailrace, downstream of the turbine or in an approach channel upstream of the system intake structure. Under proper conditions, accuracies of about 2% can be obtained. Limitations include the need for a properly constructed and placed weir, and the need for smooth flow (without air bubbles or flow disturbances) in the approach canal.

10. Magnetic flowmeters

Magnetic flowmeters use the principle of electromagnetic induction by a conductor moving through a magnetic field (based on Faraday's law). Water flowing through a magnetic field induces a voltage that is proportional to the velocity of the fluid. Single-point

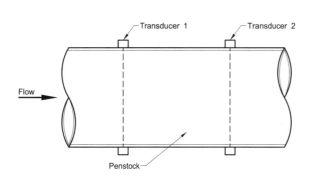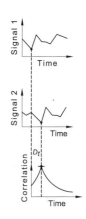

Fig. 15-2. Schematic representation of velocity measurement by acoustic scintillation

insertion types are available, as well as full pipe systems. Single-point insertion types must be traversed across the pipe diameter to obtain the overall flow rate (similar to a pitot tube). For the insertion type, accuracies of about 1.5% can be obtained under proper conditions. Limitations (similar to those for the pitot tube) include difficulties in traversing across a large pipe diameter (such as support problems or vibrations), the need to obtain a good measurement of the penstock inside diameter, and the need for a good flow profile (at least 10 penstock diameters of undisturbed, straight pipe upstream of the meter). For the full pipe system, accuracies better than 1% can be obtained.

11. Acoustic scintillation method

Acoustic scintillation drift is a technique for measuring flow in a turbulent medium by analyzing the variations in ultrasonic pulses that have been transmitted through the medium.

The acoustic scintillation flowmeter (ASFM) uses this technique to measure the velocity of the water flowing through a conduit (e.g., an intake to a hydroelectric turbine) by using the natural turbulence embedded in the flow.

Three transmitters are placed at one side of the conduit, three receivers at the other. The signal amplitude at the receivers varies randomly in time as the distribution of turbulence along the propagation paths changes with time and the flow. If the paths are sufficiently closely spaced, the turbulence may be regarded as being embedded in the mean flow, and then the pattern of these variations (known as "scintillations") at the downstream receiver is almost identical to that at the upstream receiver, except for a time delay, D_t (Fig. 15-2). If these scintillations are examined over a suitable time period, this time delay can be determined. The mean flow velocity perpendicular

to the acoustic beams is then D_x/D_t, where D_x is the separation between the paths. Using three receivers allows both the magnitude and inclination of the velocity to be measured.

The average velocity is measured at several preselected measurement levels. Total flow rate is calculated by integrating the average horizontal component of the velocity at each level over the total cross-sectional area of the conduit.

With the exception of the dye dilution method, these devices and methods require approximately 10 to 20 pipe diameters of undisturbed flow (with no change in diameter or bends). The selected device or method should be consistent with the required accuracy.

15.2.1.2 Operational Tests   The following are typical hydraulic tests performed during operational testing:

1. Speed no load (SNL) rejection (usually a critical pressure rise case for both Francis and impulse machines);
2. 25% rated load rejection and load acceptance;
3. 50% rated load rejection (usually a critical pressure rise case for a Francis machine) and load acceptance;
4. 75% rated load rejection and load acceptance;
5. 100% rated load rejection and load acceptance; and
6. 100% gate or needle rejection and load acceptance (usually a critical pressure drop case for an impulse machine).

15.2.2 Instrumentation

To monitor the pressure rises generated during the testing described in Section 15.2.1, sufficient and accurate instrumentation must be installed. The monitored pressure data are used to ensure that the pressure rises and/or drops generated by unit operation are in compliance with the penstock design criteria. Acceptance criteria and allowable deviations must be established before the testing program. (Note: Pressure rise and/or drop data must be reviewed and evaluated at the time of the test and not taken merely for the purpose of later evaluation or future reference.)

The following must be considered in using instrumentation equipment to monitor pressure:

1. Strain gauges, if necessary, must be located at the most critical areas of the penstock system. See Section 14.4 for information on hydrostatic testing.
2. Flow measuring devices can be used for unit flow and corresponding penstock head losses.
3. Pressure transducers and analog or digital pressure charts should be used for measuring pressures and water surfaces. Pressure gauges do not provide sufficient speed response and accuracy. The type of pressure de-

vice should be compatible with the speed of the transient pressure wave. Analog pressure devices are adequate for surge-tank fluctuations and slow-moving valves. However, pressure transducers should be used for cases such as those involving wicket gate closures, PSV operations, or needle operations. Long lengths of tubing should not be used for transient conditions because they can cause amplification or attenuation of data. They should be used only for steady-state conditions.

REFERENCES

AWWA. (2010). "Transit-time flowmeters in full closed conduits." *AWWA C750-10,* Denver.

CHAPTER 16

Documentation and Certification

Source: Photograph courtesy of American Cast Iron Pipe Co.; reproduced with permission from American SpiralWeld Pipe LLC.

16.1 CERTIFIED MILL TEST REPORT

The penstock documentation package must contain a copy of the certified mill test report, which includes records for the following:

1. Steel plate, forgings, structural shapes, pipe, and flanges;
2. Welding consumables;
3. Connecting devices, such as bolts and gaskets;
4. Valves, expansion joints, and couplings; and
5. Other applicable items.

Suppliers of these items must provide records for each item for inclusion in the certified mill test report. Certified mill test report records from items 1 and 2 must be traceable to heat and lot numbers of material used and must comply with the requirements outlined in the project specifications.

16.2 PERSONNEL CERTIFICATION

16.2.1 Designers

Both in-house and outside contracted designers performing calculations on penstock systems, and personnel checking those calculations must have sufficient formal education, experience, and training to perform the calculations. For outside contracted design, personnel qualifications should be left to the discretion of the design supplier, with the approval of the owner. For outside contracted design, current resumes of personnel working on the design team must be included for review by the owner. Final design packages, including project specifications, however, must be signed off and certified by a registered professional engineer.

16.2.2 Welders, Welding Operators, and NDE Personnel

Welders and welding operators performing welding on pressure- or load-carrying components, both in the shop and field, must be qualified by testing to the types of welding being performed on the penstock.

Certain documentation must be submitted to the owner for approval.

1. Fabrication, welding, and inspection procedures
    Before the start of welding, copies of the following must be submitted: proposed welding procedure specification (WPS), procedure qualification record (PQR), welding sequences, repair procedures, and welding rod control. Documentation must include a weld map or table identifying each type of weld joint, the assigned WPS, the parts being joined, the material thickness, and any requirements for preheat and postweld heat treatment.
    Welding procedures must be qualified in accordance with ASME (2010b).
    Procedures for welding and nondestructive examination (NDE) must be submitted before the start of work.
2. Personnel qualification records
    Welders and welding operators must be qualified in accordance with ASME (2010b), with the WPQ recorded on ASME data report forms QW 482–484 (ASME 2010b).
    NDE personnel must be qualified under ASNT (2011) (visual examination excepted). Qualification records of NDE personnel must be submitted to the inspector.

16.3 PROCESS CERTIFICATION

When NDE is required, the fabricator or installer must designate in writing suitable examination procedures, such as those given in ASME (2010a).

Nondestructive testing technicians and operators must be qualified at NDT level II or level III, as defined in ASNT (2011). Weld inspection work must be performed by a certified welding inspector (CWI) certified in accordance with AWS (2007) provisions. NDE personnel must show sufficient experience and training and must be certified in accordance with ASNT (2011).

Weld examination and inspection reports must include detailed records showing evidence of weld quality. For each section of weld inspected, a report form is required. The report must identify the work and show the welder's identification, the area of inspection, the acceptance of the welds, and the inspector's approval signature. Each report must be completed at the time of inspection. A complete set of forms must be supplied upon completion of the work. For radiographic examination (RT), in addition to the report forms, a complete set of radiographs must be supplied.

16.4 DOCUMENTATION RECORDS

16.4.1 Coatings and Linings Documentation

Chapter 10 describes requirements for penstock coatings and linings. The following documents must be submitted and approved before coating and lining application:

1. List of materials: A list of coating materials used in the work, with specific products identified by manufacturer and catalog number for each coating;
2. Written coating procedures: Written procedures for storage, handling, surface preparation, environmental control, application, touch-up and repair, curing, and inspection of each coating system; and
3. Quality control inspection reports: Reports of quality control inspections performed in accordance with the project specifications.

16.4.2 Shop and Installation Drawings

A copy of shop fabrication and field installation record drawings must be included in the documentation records. These record drawings must include traceability of the material identification, by heat and slab numbers, for each piece of material and the physical location of each item.

16.4.3 NDE Records

The documentation records must include checklists showing NDE, along with traceability to each area examined and the identity of the individual making the examination.

16.4.4 Owner's As-Built Drawings

A copy of the applicable owner's as-built drawings must be included in the documentation records.

16.4.5 Operations and Maintenance Manuals

Fabricators, installers, suppliers, and/or constructors must provide the owner with operations and maintenance manuals for pertinent parts, components, and systems within their scope, such that the owner can operate and maintain the penstock system as originally designed.

16.4.6 Document Retrieval

These documents must be submitted to the owner and logged and indexed by the owner to facilitate identification and retrieval.

16.5 DATA REPORT

Completed steel penstocks must be documented by means of a data report completed by the fabricator. The data report must include a description of the penstock, the mate-

rial used, and the applicable design parameters, including design pressures, to the extent necessary to provide adequate identification.

16.6 NONCONFORMANCE REPORT

During fabrication, installation, and start-up testing, any deviation from project specifications and/or design drawings, including the specific final disposition of the deviation, must be documented in a nonconformance report. Nonconformance reports must be retained in the final documentation package.

REFERENCES

ASME. (2010a). "Boiler and pressure vessel code, section V: Nondestructive examination." New York.

ASME. (2010b). "Boiler and pressure vessel code, section IX: Welding and brazing qualifications." New York.

ASNT. (2011). "Personnel qualification and certification in nondestructive testing." *Recommended Practice No. SNT-TC-1A,* Columbus, Ohio.

AWS. (2007). "Standard for the AWS certification of welding inspectors." *AWS QC1,* Miami.

CHAPTER 17

Maintenance

Source: Photograph courtesy of National Welding Corp.; reproduced with permission.

Steel penstocks are highly durable structures, as evidenced by numerous penstocks more than 80 years old still in operation. Although penstocks do not have the obvious maintenance activities of other mechanical systems, such as rotating machinery, there are numerous examples of catastrophic failure of aging penstocks caused by lack of maintenance that have had significant effects. A proper penstock maintenance program cost effectively extends the safe and reliable service life of any penstock.

The primary elements of a penstock maintenance program are

1. Monitoring,
2. Periodic maintenance activities,
3. Condition assessment, and
4. Rehabilitation or replacement, if needed.

This chapter describes these elements as they relate to in-service steel penstocks.

17.1 MONITORING

The three components of a monitoring program are (1) determining the critical elements of a specific penstock system, (2) formulating an inspection program that effectively determines the condition of those elements, and (3) effectively implementing the program. In addition to what is presented here, refer to ASCE (1998) for further details.

17.1.1 Determining the Critical Aspects of a Penstock Monitoring Program

The frequency, complexity, and detail of in-service penstock inspections must be tailored to each individual hydroelectric facility and should consider the following factors:

1. Consequences of a penstock failure, including
   a. The risk to public safety,
   b. Potential for environmental impact,
   c. Extent of possible property damage, and
   d. Economic losses, such as the value of generation and ancillary services, and impact on the electrical grid.
2. Site conditions and penstock loads, including
   a. Maximum and minimum heads and the resulting positive and negative pressures for operating conditions (Obviously, higher head penstocks present greater risk. Applied pressures may change over time, and it is important to determine if the actual conditions of the penstock presently match the conditions used to determine the penstock's condition. For example, the closing times of a turbine's wicket gates or needle valves can be shortened, resulting in higher transient pressures, either intentionally by maintenance personal or unintentionally by the aging of governing control

equipment. Pertinent load cases must be evaluated, such as load rejection and emergency closure caused by an equipment malfunction that disables any cushioning of the servomotor stroke);
   b. Volume of water in an uncontrolled release;
   c. Climatological and environmental conditions;
   d. The water quality of the penstock's discharge;
   e. Aggressiveness of soil for buried penstocks;
   f. Amount of sediment in the water flowing through the penstock;
   g. Frequency of penstock watering-up and dewatering; and
   h. Accessibility for inspection.
3. Penstock characteristics:
   a. Penstock age (this parameter may help evaluate the safety of the penstock);
   b. Method of penstock manufacture (e.g., forge-welded, riveted, welded, or mechanically coupled; history has shown that forge-welded penstocks have had more failure problems than riveted penstocks);
   c. Whether the penstock has a headgate or valve that can be closed to stop an uncontrolled discharge under emergency conditions (such systems reduce damage exposure);
   d. Prior history of the penstock and any previous failure (i.e., does the penstock show evidence of previous problems?).

17.1.2 Inspection Types and Content

As discussed in Chapter 2.2.2 of ASCE (1998), different frequencies are recommended for different types of inspections.

Cursory inspections are visual observations of the penstock for obvious signs of distress, such as leakage, displacements and distortions, sudden changes in instrumentation readings, or unexpected operation performance. These inspections have at least a monthly suggested frequency.

Comprehensive inspections are more rigorous than cursory inspections and include preinspection, an initial visual inspection, and subsequent nondestructive examination and destructive testing as needed. Comprehensive inspections have a suggested minimum frequency of 5 to 10 years, or as dictated by abnormalities, such as observations during cursory inspections or special events, such as an earthquake or flood.

17.1.3 Inspection Safety

Safety is an important component of any penstock inspection. Safety includes properly safeguarded access and/or appropriate personal fall protection for areas that personnel visit. Safety is particularly critical for interior inspections, which are in confined spaces. Interior inspection procedures should include tagged and signed locks for gate operators by personnel entering the penstocks, open vents at the top, bottom, and at intermediate locations along the pipe to provide adequate pipe air flow and ventilation, air quality monitoring if applicable, primary and backup lights, appropriate platforms and/or rigging for steep sections, and emergency communication equipment. Most penstock owners have specific confined space access procedures that need to be followed. This may be in addition to what is required by OSHA and other regulators. See www.osha.gov for more information on OSHA confined space requirements.

17.1.4   General Monitoring Inspections

The general monitoring of penstocks is considered the cursory inspection, as described in ASCE (1998). This monitoring is primarily a regular and consistent external visual observation of the penstock by the station's operating and maintenance personnel for any abnormality or change in condition. Specific observations in a cursory inspection of a steel penstock include the following:

1. The foundation of the penstock should be inspected for unusual movement, particularly on a hillside and at points of restraint, such as anchor blocks. Small cracks in the penstock saddle could be indicators of penstock or foundation settlement. Similarly, expansion joints between concrete foundation slabs may open or close, and they suggest penstock or hillside movement.

2. The penstock should be inspected for excessive vibration or changes in vibration.

3. Appurtances, such as expansion joints, manholes, and drains should be inspected for leakage. Leaks from aboveground penstocks are relatively easy to observe. However, for buried penstocks, leaks are not quite as obvious. Here, one needs to look for new green vegetation growing near the penstock and new water flow in nearby drainage ditches.

4. Changes in the shell condition, such as new areas of corrosion, pitting, deterioration, and loss of rivet head cross section, if the interior is readily accessible, should be observed.

5. Deterioration of the shell coatings and linings should be observed. This would include paint blistering, delamination, rusting, or staining.

6. If dimpling of the steel shell, particularly around the horns of saddles, is observed, it may indicate shell overstressing. The horn of a saddle is the inner concrete or steel saddle top edge that comes into contract with the penstock shell. As penstocks are filled and pressurized, they deflect over this horn area and generate large concentrated bending stresses.

7. The penstock should be checked for expansion joints and coupling deterioration, such as missing or corroded bolts. Also, expansion joints should be observed for any indications that their full movement is being restrained. Conversely, any signs of unusual or additional expansion joint movement should be noted.

8. The penstock should be checked for damage to saddles or supports, such as broken horns of concrete saddles, damaged structural members, or missing bolts on shell support structures.

9. The growth of any trees or brush above a buried penstock may produce roots that could disturb the pipe and backfill.

10. The correct operation of air–vacuum valves should be confirmed by affirming that the springs are operational and, in colder climates, that any heaters required to prevent freezing are correctly functioning.

11. Site-specific instrumentation should be verified to be operating correctly.

12. Confirm that air vents behind gates are not plugged and, in colder climates, any devices to prevent ice plug formation are operating correctly.

The more detailed comprehensive inspection programs, as described in ASCE (1998), are part of the periodic condition assessments, and these are discussed in Section 17.3.

17.2   PERIODIC MAINTENANCE ACTIVITIES

The following sections describe common penstock maintenance activities.

17.2.1   Mechanical Expansion Joints

Under normal operating conditions, mechanical expansion joints require little maintenance. Packing chamber leakage may be caused by loosening of the bolts in the end ring or by drying out of the packing material located near the surfaces exposed to the sun. If this leakage occurs, the bolts should be backed off slightly and then retightened using the recommended torque settings furnished by the manufacturer. If further tightening does not stop the leakage, it may be necessary to replace the packing in the packing chamber. The following procedure may be used:

1. After dewatering the penstock, loosen and remove all bolts.

2. Using a mechanical means, such as a come-along attached to the end-ring section at three places equidistant around the circumference, pull the end-ring section out of the packing chamber.

3. After the packing chamber area has been exposed, pry out the alternate rubber and lubricating rings. High-pressure jet streams are effective in removing gaskets on older expansion joints.

4. Install new packing rings according to the manufacturer's instructions, making sure to alternate the rubber with the 1 or 2 flax rings, and successively alternate the splices 120 degrees for each material. There are usually

five rubber rings and four flax rings. The rubber rings are always the end pieces that come into contact with the metal parts. The engineer should consult with the manufacturer in regard to what durometer rubber should be used for the replacement rings.

5. After the packing rings have been properly installed, push the end ring into place and follow the installation instructions to complete the maintenance.

6. After watering-up the penstock, check the joint for leakage and adjust bolts accordingly, as necessary, to attain a watertight joint.

17.2.2 <u>Bolted Sleeve Type Couplings</u>

With the penstock pressurized, axial movement or deflection may cause bolted sleeve type couplings (BSTCs) to be slightly displaced from their original position, and leakage may occur. The following leakage repair procedure is recommended:

1. Monitor the position of the end rings and center sleeve in relation to the outside circumferential gap between the coupling parts and the penstock shell.

2. Where gaps that exceed the spacing allowed during original installation are observed, mark these areas to properly identify the space and gap.

3. At areas where the gaps exceed the spacing allowed by the installation instructions, loosen the nuts a few turns to allow the gasket to creep.

4. Where minimum gaps are observed, even at steel-to-steel interfaces, commence tightening in these areas to force the spacing to increase, thus diminishing spacing where it is excessive. The bolt torquing action causes cold flow of the rubber gaskets to equalize the gaps.

5. After the gaps have been made uniform around the entire circumference, follow the manufacturer's recommended torquing procedure to again seat the parts. If leakage continues after following this procedure, it may be necessary to replace the BSTC gaskets after dewatering the penstock. This replacement is accomplished by the following procedure:

    a. Remove the bolts.

    b. Separate the end rings from the center sleeve and remove the gaskets by cutting at two or three places around the circumference. High-pressure jet streams are effective in removing gaskets from older BSTCs.

    c. Slide one end ring completely away from the center sleeve to allow the sleeve to be displaced from the original position, thereby exposing the pipe gap.

    d. Using replacement gaskets from the manufacturer, insert them one at a time, feeding them through the pipe gap. The generally accepted practice is to stretch the gasket over the top of the sleeve to

position it in the proper space between the end ring and sleeve on one side. If gaskets need to be replaced and the pipe gap is not available to the maintenance personnel, use an unspliced or nonvulcanized gasket supplied by the manufacturer.

    e. Take the proper measurements, cut the gasket to the manufacturer's recommendations, and bond the ends together to hold the gasket in shape.

    f. After all parts are properly positioned, follow the installation instructions furnished by the manufacturer.

The results of visual inspections, evaluations, and repairs should be entered accurately into the plant maintenance log for future reference.

17.2.3 <u>Maintenance of Air Valves</u>

Air-release valves allow air trapped inside a penstock to escape during penstock watering-up or operation to prevent the trapped air from producing unsteady flow or excessive shell internal positive pressures. Combination air-release–vacuum valves can either release air as an air-release valve or allow air to enter the penstock when the penstock's internal pressure falls below atmospheric air pressure during dewatering. This feature prevents the penstock's shell from experiencing unacceptable negative (vacuum) pressures that could buckle thin penstock shells. For convenience, both air-release and combination air-release–vacuum valves are referred to as air valves.

17.2.3.1 Maintenance Operations   It is recommended that each air valve and appurtenance feature be checked for proper operation and for operational damage at least once each year. A maintenance procedure should be established that includes the following:

1. Inspection of vault or manhole structures for cleanliness and ease of access;

2. Inspection of vent pipes to ensure proper ventilation of air-valve enclosures;

3. Inspection of inside working parts of air valves (valve shafts, float stems, and lever mechanisms should not be struck with a hammer);

4. Repair or replacement of defective or nonworking parts, including bolts and gaskets;

5. Flushing air valves after maintenance to ensure a proper seal and operation;

6. Evaluation of protective coatings and linings;

7. Checking the operation of isolation valves and also checking that valves are left open after completion of maintenance;

8. Protecting air valves from freezing and discontinuation of the protection when it is not required; and

9. Including scheduled maintenance during normal dewatered outages if possible.

17.2.3.2 Maintenance Records It is recommended that air-valve maintenance records be kept and include the following items:

1. Date of inspection or maintenance;
2. Location of each air valve assembly, including reference point measurements;
3. List of items repaired in the vault or manhole structure;
4. List of replaced or repaired air-valve parts;
5. Checklist of items observed;
6. Environmental conditions, including groundwater elevation;
7. Recommendations for additional repairs and improvements; and
8. Name of person completing the inspection or maintenance.

17.2.4 Leakage Repair

Preventative maintenance, such as bolt tightening, at couplings and expansion joints must be performed as needed to help prevent leakage. Usually this can be done at minimal cost. Whenever leakage is identified, it should be evaluated immediately by the appropriate personnel and, based upon the severity of leakage, repaired at the first appropriate opportunity.

One method currently used for the repair of nonstructural leaks at steel-to-concrete interfaces or temporary repair of other appurtenances, such as couplings and expansion joints, is the pressurized injection of expandable hydrophilic or hydrophobic chemical grout that expands when exposed to water. It is important that the grout's flexibility match the movement, such as thermal seasonal expansion and contractions, that the repair area experiences and also has sufficient adhesion to the base materials. This method is relatively inexpensive and usually does not require a plant shutdown or penstock dewatering during the repair process.

17.2.5 Coatings and Linings

A wide variety of coatings and linings is available with steel penstocks. For this discussion, coating refers to the exterior and lining to the interior penstock surfaces. Generally, coatings are divided into two categories: those for buried conditions and those for exposed or atmospheric conditions.

Coatings for buried penstocks, listed in Chapter 10, should provide long-lasting performance with minimum, if any, maintenance. Should the coating of a buried penstock show signs of deterioration, the application or replacement of a cathodic protection (CP) system is usually the most appropriate maintenance practice. It is important that CP monitoring stations be tested annually to ensure that the CP system is working properly and the penstock is adequately protected.

A number of coating systems have been used wherein different coatings are layered atop each other to provide the finished system with the benefits of each material, for instance, a substrate layer of inorganic zinc for its corrosion protection properties, followed by a layer of epoxy for its resistance to permeation and abrasion, and completed with a layer of aliphatic polyurethane for its resistance to UV degradation. Any coating exposed to the atmosphere is likely to need maintenance within 20 years.

Linings generally present operational consideration. One of the best protective linings for steel pipe, providing long-term protection, is cement-mortar in accordance with AWWA (2007). However, because of concerns that this material may have small particles dislodge, particularly at the hand-plastered joints, this material sees limited application on hydoelectric station penstocks. This limitation is due to concern for potential damage to the turbine blades caused by portions of failed lining impacting the blades. Cement-mortar lining should also be carefully scrutinized if the flow exceeds 20 ft/sec because the lining may abrade, depending on the level of suspended solids in the water. However, if cement-mortar lining has been applied because it was an appropriate choice, it should provide the longest successful service without any maintenance, generally on the order of magnitude of 100 years or longer.

Polyurethane and epoxy linings are currently used extensively. These materials should provide acceptable service for at least 20 years before maintenance may be required. Be aware that the completed lining at welded joints is field-applied and therefore susceptible to poor application practices. Particular attention must be paid to these areas during inspection.

Coal tar enamel coating and lining conforming to AWWA (2008) was used extensively until about 1980. This material was used for both buried and exposed penstocks and generally provided excellent service. Health concerns have largely eliminated this product's use, at least in the United States. Coal tars and coal tar pitches are known to be human carcinogens, based on sufficient evidence of carcinogenicity in humans (IARC 1985, 1987). Coatings were usually reinforced with an asbestos or fiberglass mat and coated with a kraft paper or whitewash. This material was not intended for exposed application in environments where temperatures could reach 0°F. Where properly installed, these coatings and linings could provide 50 years or more of service without maintenance.

For the repair of any coating or lining, the manufacturer's recommendations should be followed. However, in general, they will include surface preparation, material selection, and inspection.

17.2.5.1 Surface Preparation Surface preparation and keying into the existing coating and lining material is necessary to ensure that the recoating or relining provides performance capabilities equivalent to the original material.

Existing coating or lining materials surrounding the area to be recoated or relined should be checked for adhesion to the penstock. Loose coating or lining materials must be removed back to the point that the material exhibits sound adhesion to the penstock. Adhesion of adjacent or remaining materials should be evaluated by reference to the manufacturer's recommendations or the original project specifications. Edges of adhering material adjacent to the area of the recoat or reline should be feathered to provide a smooth, even surface. These procedures are not applicable to cement-mortar lined systems.

Surface preparation of the area to be repaired should be in accordance with the requirements of the coating and lining manufacturer. In addition, the area to be repaired should be cleaned in accordance with SSPC-SP 1 (SSPC 2004a) to remove oil and grease. Only clean rags and a solvent that does not leave a residue or have detrimental health effects should be used. Follow OSHA requirements for required ventilation and allowable flash-point requirements. After solvent cleaning, the repair area and coated or lined areas adjacent to the repair area should be abrasive- or water-blasted to remove loose materials, rust, and other foreign residue (cement-mortar coated or lined penstocks should not be water-blasted). Water for the blasting must be clean and free of contaminants that may leave deposits.

Before coating or lining, all surfaces (bare or coated) must be abrasively blasted in accordance with SSPC-SP 5 (SSPC 2007a) or SSPC-SP 10 (SSPC 2007b) or cleaned with a power tool in accordance with SSPC-SP 3 (SSPC 2004b) to provide the required surface profile in accordance with the requirements of the coating and lining manufacturer or the project specifications.

Applied coating and lining materials should be checked for adhesion either by reference to the manufacturer's recommendations or the project specifications.

17.2.5.2 Material Selection The repair materials should be the same as or compatible with the penstock's existing coating and lining materials, and at the same thickness.

17.2.5.3 Inspection Repaired areas should be inspected to confirm that the new coating has the minimum required dry film thickness (DFT) and must be holiday free, in accordance with the ASTM D 5162 (ASTM 2008). Holidays in linings must be repaired before watering-up the penstock.

17.2.6 Safety

Maintenance work should be performed in a manner that is safe for personnel and the environment. Primary considerations for personnel safety are appropriate access, proper equipment, personnel training, and defined emergency action plans. Environmental safety includes the proper use and disposal of potentially hazardous material and preventing pollution during completion of the work. The material safety data sheet (MSDS) and manufacturer's recommended procedures for potentially hazardous material should be approved before being brought on site, and once the material is on site, the MSDS and procedures must be posted in a readily accessible location.

17.3 CONDITION ASSESSMENT

A penstock condition assessment should provide a thorough evaluation of the condition of an existing steel penstock. An important aspect is to estimate the potential failure modes applicable to the specific penstock being evaluated. Further details on penstock condition assessment are presented in ASCE (1995).

17.3.1 General

The key general items involved in evaluating existing penstocks include the following:

1. Obtaining pertinent engineering documents, including but not limited to drawings, material property or classification, age, type of fabrication, fabrication and/or construction records, and operating records (including any operational test data);
2. Inspecting the penstock to identify any defective areas and to determine if such defective areas are localized or global;
3. Determining the cause of the defect;
4. Determining if the defect is self-limiting or may tend to propagate further;
5. Assessing the severity of the problem and any effect on the penstock's structural integrity and remaining service life; and
6. Evaluating nontechnical requirements or concerns that may pose additional constraints. Some common requirements include special operational and/or maintenance requirements, accessibility, proximity to other structures or features (especially underground structures, cables, or pipes), and annual hydropower plant outage times and durations. This information should be provided by the facilities hydropower operations department.

Additional items that may be performed to follow up any initially identified problems are the following:

1. Developing alternative schemes at a conceptual level, including an examination of the pros and cons associated with each and
2. Establishing a time frame for implementing any repair or replacement schemes.

17.3.2 Penstock Elements

This section addresses some of the essential elements necessary to better understand a penstock to allow a thorough and complete inspection and evaluation.

1. Review available previous penstock drawings, data, and calculations.
2. Develop an understanding of the facility, including:
   a. Age,
   b. Head,
   c. Flow,
   d. Penstock diameters,
   e. Thicknesses,
   f. Type of construction,
   g. Strengths of steels,
   h. Support types,
   i. Coating and lining types,
   j. Transient operating pressures,
   k. Wicket gate and needle valve timings,
   l. Inclinometer data,
   m. Inspection dates,
   n. Previous test dates,
   o. Inspection record reference,
   p. Records of previous problems and repairs, and
   q. Drawing and design references.
3. Review available information to prepare a systematic and comprehensive inspection plan for the penstock.
4. If site conditions warrant, a geologic evaluation may also be performed. This evaluation may be based on observations of movement in the penstock and supports.
5. These inspections also may include mechanical testing (material properties) of penstock shell, and weld nondestructive examination (NDE) to help in evaluating the strength parameters of the penstock.

17.3.3 Inspection

17.3.3.1 Visual Inspection
This section addresses some of the common elements of a penstock that require a visual inspection.

1. A condition assessment visual inspection should first be performed on exposed penstock surfaces and on the ground above buried penstocks when there is adequate access. Visual inspection is particularly effective for detecting surface defects and potential subsurface defects, which can then be subsequently further examined by other methods.
2. Inspect rivet heads on circumferential and longitudinal joints and straps, particularly if there is any leakage.
3. A representative portion of structural welding on the inside and outside of the penstock should be examined visually for signs of significant rusting, pitting, or other structural defects. If justified by visual inspection, structural weld integrity then should be evaluated by NDE methods.
4. Inspect for rust blisters, which may indicate pinhole leaks from pitting.
5. Inspect for linear indications, particularly along the longitudinal axis, which may reveal cracks in the penstock shell.

6. Examine the penstock shell for excessive corrosion, pitting, and deterioration, as well as for coating and lining deficiencies as applicable. This step should be done both internally and externally. When a visual interior inspection is performed, use caution because of limited access, poor lighting, and ventilation problems. Also, the penstock must be dewatered. However, this type of inspection usually can be scheduled during normal plant outages. Review OSHA confined space requirements.
7. An increasingly common visual inspection method is the use of remote operated vehicles (ROVs) to visually inspect penstocks that are difficult to dewater and/or access. These devices can perform inspections rapidly and safely inside slippery, steeply inclined penstocks under either watered or dewatered conditions. The shell's internal surface condition, such as pitting, paint, rust, and erosion conditions, can be observed in real time by using a camera mounted on the inspection unit. The video can be recorded to document current conditions for future reference. Technology changes quite rapidly in this field, so a qualified inspection company should be consulted in regard to abilities and limitations.

17.3.3.2 Nondestructive Examination
This section provides some of the basic inspection techniques used in nondestructive examination of penstocks.

1. Use ultrasonic examination (UT) to determine the thickness of a penstock shell. Record penstock shell thickness measurements at selected locations along the penstock. A history of these readings gives an indication of the expected yearly decrease in shell thickness. These readings can be taken easily on the outside of an exposed penstock without the need for dewatering. For a buried penstock without a cement-mortar lining, shell thickness readings can be taken from the inside during dewatered periods or with selective excavation, depending on the depth of burial. These readings should be compared with the original shell thickness specified in the design to determine if any corrective action is needed. Generally, ultrasonic thickness testing is performed as spot readings taken on a representative grid profile. If grinding to sound metal is required, then the coating and lining should be repaired upon completion of the work.
2. Use magnetic particle examination (MT) and/or liquid penetrant examination (PT) to follow up on surface indications that were previously detected by visual inspection.
3. Use radiographic examination (RT) to evaluate interior pitting or plate–weld cracking and corrosion between steel shell and concrete saddles. The film should be placed between the steel and concrete, with the source on the other side of the penstock. The penstock should be dewatered; otherwise exposure time becomes excessive.

17.3.4 Documentation

The inspection program should be well documented and implemented by competent, well-trained personnel using calibrated equipment. A log should be set up at the plant to record the date, type of inspection performed, and results of inspections performed on penstocks. Inspection results should be forwarded to the engineering staff or other appropriate personnel for review and evaluation. These records should be retained for future reference.

The penstock evaluation documentation should include a chronology of inspections, results, evaluations, and repairs that can help identify the development of any adverse trends. This chronology is essential for the proper maintenance of safe penstocks.

17.4   REHABILITATION

17.4.1   Decrease in Shell Thickness

A reduction in wall thickness usually is attributed to uniform corrosion, localized corrosion or pits, cavitation at sharp boundary edges, or scour and abrasion from hydraulic transport of sediment. Corrosion and pitting can occur on both the exterior and interior surfaces. Uncoated penstocks can experience corrosion from anodic–cathodic reactions.

To determine the extent of material loss, wall thickness measurements should be obtained. From these data, the associated stress level can be determined and compared against the maximum allowable stress. Depending upon the extent and the effect of reduced wall thickness, four restoration options are available: relining, recoating, repair, and replacement.

17.4.1.1   Relining or Recoating   Usually coatings and linings are intended to prevent wall thinning and/or pitting and not to act as structural replacement systems. Therefore, this method should be considered only when the current stress levels are within an acceptable limit. Coating and lining information is discussed in greater detail in Chapter 10.

Coating or lining concerns include material life (which varies according to material product and method of application); defects associated with lack of complete adhesion or holidays; proper removal, handling, and disposal of the original coating or lining; and environmental application regulations.

17.4.1.2   Repairs   Common methods of repairing steel penstocks include installation of one of the following:

1. Patches or wedding bands: Depending on the chemical composition and weldability of the penstock material, patches or "wedding bands" (circumferential banding) can be welded integrally to the penstock shell. Welding to cast sections requires special precautions and may not provide the full strength of the weld filler or patch material. Patches can be installed over the affected area to provide additional reinforce-

ment, or the affected area can be cut out and replaced. Whichever method is selected, corners or sharp edges should be rounded to reduce the associated stress concentrations, and welding should be performed with the penstock dewatered to reduce the possibility of thermal crack development in the welds' or plate's heat-affected zone (HAZ). Welding should be performed under the rules of the appropriate code (ASME, AWS, API, or other acceptable code) and by certified welders as set forth in Chapter 11 of this manual. In addition, all surfaces should be coated or lined to prevent further corrosion.

2. Concrete encasement: Full concrete encasement is used to transfer internal pressure loads to the concrete so that the steel shell functions only as a waterproof membrane. Concrete should be placed with the penstock completely dewatered and with circumferential rebar reinforcement, with rebar joints adequately transferring the hoop reactions, to prevent the internal penstock pressure from cracking the concrete.

3. Carbon fiber wrapping: This type of wrapping is a recently developed technique that provides additional circumferential shell rehabilitation reinforcement.

4. As background, one type of carbon fiber wrapping system available includes a bidirectional weave of carbon fiber and a 100% solid epoxy, forming a composite system that is stronger than steel. The structural system forms a pipe around a pipe, and each successive wrap increases the pressure rating. Its bidirectional weave allows for strength in both the hoop and axial directions, experiencing little or no creep properties over time and ensuring that there is no reduction in strength. Because of its low profile, carbon fiber wrapping can be installed on tees, elbows, straight runs of pipe, and irregular surfaces that require structural reinforcement or leak containment. The adhesive properties of the epoxy allow it to be applied to most substrates.

17.4.1.3   Replacement   Replacement sections should be designed and manufactured in accordance with Chapters 2, 3, and 12.

17.4.2   Supports

This section addresses some of the more common problems and remedies associated with the movement of penstock supports.

17.4.2.1   Causes of Movement   The following are some of the common causes of penstock support movement:

1. Undersized supports: If all loads were not considered during the initial design, it is possible for supports to move. If there is movement in a support, the design should be reviewed to determine that the support is adequate.

2. Damaged supports: A support may be damaged (because of a variety of causes) such that the penstock is not being restrained or supported by the full mass or area of the support.

3. Improper sliding resistance coefficient: If the actual surface condition of the foundation that the concrete support is placed against differs from the condition assumed during initial design, the support may move.

4. Unsound rock: If there was improper inspection during construction, rock that was assumed sound during initial design actually may be heavily fractured or unsound.

5. Deteriorated foundation: When insufficient consideration is given to runoff or erosion, both rock and soil foundations can deteriorate, resulting in lower sliding resistance coefficients.

6. Slope instability: Because of the steepness of many penstock installations, the slope on which the penstock supports are placed can move, causing the supports to move with the slope or an increased loading on the supports.

17.4.2.2 *Repair and Replacement of Supports* The following are some remedies for the problem of penstock support movement:

1. Support replacement: If the support has settled or is undersized or damaged beyond repair, it may be necessary to replace it with a properly designed support system. For replacing saddle-supported penstocks, a common approach is to construct new foundations with saddle supports in between the existing saddle supports to minimize station interruptions.

2. Support repair: If the supporting foundation is in acceptable condition and the support is not severely damaged or undersized, it may be supplemented by adding a system of grouted rock anchors and structural steel plates. Or additional concrete surrounding the existing support may be used to repair the support to decrease bearing pressure or increase overturning capacity.

3. Foundation improvement: If the foundation is unstable or weathered, the foundation may need to be improved. Unsound rock can be consolidated using pressure grouting.

4. Slope stability: Unstable slopes can be repaired using rock buttresses, a system of tie-back retainers, or in some instances, by removal and replacement using geotechnically reinforced soil slopes. Geotechnical engineers should be used for these types of problems.

17.4.3 Conditions That May Adversely Affect Steel Properties

When evaluating the maintenance of existing penstocks, certain material properties should be considered. Changes in steel mechanical properties or fracture characteristics can be attrib-

uted primarily to (1) fatigue-type loading caused by excessive vibration (see Section 1.4 for a discussion of vibration prevention); (2) in-service operating temperature range relative to the material's ductile-to-brittle transition zone; (3) excessive heat imparted to the steel, such as during welding or stress relieving; and (4) an abnormally high transient pressure rise that exceeds the yield point of the material or joints.

17.4.4 Strength Reduction Caused by Joints

Causes of in-service joint decreases in strength are unique to the type of joint or connection. The most common types of connections and associated problems are the following:

1. Forge-welded connections: Problems with forge-welded joints include variable joint efficiency and accelerated corrosion at the joint (see Section 17.5.6). To verify the efficiency of a forge-welded joint, "coupons" should be removed and tested. From these data, an acceptable joint efficiency can be determined and used in ascertaining if repair or replacement schemes are warranted. If the tested joint efficiencies are determined to be too low, a fracture mechanics evaluation may be performed to demonstrate that certain flaw sizes and orientations are acceptable for the operating service loads. If the design steel specification is unknown, a metallurgical evaluation should be performed to determine the material's chemical composition and ascertain if weld repairs may be appropriate. If the stresses in the joint are deemed unacceptably high, replacement may be the preferred alternative. However, some repair options exist. The most common joint repair options include the following:
   a. Weld repair of the joint: This approach is highly dependent on the chemical and mechanical properties of the steel and may require a postweld heat treatment to relieve stress in the weld zone (especially where plate thickness exceeds 1 1/2 in. and carbon content is too high). Refer to Chapter 11 for more details on welding.
   b. Installation of structural banding: Structural banding may be installed, but it may be difficult to install and may not provide sufficient long-term reliability.

2. Riveted connections: Design joint efficiencies for this type of connection usually are in the range of 46% to 95%. See Sections 17.5.1 and 18.7 for methods of determining riveted joint efficiencies. Reductions in the design joint efficiency, or conversely, an increased stress distribution across the joint can be caused by improper installation of the rivet or corrosion of the rivet shanks. Scour or abrasion of a rivet head is indicative of a loss of plate thickness (see Section 17.4.1). If the efficiency of the joint is determined to be unacceptable, common repair options include the following:

a. Replacing rivets with bolts: Obtaining full plate contact of both the bolt and nut by applying sufficient torque to the bolts is of primary importance. If a cluster of rivets is to be replaced, the rivets should be replaced one at a time to prevent the joint from springing open. Removal of rivets can be accomplished in several ways without damaging the surrounding plate. One method is to directly heat one head of the rivet and then quickly blow out the rivet with compressed air (Blodgett 2006). Another method is to cut off the heads and core the shanks. It is also sometimes possible to cut off the heads and press out rivets, although if the rivet was installed in a hole where the plies were not perfectly aligned, the rivet may be deformed and difficult to remove. To prevent joint slippage, replacement bolts need to either have their shanks fit tight to the holes without the typical bolt clearances or to be fully tightened (i.e., slip critical).

b. Welding both sides of the lap joint: Welding both sides of the lap joint tends to increase the joint efficiency, assuming that the rivet strength is the critical element. However, it also introduces additional residual stresses associated with the welding process and may result in leakage around the rivets, which can lead to corrosion problems. Should this method be used, welding should be performed when the penstock is dewatered to minimize the thermal effect associated with welding on the cold plate, and the repair should be monitored during repressurization to verify that there is no leakage around any rivet heads. If leakage occurs, caulking around the interior of the rivet head or installing a flexible interior lining system may be required.

3. Welded connections: Typically, joint efficiency is established during the original design phase (see Section 3.5). The welded joint usually can be repaired by air carbon arc gouging or grinding out the defect and rewelding. Depending on the steel plate thickness and other factors, preheat or postweld heat treatment may be required. In addition, weld repairs on pipe sections encased in concrete may not be effective because of the difficulty in performing the necessary preheat and postweld heat treatment.

4. Jointing new welded pipe to existing riveted pipe: There may be instances where a section of riveted pipe must be removed and replaced with a new section of welded steel pipe. This work can be done, but there are a number of precautions that need to be incorporated into the design and construction activities:

- Samples of the riveted pipe that are going to be welded to need to be tested for chemistry to develop the appropriate welding procedure.

- When the girth weld, between the new and old penstock, is welded, the heat from the welding process often causes some of the rivets, adjacent to the weld, to start leaking. Some things that have been used by owners are (1) seal-welding the rivet heads (this step may just chase the problem further), (2) developing an encompassing cover unit that is pressure-grouted with epoxy to encapsulate the welded joint, (3) caulking the rivet heads in the interior with a flexible sealant material, and (4) relining the interior in the repair area with a polyurethane lining material (probably the most effective system).

- Welding across the existing riveted joint (lap or cover plated) creates a built-in crack starter at the joint. A "crack stopper" detail should be used to prevent the crack propagation into the welded section from the riveted section.

17.4.5 Cracks

Cracks develop at areas of inherent material or weld defects or by excessive loading. Once formed, certain types of cracks tend to propagate because of in-service conditions (i.e., when subjected to cyclic loading or vibration, especially). Factors such as chemical composition, material properties, fracture toughness, crack geometry, and applied loading (and frequency) determine whether a crack tends to propagate. To determine if repairs are required or even possible, evaluations for both fracture mechanics and metallurgy should be performed. Such evaluations should indicate the maximum crack geometry that may tend to be self-limiting and not propagate further, providing that the mechanism responsible for initiating the crack can be arrested. In addition, this type of evaluation determines if repairs are feasible and, if so, indicates the type of repair procedure to be implemented.

Crack repairs typically consist of grinding, air carbon arc gouging, or plasma arc cutting to remove the crack for its entire length and depth and result in a semicircular gouge termination. Use an appropriate weld procedure to prepare and replace the removed material. Depending upon the plate thickness and chemical composition, preheat or postweld heat treatment may be required. One cautionary note in performing weld repairs is that, depending upon the chemical composition of the steel, the heat associated with the welding process may tend to propagate the crack further. Therefore, weld repair of cracks should be performed only under the careful supervision of a qualified metallurgist or welding engineer. In addition, if the penstock section in which cracks have been identified is to be left in service (whether or not it has been repaired), periodic inspections and NDE should be performed to confirm that no new cracks have developed and that unrepaired cracks are not propagating.

17.4.6 Casting Defects

Old steel castings typically are associated with penstock elbows, tees, transitions, and wye branch sections. Casting defects range from superficial delineation and blemishes to structural defects, such as cracks associated with shrinkage during cooling near discontinuities, inclusions of foreign material in the casting, and poor chemical or crystalline grain structure.

Repair of casting defects is dependent on the chemical composition, weldability, and geometry of the casting, whether the casting is encased in concrete or exposed, and the type of defect. Most repairs consist of air carbon arc gouging, plasma arc cutting, or mechanically grinding out the defective area and may not require performing a weld repair. For shallow cracks or defects, it may be sufficient to grind out the defect and feather back all sharp edges to minimize stress concentrations. For larger cracks or defects, weld repairs with preheat and potential postweld heat treatments may be required. Successful repair of casting defects requires special precautions and should be performed with carefully prepared welding procedures and preferably under the careful supervision of a qualified metallurgist or welding engineer. For detailed information on assessing casting defects, the engineer should consult with a certified welding engineer.

It should be noted that not all types of casting defects require repairs or can even be repaired. Small cracks and defects may be acceptable without repair, pending a review using fracture mechanics techniques to determine the maximum acceptable defect size (length and depth) that is considered self-limiting. Shrinkage-type defects generally are not considered repairable but may be acceptable depending on defect size, relative location, material toughness, and relative state of stress. In addition, depending upon the weld procedure, weld repairs on castings encased in concrete may not be effective because of the difficulty in obtaining the necessary preheat and postweld heat treatment.

17.4.7 Ovalization/Out-of-Roundness

Low-pressure, thin-walled penstocks are most susceptible to losing their shape and becoming out of round. Potential negative effects include damage to coatings or linings, susceptibility to general or local buckling, and increased joint leakage, especially at mechanical couplings.

The following are some of the most common causes of penstock ovalization:

1. When the normal internal pressures are low and the diameter-to-thickness ratios are high, the penstock typically cannot maintain its circular shape. For low-head sections, the rounding effect associated with pressurizing the conduit may not be sufficient to offset the weight of the fluid acting outward and tending to flatten out the pipe. This problem may be of particular concern for penstocks on concrete saddles because of high stress concentration at the saddle horns (see Chapter 4).

2. Improper installation of buried or partially buried penstocks can cause the penstock to lose its shape. Typically, either improper compaction or the application of excessive surcharge loads can cause the penstock to lose its shape. Proper compaction from penstock invert to spring line is essential to proper installation. Overcompaction at the spring line can deflect the penstock sides inward, whereas undercompaction can cause the sides to splay outward. Exceeding the design surcharge or external pressure design loading (e.g., under road crossings) can also result in ovalization of the penstock. These loadings are particularly important with cement-mortar lined penstocks.

3. Penstock sections that have not been evaluated for external loads and that are to be backfilled in soil or encased in concrete can become ovalized. Penstock sections that are deeply buried or are to be subjected to large surcharge loads should be reviewed based on external loads with the penstock empty, which is typically the more severe external loading condition.

17.4.8 Voids in Backfill or Concrete-Encased Penstock Sections

For buried or concrete-encased penstock sections, voids may be present in the backfill or concrete.

Voids in backfill are typically caused by groundwater erosion of the backfill material near the invert of the penstock. Prolonged erosion of the backfill can undermine the penstock foundation, leading to differential settlement and potential failure. This type of defect usually can be detected by striking the penstock shell with a hammer at multiple locations and listening for a hollow sound. Once a void has been detected, the extent of the hollow region should be further evaluated to determine the penstock's susceptibility to buckling.

If it is determined that remedial repairs are necessary, the best method for repair is to excavate the backfill and then recompact backfill material into the void. Installation of a drainage system using a geotextile fabric also helps to minimize soil erosion at such locations. Also, the water source should be located and a system developed to channel the water away from the penstock backfill.

Voids in concrete-encased steel penstock are caused by poor consolidation of the fresh concrete during concrete placement or by the trapping of excess water in the concrete "bleed water" near the penstock invert. Typically these types of voids are localized and relatively small. However, voids that allow sufficient external pressure to build up that could cause buckling should be evaluated as outlined in Chapter 6. The detection of voids in concrete is similar to the detection of voids in backfill. The repair of unacceptable voids usually consists of drilling through the penstock shell and filling the void by contact grouting. This filling uses water, sand and cement, or water and cement, depending on the size of the void. A patch or plug is then installed on the drilled-out area of the penstock shell (see Figs. 6-16, 6-17, and 6-18). External encasement of the penstock can also be accomplished using flowable fill, which is a lean, low cement-to-sand ratio, high-slump concrete that can be placed into constricted areas without vibration.

17.4.9 Buckling

Shell buckling can be either localized or general. Localized buckling is a relatively small area of shell deformation that is typically caused by localized external loads, such as heavy overhead traffic, protruding rocks, or concentrated saddle reactions. Circumferential local buckling is most commonly associated with thermal changes that affect a section of penstock between two fixed points, such as anchor blocks. The penstock is most susceptible to thermal problems when the penstock has been dewatered and the ambient temperatures exceed the penstock's normal designed operating temperatures. Therefore, dewatering of aboveground penstocks during the warmest seasons of the year should be evaluated. See Chapter 3 for thermal considerations. For repair, the damaged penstock section should be removed and an expansion joint should be installed. If an expansion joint already is installed in the section, it should be inspected closely to determine if it is functioning properly. For circumferential buckling, the section can be cut out and replaced, or tunnel liners can be backfill grouted and rock bolted to prevent additional deformation.

General buckling commonly occurs as a longitudinal deformation and is an indication of excessive external vacuum, water pressures, or dewatered tunnel liners. The blocking of penstock vents by ice or debris is the most typical cause of penstock vacuum pressures. Shell replacement or the installation of circumferential stiffeners to hold the shell's round shape are common approaches to repairing longitudinal buckled shells.

17.4.10 Transient Pressures

A change in load rejection and acceptance conditions can potentially produce unacceptably large positive and negative pressure waves along the entire length of the penstock. The actual operating conditions need to be checked with the design transient conditions to confirm that the penstock's design is adequate.

Large positive pressure rises can produce unacceptably high tension stresses in the penstock shell. Remedies for large positive pressures, aside from replacement with pipe of higher pressure capacity, include installing a bypass valve that opens when the turbine wicket gates close, installing rupture discs that open when the pressure rise exceeds the structural capacity of the disc (lower than penstock design pressure), changing the turbine gate or needle valve closure timing, installing jet deflectors on Pelton-type turbines, and eliminating or significantly reducing the time lag associated with mechanical linkage systems. All options should be explored based on effectiveness and economics.

Large negative pressure drops occur when the elevation of the hydraulic grade line drops below the penstock crown elevation. Water-column separation may then occur. This condition can produce a negative internal pressure that can buckle the penstock and very large recombining pressures that could rupture the penstock. The previously described remedies can be used if the negative pressure drop is attributed to load rejection. However, if the negative pressure drop occurs during the "load on" condition, the wicket gate or needle valve opening timing can be increased.

17.5 PENSTOCK CONSTRUCTION TECHNOLOGIES

The types of construction most commonly used seemed to follow the development of the technology of welding and steel plate manufacturing. Prior to 1900, cast steel pipe was used as a primary material for small diameter penstocks. Larger penstocks used plate steel with riveted joints. The advent of forge welded technology made forge welded penstocks popular during the 1920s. Arc-welding technology soon became the standard mode of construction, beginning in the early 1930s and gaining popularity and refinement as the industry met the challenges of the 1940s and World War II. Are-welded penstocks became the norm by the mid-1930s, although circumferential field joints were still riveted. After 1940, certain variation on the electric arc-welded mode became popular.

Because many older existing steel penstocks were constructed with obsolete technologies, contemporary engineers may be unfamiliar with the characteristics and limitations of these techniques. The following sections review some of the technologies used in older steel penstocks. Most of this information to follow was extracted from the ASME Boiler Codes of 1914 to 1915 (ASME 1914, 1915) and 1925 (ASME 1925). There may be some variability to the presented strength values because of the lower levels of quality control as compared to that for today's steels.

Many penstocks in this era were designed using these codes. However, there may be penstocks still in existence that were designed to some other criteria or material.

17.5.1 Riveted Penstock Joints

Although current penstock design and fabrication practices no longer use riveted joints, many penstocks still in service are either fully riveted or a combination of welded and riveted.

It is important to know more about the original design and construction of these older penstocks because many of them are connected to powerhouses that are being upgraded with additional powerhouse units through repowering programs. Also, with the increase in age of these older penstocks, many are being repaired or replaced as a result of safety evaluation programs or normal maintenance programs. A detailed material, weldability and joint efficiency analysis should be performed.

For additional information, see ASCE (1995).

17.5.2   Types of Joints

Two combinations of welded and/or riveted joints that have been used for steel penstock fabrication and erection are

1. Shop- or field-riveted longitudinal and field circumferential joints and
2. Shop-welded longitudinal joints and field-riveted circumferential joints.

The combination of shop welding and riveting was common during the 1950s. These types of joints did not change until improved welding techniques were developed that made high-quality field-welding feasible. Before the 1950s and dating back into the late 1800s, most joints were fully riveted. One exception was the use of forge-welded and banded steel penstocks during the 1910s and 1920s (see Sections 17.5.6 and 17.5.8).

Usually shop-welded joints, either longitudinal or circumferential, were radiographed and stress-relieved in accordance with the ASME code of the time. In addition, it was quite common to pressure-test each welded pipe section to 150% of the normal working pressure.

Before the 1950s, there was generally no field-welding of the penstock shells, except in cases where access was a problem (tunnel liners made field-riveting impractical).

It was not considered practical to pressure-test riveted pipe in the shop because the strength could not be accurately computed and because of the difficulty of testing.

17.5.3   Material Properties

17.5.3.1   Chemical Composition   The actual chemical composition of an existing penstock can be determined by removal of a small physical sample for examination by a mass spectrometer. The results can then be evaluated for weldability and other composition considerations. Also, coupons can be removed from the penstock for detailed material property testing and weldability. After removal, the coupons must be replaced.

Generally, the two grades of steel used for penstock shells were ASTM flange and firebox grades. The chemical composition of the shell material usually conformed to the specifications given in Table 17-1.

The chemical composition of the rivet steel material usually conformed to the requirements given in Table 17-2.

17.5.3.2   Mechanical   Properties   The mechanical properties of the shell steel used usually conformed to the requirements given in Table 17-3.

17.5.3.3   Crushing Strength of Mild Steel   The 1914 edition of the ASME code allowed the crushing strength of mild steel to be 95,000 psi (for the projected area).

17.5.3.4   Strength of Rivets in Shear   Table 17-4 gives stresses for the cross-sectional area of the rivet shanks, which are used for calculating joint strengths of circumferential and longitudinal joints.

17.5.4   Design

17.5.4.1   Minimum Thickness of Plate   The minimum thickness of any plate under pressure was $\frac{1}{4}$ in., except that for shells 24 in. and less in diameter, the minimum plate thickness was $\frac{3}{16}$ in.

17.5.4.2   Minimum Thickness of Butt Straps   The minimum thickness of butt straps for double strap joints is given in Table 17-5.

For plate thicknesses exceeding $1\frac{1}{4}$ in., the thickness of the butt straps could not be less than two-thirds the shell thickness.

17.5.4.3   Allowable Stress   The maximum allowable working stress (S) was determined by

Table 17-1.   Shell Material Chemical Composition for ASTM (1954)

Element	Flange steel	Firebox steel
Carbon	—	Plates <3/4 in., 0.12%–0.25%
		Plates >3/4 in., 0.12%–0.30%
Manganese	0.30%–0.80%	0.30%–0.80%
Phosphorus (acid)	<0.05%	<0.04%
Phosphorus (base)	<0.04%	<0.035%
Sulfur	<0.05%	<0.04%
Copper	—	<0.25%

Table 17-2.   Rivet Steel Material Chemical Composition

Element	Composition
Manganese	0.30–0.50%
Phosphorus	<0.04%
Sulfur	<0.045%

$$S = 0.0125 f_y (e + 8), \text{ and } S < 0.4 f_y \text{ or } 0.2 f_t \quad (17\text{-}1)$$

where S = maximum allowable unit working tension stress (psi);

f_y = yield strength of material (psi);
f_t = tensile strength of material (psi); and
e = elongation of material in 8 in. (percent).

For example, the allowable stress for 55,000 psi tensile strength material (f_t) was 11,000 psi.

17.5.4.4   Joint Efficiencies   The efficiency of a joint is the ratio between the strength of the joint and the strength of the solid plate. In the case of a riveted joint, this ratio is determined by calculating the breaking strength of a unit section of the joint, considering each possible mode of failure separately, and then dividing the lowest result by the breaking strength of the solid plate of a length equal to that of the section considered. In computing the joint efficiency, the hole diameters are commonly assumed to be 1/16 in. larger for drilled holes and 3/16 in. larger for punched holes than the nominal rivet diameter.

17.5.4.5   Longitudinal Joints   On longitudinal joints, the distance from the center of rivet holes to the edges of the plates, with the exception of rivet holes in the ends of butt straps, was not less than 1.5 times and not more that 1.75 times the diameter of the rivet holes. This distance was measured from the center of the rivet holes to the caulking edge of the plate before caulking.

The longitudinal joints of a shell that did not exceed 36 in. in diameter and were of lap-riveted construction had a maximum allowable working pressure of 100 psi. The plate was rolled and the two ends lapped over each other and then riveted (Fig. 17-1). Generally these joints had low joint efficiencies that varied from 46% to 95% and therefore were used primarily for low-head locations.

The longitudinal joints of a shell that exceeded 36 in. in diameter were to be of butt and double-strap construction. The butt joint had outer and/or inner cover plates that were riveted together (Fig. 17-2). These joints had a much higher joint efficiency (77% to 95%) and were used for higher head applications.

17.5.4.6   Circumferential Joints   The strength of circumferential joints of penstocks was assumed to be at least 50% the strength of the longitudinal joints of the same structure.

17.5.4.7   Butt Straps   Butt straps and the ends of shell plates forming the longitudinal joints were rolled or formed by pressure, not blows, to the proper curvature.

17.5.4.8   Rivets   The following outlines the fabrication elements associated with riveted joints:

1. Centerline distances: The distance between the center lines of any two adjacent rows of rivets, or the "back pitch" measured at right angles to the direction of the joint, was to be at least twice the diameter of the rivets and also had to meet the following requirements:
   a. Where each rivet in the inner row came midway between two rivets in the outer row, the sum of the two diagonal sections of the plate between the inner rivet and the two outer rivets was to be at least 20% greater than the section of the plate between the two rivets in the outer row.

Table 17-3.   Mechanical Properties of Shell Steels for ASTM (1954)

Property	Flange steel	Firebox steel	
		Grade A	Grade B
Tensile strength, psi	55,000–65,000	55,000–65,000	48,000–58,000
Yield point, minimum, psi	30,000	30,000	26,000
Elongation in 8 in., minimum, %	24[a]	25[a]	27[a]
Elongation in 2 in., minimum, %	28	29	30

[a]For material less than 5/16 in. thick, a deduction from the percentage of elongation in 8 in. of 1.25% shall be made for each decrease of 1/32 in. of the specified thickness below 5/16 in. For materials more than 3/4 in. thick, a deduction from the percentage of elongation in 8 in. of 0.5% shall be made for each increase of 1/8 in. of the specified thickness above 3/4 in. This deduction shall not exceed 3%

Table 17-4.   Stresses for Cross-Sectional Area of Rivet Shanks

Rivet type and stress condition	Allowable stress (psi)	Ultimate strength (psi)
Iron—single shear	—	38,000
Iron—double shear	—	76,000
Steel—single shear	11,000	44,000
Steel—double shear	22,000	88,000
Steel—bearing	23,750	95,000
Steel—tension	13,750	55,000

Source: Reprinted from ASME (1915), by permission of the American Society of Mechanical Engineers. All rights reserved.

b. Where two rivets in the inner row came between two rivets in the outer row, the sum of the two diagonal sections of the plate between the two inner rivets and the two rivets in the outer row was to be at least 20% greater than the difference in the section of plate between the two rivets in the outer row and the two rivets in the inner row.

c. The maximum spacing of holes along caulked edges was governed by the formula $P = 2.5t + d$ (not to exceed 1 1/2 in.), where P = pitch, t = plate thickness, and d = diameter of the rivet hole.

2. Edge distance: On longitudinal joints, the distance from the centers of rivet holes to the edges of the plates, with the exception of rivet holes in the ends of butt straps, was not to be less than 1.5 times the diameter of the rivet holes.

3. Rivet holes: Rivet holes were drilled full size, with plates, butt straps, and heads bolted in position, or they were punched not to exceed 1/4 in. less than full diameter for plates more than 5/16 in. thick, and 1/8 in. less than full diameter for plates not exceeding 5/16 in. thick, and then drilled or reamed to full diameter with plates, butt straps, and heads bolted in position. After having rivet holes drilled, the plates and butt straps were separated and the burrs were removed.

4. Rivet length: Rivets were of sufficient length to completely fill the rivet holes and form heads at least equal in strength to the bodies of the rivets.

Table 17-5.   Minimum Thickness of Butt Straps for Double Strap Joints

Thickness of steel plate (in.)	Minimum thickness of butt straps (in.)
3/16	3/16
1/4	1/4
9/32	1/4
5/16	1/4
11/32	1/4
3/8	5/16
13/32	5/16
7/16	3/8
15/32	3/8
1/2	7/16
17/32	7/16
9/16	7/16
5/8	1/2
3/4	1/2
7/8	5/8
1	3/4
1 1/8	3/4
1 1/4	7/8

Source: Reprinted from ASME (1915), by permission of the American Society of Mechanical Engineers. All rights reserved.

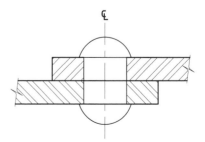

Fig. 17-1.   Riveted lap joint
Source: Reprinted from ASME (1915), by permission of the American Society of Mechanical Engineers. All rights reserved.

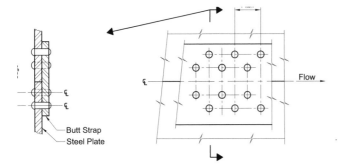

Fig. 17-2. Butt strap riveted joint
Source: Reprinted from ASME (1915), by permission of the American Society of Mechanical Engineers. All rights reserved.

17.5.4.9 Manholes The following outlines the design and detailing elements that were used for riveted manholes on penstock components:

1. Size: An elliptical manhole opening was not less than 11 in. by 15 in. or 10 in. by 16 in. in size. A circular manhole opening was not less than 15 in. in diameter.
2. Reinforcing ring: When a manhole reinforcing ring was used, it was of steel or wrought iron at least as thick as the shell plate. When reinforcing rings were required, they were to have the proper curvature. Shells more than 48 in. in diameter were riveted to the shell with two rows of rivets, which could be pitched as shown in Fig. 17-3. The strength of the rivets in shear on manhole frames and reinforcing rings was at least equal to the tensile strength of that part of the shell plate removed, on a line parallel to the longitudinal axis of the shell, through the center of the manhole or other opening.

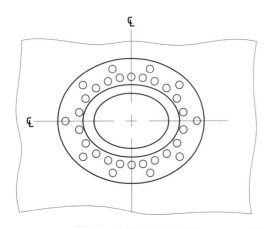

Fig. 17-3. Manhole rivet connection
Source: Reprinted from ASME (1915), by permission of the American Society of Mechanical Engineers. All rights reserved.

The following formulas determined the proportions of manhole frames and other reinforcing rings necessary to conform to the above specifications. It was assumed that the rings should have the same tensile strength as, and have a thickness equal to or greater than, the shell plate removed for the opening.

For a single-riveted ring

$$W = 1/2(t_1/t) + d \qquad (17\text{-}2)$$

For a double-riveted ring

$$W = 1/2(t_1/t) + 2d \qquad (17\text{-}3)$$

For two single-riveted rings

$$W = 1/4(t_1/t) + d \qquad (17\text{-}4)$$

For two double-riveted rings

$$W = 1/4(t_1/t) + 2d \qquad (17\text{-}5)$$

where W = least width of reinforcing ring (in.);

t_1 = thickness of shell plate (in.);
d = diameter of rivet when driven (in.);
l = length of opening in shell in direction parallel to axis of shell (in.); and
t = thickness of reinforcing ring (not less than thickness of the shell plate) (in.).

The following formula was used to determine the number of rivets for a single or double reinforcing ring:

$$N = 5.1Ta/Sd^2 \qquad (17\text{-}6)$$

where N = number of rivets;

T = tensile strength of the ring (lb/in.^{2} of section);
a = net section of one side of the ring or rings (in.^{2});
S = shearing strength of rivet (lb/in.^{2} of section); and
d = diameter of rivet when driven (in.).

3. Covers: Manhole cover plates were to be of wrought steel or steel castings.
4. Bearing surface: The minimum width of a bearing surface for a gasket on a manhole opening was 1/2 in. No gasket used on a manhole or handhole of any penstock was to have a thickness greater than 1/4 in.

17.5.5   Fabrication

17.5.5.1   Rivet Application Rivets were machine driven wherever possible, with sufficient pressure to fill the rivet holes, and allowed to cool and shrink under pressure.

17.5.5.2   Caulking The caulking edges of plates, butt straps, and heads were beveled to an angle not sharper than 70 degrees to the plane of the plate. Every portion of the caulking edges of plates, butt straps, and heads was planed, milled, or chipped to a depth of not less than 1/8 in. A round-nosed tool was used for caulking.

17.5.6 Forge-Welded Penstocks

Although current penstock design and fabrication practices no longer use forge-welded joints, many penstocks of this type are still in service.

It is important to know more about the original design and construction of these older penstocks because many connect to powerhouses that are being upgraded with additional units through repowering programs. Also, with the increase in age of these older penstocks, many are being evaluated for acceptance, repair, or replacement as a result of safety evaluation programs or normal maintenance programs.

The steels used for forge-welded penstocks were the same flange and firebox steels used for riveted penstocks, but they were open-hearth steels, described herein.

17.5.6.1 Chemical Composition of Open-Hearth Steels The following are specifications for the chemical composition of the open-hearth steels used for forge-welded penstocks:

1. Carbon (plates 3/4 in. and less thick): Not to exceed 0.18%;
2. Carbon (plates more than 3/4 in. thick): Not to exceed 0.20%;
3. Manganese: 0.40%–0.60%;
4. Phosphorus: Not to exceed 0.04%; and
5. Sulfur: Not to exceed 0.05%.

Preferably, the steel for forge-welded plates was free of silicon, nickel, and chromium. Where these elements were present, the maximum content of any one element was not to exceed 0.05%.

17.5.6.2 Tensile Strength of Open-Hearth Steels The open-hearth material used for forge-welded penstocks conformed to the following requirements for tensile properties:

1. Tensile strength, minimum: 55,000 psi;
2. Yield point, minimum: 24,000 psi; and
3. Elongation in 8 in., minimum, %: 1.5×10^6/(tensile strength).

17.5.6.3 Design The following outlines the criteria that were used for the design of a forge-welded penstock:

1. Minimum thickness: The minimum thickness for any plate was ¼ in., but in no case could the shell plate thickness be less than the diameter of the penstock, in in., divided by 200.
2. Allowable stress: The allowable stress for steels using the forge-welding process was 12,500 psi.
3. Joint efficiency: The efficiency of the joint when properly welded by the forge-welding process was taken as 85% of the minimum ultimate strength for flange and firebox steels and 95% for open-hearth steels.

Joint efficiency is directly related to the quality of the manufacturing of the joint. Because the plates are heated to approximately 2,000°F, high-temperature oxides develop on the joint.

If these oxides are not removed, adequate bonding does not occur during the pressure welding process. As a result, joint efficiency can drop considerably. This lack of fusion acts like a flaw and therefore reduces the joint efficiency.

In reevaluating forge-welded penstocks, it is recommended that coupons be removed for destructive testing. These tests should include tensile tests in the joint and parent metal (for determination of yield strength, tensile strength, and joint efficiency) and Charpy tests. This material generally has poor fracture toughness and exhibits low Charpy test results. Testing has shown that they can have values on the order of 5 ft-lb at 32°F.

Alternatively, ultrasonic examination (UT) imaging inspection (A-, B-, and/or C-scan) and a fracture mechanics evaluation can be performed to determine the allowable joint strengths.

17.5.7 Fabrication

17.5.7.1 Forge Welding The edges to be welded together were lapped a distance at least equal to the thickness of the plate to be welded. Plates 1 in. and less thick were welded without scarfing. Plates exceeding 1 in. thick were scarfed, if desired by the manufacturer. The scarf started at least one-half the thickness of the plate from the side next to the weld. Fig. 17-4 shows examples of joint details; details (A) through (C) show how various penetration systems are attached to the shell with forge welds; detail (D) shows the joint preparation before the welding process and also the welded joint.

17.5.7.2 Decarbonization Because the forge-welding process requires heating the steels to approximately 2,000°F, decarbonization (loss of carbon) of the weld joint occurs. Because of this, the joints corrode at a more accelerated rate than the base metal. If the interior lining and exterior coating do not provide adequate protection, severe wall corrosion can occur.

After the material was brought up to the proper welding temperature, it was placed (1) between an anvil and a hammer, (2) between rolls, (3) between a mandrel and a roll, or (4) between mandrels. The plates were welded together by pressure (applied by the hammer, rolls, or mandrels that actually displaced the material during welding). The metal in and adjacent to the weld was not worked at what was termed the critical blue heat temperature of the steel (between 400°F and 800°F).

The minimum thickness (t_{min}) of the weld for longitudinal and circumferential seams and special welds was the actual thickness of the shell. The maximum thickness (t_{max}) was 1.1 times the actual thickness of the shell. The engineer should pay particular attention to the actual thickness at these joints because many joints have been shown to be thinner than the normal plate thickness.

The contact line of completed forge welds was equal to at least 2.5 times the thickness of the plate thickness.

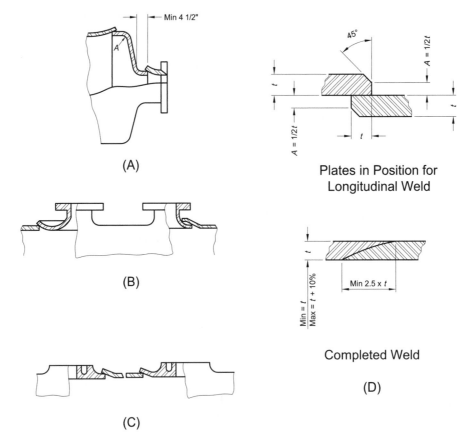

Fig. 17-4. Forge-welded joint details
Source: Reprinted from ASME (1915), by permission of the American Society of Mechanical Engineers. All rights reserved.

17.5.7.3 Annealing Longitudinal and circumferential welds were annealed by heating to the proper temperature to relieve strains and then were allowed to cool slowly. Longitudinal welds on the cylindrical pipes were heated in the area within 8 in. of the center of the weld or the entire shell, after which they were rolled to a commercially true cylindrical form. Any vessel distorted out of shape was re-formed and then annealed or re-formed at the proper annealing temperature. For a finished cylindrical shell, variations in diameter did not exceed 1% of the mean outside diameter when measured at any section. When a straight edge, two diameters long, was laid longitudinally along the outside of the shell, it had to be possible to set the straight edge such that no part of the edge exceeded a distance 1% of the mean outside diameter from the outer surface of the shell.

17.5.7.4 Hydrotest The 1914–1915 and 1925 ASME code specified that all penstocks with seams or joints fabricated by the forge-welding process were hydrotested. They were tested to 50 psi over the working pressure when that pressure did not exceed 100 psi and to 1.5 times the working pressure for pressures higher than 100 psi. While under the hydrotest pressure, a thorough hammer or impact test was performed. It consisted of striking the shell on both sides of the welded seam with a sharp vibratory blow (hammer test) using a 2- to 6-lb hammer. The blows were struck 2 to 3 in. apart and within 2 to 3 in. of and on each side of the seam. The blows were as rapid as a man could conveniently strike a sharp, swinging blow and as hard as could be struck without indenting or distorting the metal of the shell. During this test, the penstock was filled completely with water and pressurized.

Welded seams or joints that did not pass this test (because of leaks, distortion, or other signs of distress) were not accepted until the defects were repaired and the test was passed. Defective sections of welded seam were cut out and rewelded, provided the shell integrity was not lowered. When in question, a coupon was cut out across the weld at points in question and subjected to microscopic or other examination.

17.5.8 Banded Steel Penstocks

Although current penstock design and fabrication practices no longer use the banded steel pipe process, many penstocks of this type are still in service.

It is important to know more about the original design and construction of these older penstocks because many connect to powerhouses that are being upgraded with additional units through repowering programs. Also, with the increase in age of these older penstocks, many are being repaired or replaced as a result of safety evaluation programs or normal maintenance programs. It is also important to understand because this design was used for high-head applications.

17.5.8.1   General   Banded steel penstock fabrication began around 1910 and continued for approximately 50 years. These penstocks were fabricated primarily in France, Germany, and Poland. They were used when the design pressures required plate thicknesses greater than approximately 1¼ in. The lack of technology prohibited forge welding of plates exceeding this thickness.

There were two methods of fabricating banded steel penstocks. The first was to shrink the bands over the cylinders. The second was to install oversized rings on the pipe and expand the shell such that it pinched against the bands.

17.5.8.2   Band Shrinking   The banded pipe consisted of a cylindrical pipe manufactured in the same manner as for plain forge-welded pipe. The forge-welding process, using water gas to prevent oxidation of the metal, was used. The metal was reheated and circularly rolled. Any scale formed during this stage of the process was removed carefully using steel brushes. Areas over which the bands were to be shrunk were cleaned. The circumference was measured carefully to determine the proper size to which the bands were to be machined. The bands, which were seamless, were pressed and rolled and finally machined to the correct size. Then they were heated sufficiently, without altering their mechanical properties, to allow them to fit. When they had cooled to the ambient temperature, they formed a single unit with the cylinder.

When the cylindrical part of the vessel was fabricated by this method, the end flanges were attached and the whole unit was stress-relieved. Residual stresses caused by welding, as well as the prestressing resulting from the hoop fitting process, were practically eliminated. The intervals for the bands were determined in accordance with the pressure for which the pipe was designed.

The amount of shrinkage was computed such that the bands and the cores were stressed to an equal factor of safety, thereby preventing the bands or the cores from being overstressed. As a result, the shell and the hoops were stressed such that the hoops were in tension and the shell in compression.

17.5.8.3   Expansion of Shell   The hoops, with a slightly larger inside diameter than the outside diameter of the pipe, were equally spaced. The pipe was placed between two end plates of a hydrotesting machine, and the cylinder was expanded. It was pressurized until it reached its elastic limit, after which the cylinder expanded until it pressed against the hoops. Once the internal pressure was reduced to zero, the cylinder and hoops formed a single unit.

17.5.8.4   Advantages   Banded steel pipes had a number of advantages, as discussed here:

1. The weight, at the same factor of safety, was considerably lower than that of plain forge-welded or riveted pipes of comparable pressure ratings. As a result, overall costs were less.
2. The factor of safety was accurately known because the bands (which represented approximately one-half of the material used) had a 100% joint efficiency (being solid rolled).
3. Because the shell was fabricated much thinner than an equivalent plain forge-welded pipe, better notch toughness properties could be obtained for the banded pipe.
4. Should a failure occur (during an emergency or exceptional condition), any rupture would occur only at the weakest point (i.e., in the core between two bands). Thus, leakage would be limited to the length of pipe between two bands, ensuring the facility against a disaster. Banded pipes had a decided advantage when compared with seamless pipes; the combined use of materials and designs of different strengths resulting in banded pipes acting as an automatic relief in case of a failure.
5. Erection was simplified considerably, resulting in decreased costs, because the core pipes had practically one-half the thickness of a plain forge-welded pipe of comparable pressure ratings.

17.5.8.5   Material Properties   The material used in the manufacture of plain forge-welded and core pipe for banded steel pipes was a high-grade, open-hearth steel with the physical properties given in Table 17-6.

17.5.8.6   Design Stresses   The determination of the thickness was based upon the following criteria:

1. Working stress in shell: Working stress at static pressure in the full section of a plain forge-welded pipe was not to exceed 10,000 psi (equal to the yield strength of 27,800 psi times the joint efficiency of 90% divided by 2.5).
2. Maximum stress in shell: The maximum stress was not to exceed 18,800 psi. (Static head plus water hammer was not to exceed 75% of the yield strength of 27,800 psi times the joint efficiency of 90%.)
3. Working stress in bands: The working stress at static pressure in the steel bands was not to exceed 13,300 psi.
4. Maximum stress in bands: The maximum stress at static pressure plus water hammer in the steel bands was not to exceed 25,000 psi.
5. Working stress in combined section: The average combined stress used for design of the bands and the shell was 13,200 psi. At least 50% of the pressure was to be taken by the bands. The shell thickness and dimensions, spacing between bands, and shrinkage in the bands were selected to ensure that the longitudinal stresses in the shell (caused by beam action between bands) when combined with the circumferential tension

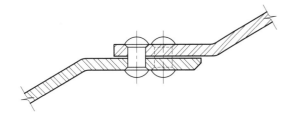

Fig. 17-5. Double bump joint
Source: Reprinted from ASME (1915), by permission of the American Society of Mechanical Engineers. All rights reserved.

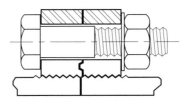

Fig. 17-7. General joint for non-encased specials
Source: Reprinted from ASME (1915), by permission of the American Society of Mechanical Engineers. All rights reserved.

did not result in extreme fiber stresses in excess of these values.

6. Test pressure: The pipes usually were hydrotested to 175% to 200% of their working pressures.
7. Factor of safety: The following is a comparison of the factor of safety (*FS*) of plain forge-welded steel pipe with that of pipe for pressures of static plus 25%.
   a. Forge-welded pipe:

$$FS = \frac{(48,500)(0.90)}{10,000} = 4.36 \qquad (17\text{-}7)$$

   b. Banded pipe:

$$FS = \frac{48,500 + 64,500}{(2)(13,200)} = 4.30 \qquad (17\text{-}8)$$

17.5.8.7 Joints Generally, for pipe shells up to 1 in. thick, single and double bump joints were used (Fig. 17-5). For shells with greater thicknesses, riveted band joints were provided (Fig. 17-6).

For banded pipe connection to specials, such as cast elbows, gasketed flanged joints were used (Figs. 17-7 and 17-8).

17.5.8.8 Anchorage Rings Anchorage rings usually were cast steel and manufactured in two parts for mounting on the bends (forge-welded or cast bends) or on adjacent pipes, thereby conveying the thrust to the penstock by means of rivets.

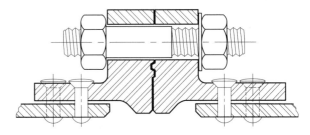

Fig. 17-8. General joint for concrete-encased specials
Source: Reprinted from ASME (1915), by permission of the American Society of Mechanical Engineers. All rights reserved.

17.5.8.9 Expansion Joints Expansion joints were provided immediately below each angle anchor block and where casings and sleeves were forge-welded for low pressures. For higher pressures, the casing was cast steel, and the sleeve was either welded or cast steel.

REFERENCES

ASCE. (1995). *Guidelines for evaluating aging penstocks.* American Society of Civil Engineers, New York.

ASCE. (1998). *Guidelines for inspection and monitoring of in-service penstocks.* American Society of Civil Engineers, Reston, VA.

ASME. (1914). "Report of the Boiler Code Committee of the American Society of Mechanical Engineers." American Society of Mechanical Engineers, New York.

ASME. (1915). *Boiler and pressure vessel code, section I.* New York.

ASME. (1925). *Boiler construction code, section VIII: Unfired pressure vessels.* New York.

ASTM. (1954). "Specification for boiler and firebox steel for locomotives." *ASTM A30,* West Conshohocken, PA (withdrawn 1964).

ASTM. (2008). "Standard practice for discontinuity (holiday) testing of nonconductive protective coating on metallic substrates." *ASTM D5162-08,* West Conshohocken, PA.

AWWA. (2007). "AWWA standard for cement-mortar protective lining and coating for steel water pipe—4 in. (100 mm) and larger—Shop-applied." *AWWA C205-07,* Denver.

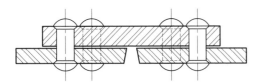

Fig. 17-6. Riveted band joint
Source: Reprinted from ASME (1915), by permission of the American Society of Mechanical Engineers. All rights reserved.

AWWA. (2008). "AWWA standard for coal-tar protective coatings and linings for steel water pipelines—Enamel and tape—Hot applied." *ANSI/AWWA C203-08,* Denver.

Blodgett, O. (2006). "How to Blow Out a Rivet." *Modern Steel Construction,* 46(7), 63–64.

IARC. (1985, 1987). "IARC monographs on the evaluation of carcinogenic risks to humans." 35, 83 (1985), Supplement 7, 174 (1987),

International Agency for Research on Cancer, Lyon, France.

SSPC. (2004a). "Solvent cleaning." *SSPC-SP 1,* Pittsburgh.

SSPC. (2004b). "Power tool cleaning." *SSPC-SP 3,* Pittsburgh.

SSPC. (2007a). "White metal blast cleaning." *SSPC-SP 5/NACE No. 1,* Pittsburgh.

SSPC. (2007b). "Near-white blast cleaning." *SSPC-SP 10/NACE No. 2,* Pittsburgh.

CHAPTER 18

Examples

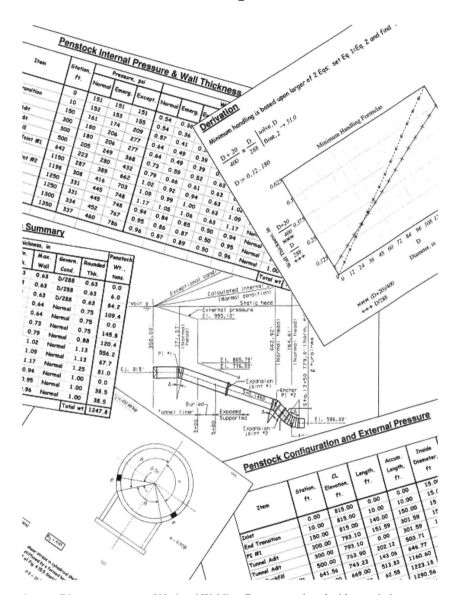

Source: Diagram courtesy of National Welding Corp.; reproduced with permission.

The following 10 example problems are based on the data presented in Fig. 18-1. The examples shown on pages 205–289 are intended only to provide the reader guidance. These examples should not be considered inclusive of all the possible design considerations that would pertain to a specific penstock.

The examples were generated in Mathcad to confirm accuracy; hence they are presented in a different format from the previous chapters in this manual.

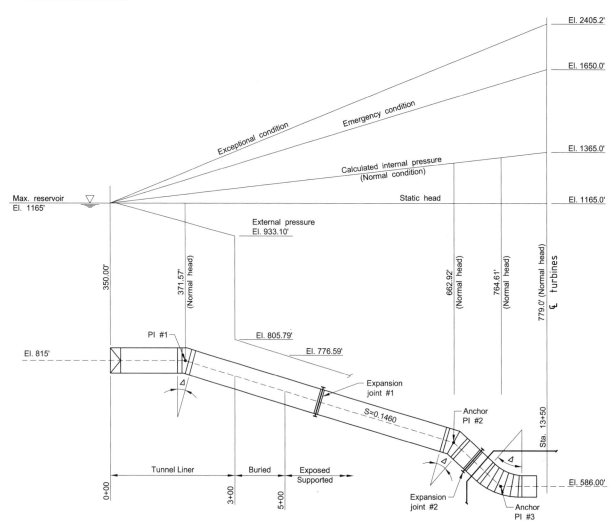

Figure 18-1. Penstock Profile Used for Example Problems

18.1 Internal Pressure

▶

Derivation

Minimum handling is based upon larger of 2 equations; set $\dfrac{D+20}{400} = \dfrac{D}{288}$ and find "D"

$$\dfrac{D+20}{400} = \dfrac{D}{288} \quad \begin{vmatrix} \text{solve}, D \\ \text{float}, 3 \end{vmatrix} \to 51.4$$

$$\dfrac{D}{288} = .0747 \quad \begin{vmatrix} \text{solve}, D \\ \text{float}, 3 \end{vmatrix} \to 21.5 \qquad \begin{array}{l}\text{14 gauge limit,} \\ \text{no mortar} \\ \text{lining}\end{array}$$

$$\dfrac{D}{240} = .0747 \quad \begin{vmatrix} \text{solve}, D \\ \text{float}, 3 \end{vmatrix} \to 17.9$$

14 gauge limit, mortar lining

$D := 0, 6 .. 180$

Figure 18-2. Minimum Handling Formulas

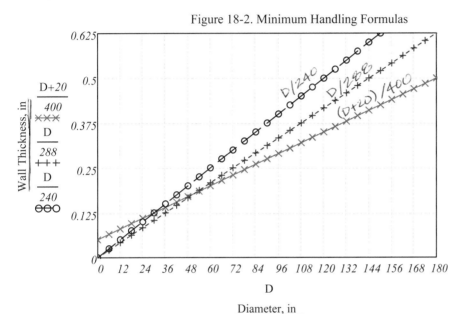

Wall Thickness, in

$\dfrac{D+20}{400}$ ✕✕✕

$\dfrac{D}{288}$ +++

$\dfrac{D}{240}$ ⊖⊖⊖

D

Diameter, in

✕✕✕ (D+20)/400
+++ D/288
⊖⊖⊖ D/240, mortar lined pipe

Conclusion

As indicated by graphing both minimum handling equations, they cross at $D = 51.4 \cdot in$
Thus for $D \geq 51.4 \cdot in$ minimum wall thickness is $t = \dfrac{D}{288}$

Governing Equations for Minimum Handling

$$a(D) := \begin{pmatrix} D < 21.5 & .0747 & \dfrac{D+20}{400} \\[2mm] 21.5 \leq D \leq 51.4 & \dfrac{D}{288} & \dfrac{D+20}{400} \\[2mm] 51.4 < D & \dfrac{D+20}{400} & \dfrac{D}{288} \end{pmatrix}$$

Matrix for Plotting Minimum Handling Equations

$b(D) := \text{lookup}\left(1, a(D)^{\langle 1\rangle}, a(D)^{\langle 2\rangle}\right)_1 \qquad c(D) := \text{lookup}\left(1, a(D)^{\langle 1\rangle}, a(D)^{\langle 3\rangle}\right)_1$

$d(D) := \max\left(.0747, \dfrac{D}{240}\right)$

INTERNAL PRESSURE

Figure 18-3. Minimum Handling Equations, AWWA & MOP 79

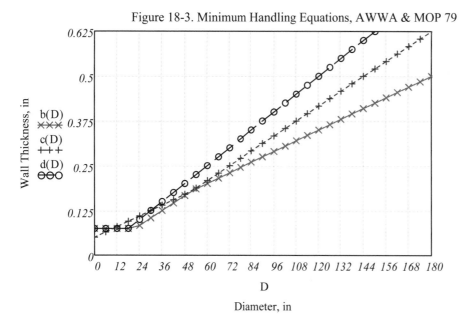

✕✕✕ AWWA M11, non mortar lined pipe
+++ MOP 79
⊖⊖⊖ AWWA M11, D/240, mortar lined pipe

Minimum Handling Equations; AWWA & MOP 79

Table 18-1. Penstock Configuration and External Pressure

Item	Station, ft	CL Elevation, ft	Penstock Segment Length, LL	Penstock. Length at Station, L ft	Inside Diameter, ft	External Pressure, ft
Inlet	0.00	815.00	0.00	0.00	15.00	350.00
End Trans.	10.00	815.00	10.00	10.00	15.00	350.00
PI #1	150.00	815.00	140.00	150.00	15.00	267.01
Adit (US)	300.00	793.10	151.59	301.59	15.00	200.00
Adit (DS)	300.00	793.10	0.00	301.59	15.00	15.00
End Backfill	500.00	763.90	202.12	503.71	15.00	15.00
Saddle Supt.	561.63	754.90	62.28	565.99	15.00	15.00
Exp. Jt. #1	641.56	743.23	80.78	646.77	15.00	15.00
PI #2	1150.00	669.00	513.83	1160.60	15.00	15.00
Exp. Jt. #2	1171.55	651.12	28.00	1188.60	15.00	15.00
PI #3 (US)	1250.00	586.00	101.96	1290.56	15.00	15.00
PI #3 (DS)	1250.00	586.00	0.00	1290.56	12.00	15.00
End Penstock	1300.00	586.00	50.00	1340.56	12.00	15.00
Center Units	1350.00	586.00	50.00	1390.56	12.00	15.00

Abbreviations:
Trans.=transition
PI=point of intersection
Supt.=support
Exp. Jt.=expansion joint
US=upstream
DS=downstream

Demonstrate calculations for normal conditions at station 11+50 (PI #2); compare penstock wall thickness with minimum handling; find the greater wall, round up to nearest 1/8 in. and find weight of the penstock segment.

Input   Steel is SA-516, Grade 70.

$F_{mts} := 70 \cdot ksi$ specified minimum tensile stress. $F_y := 38 \cdot ksi$ specified minimum yield stress.

$$\begin{pmatrix} \dfrac{F_y}{1.5} \\[2ex] \dfrac{F_{mts}}{2.4} \end{pmatrix} = \begin{pmatrix} 25.33 \\ 29.17 \end{pmatrix} \cdot ksi \qquad S := \min \left(\begin{pmatrix} \dfrac{F_y}{1.5} \\[2ex] \dfrac{F_{mts}}{2.4} \end{pmatrix} \right) = 25.33 \cdot ksi$$

allowable stress under normal conditions for SA-516, Grade 70 steel per ASME BPV Sec. VIII, Div. 2, and Sec. II, Part D, Table 5A.

$E := 1.0$ joint efficiency for double welded complete joint penetration (CJP) butt joints that receive 100% radiographic testing (RT).

$\gamma_s := 490 \cdot \dfrac{lbf}{ft^3}$ unit weight of steel.

$\gamma_w = 62.25 \cdot \dfrac{lbf}{ft^3}$ unit weight of water. $Max\_Reservoir\_El := 1165 \cdot ft$ maximum reservoir elevation.

$L := 1160.60 \cdot ft$ length of penstock at station. $CL\_EL := 669.0 \cdot ft$ centerline elevation of penstock.

$LL := 513.83 \cdot ft$ length of penstock segment. $k := 1.0$ allowable stress design increase factor, normal conditions.

$D := 15 \cdot ft$ penstock diameter.

Calculations

Normal hydraulic grade line (HGL) at station is determined by multiplying slope of HGL by length of penstock, added to maximum reservoir elevation.

$$HGL_n = \frac{1365 \cdot ft - 1165 \cdot ft}{1390.56 \cdot ft} \cdot L + Max\_Reservoir\_El = \frac{1365 \cdot ft - 1165 \cdot ft}{1390.56 \cdot ft} \cdot 1160.60 \cdot ft + 1165 \cdot ft$$

$\boxed{HGL_n = 1331.93 \, ft}$ normal conditions HGL at station 11+50.

Pressure is determined by normal HGL minus centerline elevation of penstock multiplied by the unit weight of water

$$P = \left(HGL_n - CL\_EL \right) \cdot \gamma_w = \left(\frac{1365 \cdot ft - 1165 \cdot ft}{1390.56 \cdot ft} \cdot 1160.60 \cdot ft + 1165 \cdot ft - 669.0 \cdot ft \right) \cdot 62.247 \cdot \frac{lbf}{ft^3}$$

Penstock wall thickness from internal pressure for normal conditions is determined as follows:

$$t = \frac{P \cdot D}{2 \cdot k \cdot S \cdot E} = \frac{\left(\frac{1365 \cdot \text{ft} - 1165 \cdot \text{ft}}{1390.56 \cdot \text{ft}} \cdot 1160.60 \cdot \text{ft} + 1165 \cdot \text{ft} - 669.0 \cdot \text{ft}\right) \cdot 62.247 \cdot \frac{\text{lbf}}{\text{ft}^3} \cdot 15 \cdot \text{ft}}{2 \cdot 1.0 \cdot S \cdot 1.0}$$

$\boxed{t = 1.02 \cdot \text{in}}$ normal conditions penstock, wall thickness.

↳ ASSUMES 1.0 Joint EFFICIENCY

$$t_{min} = \frac{D}{288} = \frac{15 \cdot \text{ft}}{288}$$ $\boxed{t_{min} = 0.63 \cdot \text{in}}$ minimum handling thickness.

Compare normal, emergency and exceptional conditions with minimum handling thickness and determine the greatest wall thickness (see vector results tabulated below for normal, emergency and exceptional conditions and minimum handling).

$$\begin{pmatrix} t \\ .92 \cdot \text{in} \\ .94 \cdot \text{in} \\ t_{min} \end{pmatrix} = \begin{pmatrix} 1.02 \\ 0.92 \\ 0.94 \\ 0.63 \end{pmatrix} \cdot \text{in} \qquad t' := \max\begin{pmatrix} \begin{pmatrix} t \\ .92 \cdot \text{in} \\ .94 \cdot \text{in} \\ t_{min} \end{pmatrix} \end{pmatrix} \qquad \boxed{t' = 1.02 \cdot \text{in}}$$ the greatest wall thickness occurs as a result of normal conditions.

Results

Round the wall thickness of $t' = 1.02 \cdot \text{in}$ up to the nearest 1/8 in. thickness and determine the unit weight of the penstock and the total weight of the penstock section.

$$tt' := \text{Ceil}\left(\frac{t'}{\text{in}}, \frac{1}{8}\right) \cdot \text{in} \qquad \boxed{tt' = 1.13 \cdot \text{in}}$$ penstock wall thickness (rounded up to the nearest 1/8 in.).

$$\text{unit\_wt} := \pi \cdot D \cdot tt' \cdot \gamma_s = \pi \cdot 15 \cdot \text{ft} \cdot tt' \cdot 490 \cdot \frac{\text{lbf}}{\text{ft}^3} \qquad \text{unit\_wt} = 2165 \cdot \frac{\text{lbf}}{\text{ft}} \qquad \boxed{\text{unit\_wt} = 1.08 \cdot \frac{\text{tonf}}{\text{ft}}}$$ unit weight of the penstock.

$$\text{wt} := \pi \cdot D \cdot tt' \cdot LL \cdot \gamma_s = \pi \cdot 15 \cdot \text{ft} \cdot tt' \cdot 513.83 \cdot \text{ft} \cdot 490 \cdot \frac{\text{lbf}}{\text{ft}^3} \qquad \boxed{\text{wt} = 556.16 \cdot \text{tonf}}$$ total weight of the penstock section.

▶ Programs to Calculate Tabluar Penstock Wall Thickess —————————————————————————

Table 18-2. Penstock Internal Pressure and Wall Thickness Summary

Item	Station, ft	Pressure, psi			Wall Thickness, in							Penstock Weight, tons	Length of Penstock
		Normal	Emerg.	Except.	Normal	Emerg.	Except.	Min. Thk.	Max. Wall	Govern. Cond.	Rounded Thk.		
Inlet	0	151	151	151	0.54	0.36	0.21	0.63	0.63	D/288	0.63	0.0	0.0
End Trans.	10	152	153	155	0.54	0.36	0.22	0.63	0.63	D/288	0.63	6.0	10.0
PI #1	150	161	174	209	0.57	0.41	0.30	0.63	0.63	D/288	0.63	84.2	140.0
Adit (US)	300	180	206	277	0.64	0.49	0.39	0.63	0.64	Normal	0.75	109.4	151.6
Adit (DS)	300	180	206	277	0.64	0.49	0.39	0.63	0.64	Normal	0.75	0.0	0.0
End Backfill	500	205	249	368	0.73	0.59	0.52	0.63	0.73	Normal	0.75	145.8	202.1
Saddle Supt.	562	212	263	395	0.75	0.62	0.56	0.63	0.75	Normal	0.88	52.4	62.3
Exp. Jt. #1	642	223	280	432	0.79	0.66	0.61	0.63	0.79	Normal	0.88	68.0	80.8
PI #2	1150	287	389	662	1.02	0.92	0.94	0.63	1.02	Normal	1.13	556.2	513.8
Exp. Jt. #2	1172	296	401	680	1.05	0.95	0.97	0.63	1.05	Normal	1.13	30.3	28.0
PI #3 (US)	1250	331	445	748	1.17	1.05	1.06	0.63	1.17	Normal	1.25	122.6	102.0
PI #3 (DS)	1250	331	445	748	0.94	0.84	0.85	0.50	0.94	Normal	1.00	0.0	0.0
End Penstock	1300	334	452	767	0.95	0.86	0.87	0.50	0.95	Normal	1.00	38.5	50.0
											Total wt.	1213.4	

▶ Hydraulic Profile and Pressures ————————————————————————————

18.2 Steel Tunnel Liner

The following calculations illustrate the methods for the design of tunnel liners subjected to external pressure. Tunnel liners initially are designed to withstand the internal pressure at each point along the penstock. The thickness determined for internal pressure then is checked for its capacity to accommodate the specified external pressure. The methods used for this check are given in Chapter 6. Tunnel liner thickness based on internal pressure may be adequate to accommodate the specified external pressure. If this is the case, no further analysis is required. If the liner thickness is inadequate, the following three alternate methods may be pursued.

(1) Method A

Increase the tunnel liner thickness to the point that it accommodates the specified external pressure.

(2) Method B

Use the thickness determined for internal pressure and add external stiffening.

(3) Method C

Increase the tunnel liner thickness and add external stiffening.

The following sample calculations are accompanied by an explanation of the relative advantages of each of the above methods. An economic evaluation should be made to allow the comparison of the design with the other alternatives. Initially, the designer is encouraged to select stations that represent suitable points along the tunnel where analysis should be performed.

The tunnel liner length extends from station 0 + 10 to station 3 + 00. Since the external pressure decreases and the thickness increases along the length of the penstock, a detailed description of the pressures and the thicknesses is given in Table 18-3. Data is given for intervals of 50 feet (40 feet for the first interval).

Table 18-3. Pressures and Thicknesses (Stations 0+10 to 3+00)

POINT	STATION	INTERNAL PRESSURE (psi)	EXTERNAL PRESSURE (psi)	THICKNESS* FOR INTERNAL PRESSURE (IN.)
1	0 + 10	152.3	151.7	0.588**
2	0 + 50	156.1	144.4	0.598**
3	1 + 00	160.9	128.6	0.610**
4	1 +50	165.6	115.7	0.622**
5	2 + 50	170.4	106.0	0.647
6	2 + 50	175.2	96.4	0.671
7	3 + 00	179.9	86.7	0.696

* Thickness is determined on the basis of the internal pressure using the formula $t = Pr/SE$
** Note that these thicknesses are less than the calculated minimum handling thickness of 0.625 inches based on a D/t = 288.

Method A (Increase Tunnel Liner Thickness)

Based on the data in Table 18-3, the capacity of the unstiffened tunnel liner at each point should be determined using the thicknesses determined for internal pressure. This may be done using the single lobe failure theories developed by Amstutz and Jacobsen.

The Amstutz and Jacobsen methods are presented in Chapter 6. The Jacobsen method tends to give critical collapse pressures that are somewhat lower than those calculated by the Amstutz method. Chapter 6 recommends using the Amstutz method to find an initial solution, then using the results as the starting point for the Jacobsen method. This approach is presented below in the sample calculations. A gap ratio (gap/radius = .0003) is specified in the design criteria. The factor of safety for critical buckling pressure is 1.5 for these sample calculations. Other parameters have been defined previously. Using the thickness given in Table 18-3 and the Jacobsen and Amstutz formulas shown below, the allowable external pressures can be calculated. Table 18-4 gives the results of the calculations.

Table 18-4. Allowable External Pressure (Jacobsen and Amstutz)

POINT	STATION	THICKNESS REQUIRED FOR INTERNAL PRESSURE (IN.)	REQUIRED EXTERNAL PRESSURE (psi)	ALLOWABLE EXTERNAL PRESSURE-JACOBSEN (psi)	ALLOWABLE EXTERNAL PRESSURE-AMSTUTZ (psi)
1	0 + 10	0.588	151.7	42.6	48.7
2	0 + 50	0.598	144.4	44.1	50.4
3	1 + 00	0.610	128.6	45.8	52.4
4	1 +50	0.622	115.7	47.5	54.5
5	2 + 50	0.647	106.0	51.2	59.0
6	2 + 50	0.671	96.4	54.9	63.5
7	3 + 00	0.696	86.7	58.9	68.2

(1) Amstutz formulas

$$\frac{\sigma_N - \sigma_v}{\sigma_F^* - \sigma_N}\left[\left(\frac{r}{i}\right)\sqrt{\frac{\sigma_N}{E^*}}\right]^3 \cong 1.73\left(\frac{r}{e}\right)\left[1 - 0.225\left(\frac{r}{e}\right)\frac{\sigma_F^* - \sigma_N}{E^*}\right]$$

$$p_{cr} \cong \left(\frac{F}{r}\right)\sigma_N\left(1 - 0.175\left(\frac{r}{e}\right)\frac{\sigma_F^* - \sigma_N}{E^*}\right)$$

Where:

i = $t/\sqrt{12}$

e = $t/2$

σ_v = $-(\Delta/r)E^*$

Δ/r = gap ratio, for gap ratio between steel and concrete = γ

r = tunnel liner radius = D/2

D = tunnel liner diameter

t = liner thickness

E = modulus of elasticity of liner

σ_F = yield strength

σ_N = circumferential axial stress in plate liner ring

$$\mu \;=\; 1.5 - 0.5\left[\dfrac{1}{1 + 0.002\left(E/\sigma_y\right)}\right]^2$$

$$\sigma_F^* \;=\; \dfrac{\mu\sigma_F}{\sqrt{1 - v + v^2}}$$

F = total cross-sectional area of ring between stiffeners = t

$E^* = E/\left(1 - v^2\right)$

v = Poisson's ratio = 0.3

p_{cr} = critical external buckling pressure

(2) Jacobsen formulas

The three equations that must be solved simultaneously for the three unknowns α, β, and p_{cr} are:

$$\frac{r}{t} = \sqrt{\left\{\frac{\left[\left(9\pi^2/4\beta^2\right) - 1\right]\left[\pi - \alpha + \beta\left(\sin\alpha/\sin\beta\right)^2\right]}{12\left(\sin\alpha/\sin\beta\right)^3\left[\alpha - \left(\pi\Delta/r\right) - \beta\left(\sin\alpha/\sin\beta\right)\left[1 + \tan^2\left(\alpha - \beta\right)/4\right]\right]}\right\}}$$

$$\frac{p_{cr}}{E^*} = \frac{\left(9/4\right)\left(\pi/\beta\right)^2 - 1}{12\left(r/t\right)^3\left(\sin\alpha/\sin\beta\right)^3}$$

$$\frac{\sigma_y}{E^*} = \left(t/2r\right)\left[1 - \left(\sin\beta/\sin\alpha\right)\right] + \left(p_{cr}r\sin\alpha/E^*t\sin\beta\right)\left[1 + \frac{4\beta r\sin\alpha\tan\left(\alpha - \beta\right)}{\pi t\sin\beta}\right]$$

Where:

α = one-half of the angle subtended to the center of the cylindrical shell by the buckled lobe, radians

β = one-half the angle subtended by the new mean radius through the half waves of the buckled lobe, radians

p_{cr} = critical external buckling pressure, psi

Δ/r = gap ratio, for gap between steel and concrete

σ_y = yield strength of liner, psi

r = tunnel liner internal radius, in.

t = liner thickness, in.

E = modulus of elasticity of liner, psi

E^* = modified modulus of elasticity of steel liner = $E/\left(1 - v^2\right)$, where v = Poisson's ratio for steel = 0.3

Following are sample computer calculations for both the Jacobsen and Amstutz methods. These calculations can be performed by commercially available numerical solution software.

The results of both the Jacobsen and Amstutz calculations indicate that the liner thicknesses based on internal pressure cannot accommodate the specified external pressure. It is necessary to increase the thickness of the tunnel liner until the capacity of the liner matches the required external pressure if stiffeners are not added. Table 18-5 indicates the required thickness.

Table 18-5. Thickness Required for External Pressure

POINT	STA	RECOMMENDED THICKNESS (IN.)	REQUIRED EXTERNAL PRESSURE (psi)	ALLOWABLE EXTERNAL PRESSURE- JACOBSEN (psi)	ALLOWABLE EXTERNAL PRESSURE- AMSTUTZ (psi)
1	0 + 10	1.18	151.7	152.5	184.3
2	0 + 50	1.15	144.4	145.9	175.9
3	1 + 00	1.07	128.6	128.7	154.3
4	1 +50	1.01	115.7	116.2	138.8
5	2 + 50	0.97	106.0	108.2	128.8
6	2 + 50	0.91	96.4	96.5	114.3
7	3 + 00	0.86	86.7	87.1	102.7

As discussed in Chapter 6, the use of the Jacobsen formulas results in a lower allowable external pressure for a giving thickness.

Each of the calculated allowable external pressures then is used to determine the stress level in the shell. These stress levels are based on the critical pressure, i.e., the allowable external pressure multiplied by the factor of safety. If the resulting stress is greater than 0.8 times the yield strength of the tunnel liner material, the value of E^* used in the Jacobsen or Amstutz formulas must be replaced by E_T^* using the formula:

$$E_T^* = E^* \left\{ 1 - \left[\frac{\sigma_m - 0.8\sigma_y}{0.2\sigma_y} \right]^2 \right\}$$

Where:

$$\sigma_m = \frac{(p_{cr})(r)(FS)}{t}$$

E = modulus of elasticity of liner, psi

E^* = modified modulus of elasticity of steel liner = $E/(1-v^2)$, where v = Poisson's ratio for steel = 0.3

σ_y = yield strength, psi

p_{cr} = critical external buckling pressure, psi

t = liner thickness, in.

FS = factor of safety

r = tunnel liner radius = $D/2$, in.

D = tunnel liner diameter, in.

The values given in Table 18-5 are less than 0.8 times yield strength. For example, at station 0 + 10:

$$\text{Stress} = \frac{(152.5\ psi)(90\ inches)(1.5)}{1.18\ inches} = 17{,}447\ \text{psi}$$

This value is less than 0.8 times the yield strength, σ_y.

Method B (Add Stiffeners)

Having established that the unstiffened tunnel liner thicknesses based on internal pressure are inadequate when using the wall thicknesses driven by internal pressure, the stiffener spacing needed to render the system acceptable can be determined. To do this, the designer must consider the following two types of buckling criteria: (1) rotary symmetric buckling and (2) single lobe buckling.

Rotary Symmetric Buckling

The equations for rotary symmetric buckling are presented below.

(1) Von Mises equation for rotary symmetric buckling:

$$p_{cr} = \frac{E(t/r_{NA})}{1-v^2} - \left[\frac{1-v^2}{\left((n^2-1)\left(\frac{n^2 L^2}{\pi^2 r_{NA}^2} + 1 \right) \right)^2} \right] + \frac{E(t/r_{NA})^3}{12(1-v^2)}\left(n^2 - 1 + \frac{2n^2 - 1 - v}{\frac{n^2 L^2}{n^2 r_{NA}^2} - 1} \right)$$

Where:

r_{NA} = radius to neutral axis of shell in original formulation (for practical purposes, radius to outside of shell), in.

L = length of tube between stiffeners, i.e., center-to-center spacing of stiffeners, in.

t = liner thickness, in.

p_{cr} = critical external buckling pressure, psi

E = modulus of elasticity, psi

v = Poisson's ratio = 0.3

n = number of lobes or waves in the complete circumference at collapse

(2) Donnell equation for rotary symmetric buckling:

$$p_{cr} = \frac{EI_S}{r^3} - \left[\frac{(n^2 + \lambda^2)^2}{n^2} \right] + \frac{Et}{r}\left[\frac{\lambda^4}{n^2(n^2 + \lambda^2)^2} \right]$$

Where:

p_{cr} = critical external buckling pressure, psi

E = modulus of elasticity, psi

I_S = shell bending thickness = $\dfrac{t^3}{12(1-v^2)}$

v = Poisson's ratio = 0.3

r = tunnel liner internal radius, in.
t = liner thickness, in.
λ = $\pi r/L$
L = length of panel (stiffener spacing), in.
n = number of circumferential lobes

To present a suitable comparison of methods, this sample calculation is based on a single stiffener cross section, thereby reducing the number of variables to the shell thickness and stiffener spacing. The stiffener size selected is 6 inches high by 7/8 inch thick. The stiffener spacing is the remaining variable.

Both the von Mises and Donnell formulas require that successive values of n be used to find the minimum value of p_{cr}. The value of n used in the von Mises formula may also be determined by formula. The values of n may not be the same for the two formulas.

Table 18-6 gives allowable external pressures based on a maximum stiffener spacing shown for each thickness.

Table 18-6. Allowable External Pressures for Stiffener Spacing

PT	STA	REQUIRED EXTERNAL PRESSURE (psi)	SHELL THICK. (IN.)	STIFFENER SPACING (IN.)	ALLOW. EXTERNAL PRESSURE- VON MISES (psi)	ALLOW. EXTERNAL PRESSURE- DONNELL (psi)
1	0 + 10	151.7	.588	48	152.6	147.3
2	0 + 50	144.4	.598	50	151.5	146.0
3	1 + 00	128.6	.610	58	132.4	128.9
4	1 +50	115.7	.622	66	118.5	116.9
5	2 + 50	106.0	.647	76	110.5	109.5
6	2 + 50	96.4	.671	90	99.6	99.1
7	3 + 00	86.7	.696	108	89.2	89.1

Single Lobe Buckling

Shell thicknesses, stiffener size, spacing, and the other parameters previously described are used to determine the single lobe buckling capacity using the Jacobsen formulas shown below.

$$\frac{Fr_{NA}^{2}}{I_{F}} = \frac{\left[\left(9\pi^{2}/4\beta^{2}\right)-1\right]\left[\pi-\alpha+\beta\left(\sin\alpha/\sin\beta\right)^{2}\right]}{\left(\sin\alpha/\sin\beta\right)^{3}\left[\alpha-\left(\pi\Delta/r_{NA}\right)-\beta\left(\sin\alpha/\sin\beta\right)\left[1+\tan^{2}\left(\alpha-\beta\right)/4\right]\right]}$$

$$p_{cr}/E = \left[\left(9\pi^{2}/4\beta^{2}\right)-1\right]\left[\frac{\left(I_{F}\right)\sin^{3}\beta}{r_{NA}^{3}\sin^{3}\alpha}\right]$$

$$\sigma_{y}/E = \frac{h}{r_{NA}}\left(1-\frac{\sin\beta}{\sin\alpha}\right) + \frac{p_{cr}r_{NA}\sin\alpha}{EF\sin\beta}\left[1+\frac{8\beta hr_{NA}F\sin\alpha\tan\left(\alpha-\beta\right)}{12\pi I_{F}\sin\beta}\right]$$

Where:

α = one-half of the angle subtended to the center of the cylindrical shell by the buckled lobe

β = one-half the angle subtended by the new mean radius through the half waves of the buckled lobe

p_{cr} = critical external buckling pressure, psi

r = tunnel liner internal radius, in.

Δ/r = gap ratio, for gap between steel and concrete

r_{NA} = tunnel liner radius to the centroid of the stiffened section used to calculate I_F, in.

E = modulus of elasticity of steel liner and stiffener material, psi

I_F = moment of inertia of the stiffener and $1.56\sqrt{rt}$ of the cylindrical shell, in.^{4}

F = area of the stiffener and the cylindrical shell, in.^{2}

h = distance from the centroid of the stiffened section and the extreme fiber of the stiffened section, in.

These formulas may be readily solved using commercially available numerical solution software. The solutions are given in Table 18-7.

Table 18-7. Single Lobe Buckling Capacity

PT	STA	SHELL THICK. (IN.)	STIFFENER SPACING (IN.)	CALCULATED EXTERNAL ALLOWABLE PRESSURE (psi)	REQUIRED EXTERNAL PRESSURE (psi)	COMMENTS
1	0 + 10	.588	48	149.3	151.7	MARGINAL
2	0 + 50	.598	50	148.7	144.4	ACCEPTABLE
3	1 + 00	.610	58	140.7	128.6	ACCEPTABLE
4	1 +50	.622	66	134.7	115.7	ACCEPTABLE
5	2 + 50	.647	76	130	106	ACCEPTABLE
6	2 + 50	.671	90	122	96.4	ACCEPTABLE
7	3 + 00	.696	108	114	86.7	ACCEPTABLE

The above table is based upon the following data at station 0+10:

Shell thickness / Stiffener size = 0.588 / (6 x 7/8)
I_F = 48.88 in.^{4}
F = 33.474 in.^{2}
H = 4.904 in.
r = 91.684 in.
σ_y = 38,000 psi
E = 30,000,000 psi
Δ/r = Gap ratio = 0.0003
L = 48 in.
FS = 1.5

The calculated critical buckling pressure (P_{cr}) is 224 psi. Using the suggested safety factor of 1.5, the allowable buckling pressure is 149.3 psi.

In addition, a calculation of the average compression stress (σ_{ave}) is used to determine the validity of the value of E for the above scenario is as follows:

$$\sigma_{ave} = \frac{(P_{all})(FS)(r)(L)}{(t_{shell})(1.56\sqrt{(r)(t_{shell})})+A_S}$$

18.3 Buried Penstock

▶ Define Units

Deflection of Buried Penstock   Use the modified Iowa formula at station 3+00.

Modified Iowa formula: where $W = LL + DL\_on\_pipe$, $r = pipe\_radius$, $E' = soil\_modulus$,
$I = moment\_of\_inertia\_pipe\_wall$, $E = steel\_modulus$, $K = bedding\_constant$ and $D_l = deflection\_lag\_factor$.

$$x = D_l \cdot \left(\frac{K \cdot W \cdot r^3}{E \cdot I + .061 \cdot E' \cdot r^3} \right)$$

modified Iowa formula for horizontal deflection, in. per AWWA (2004), Equation 6-5.

$$\Delta\% = \frac{D_l \cdot \left(\dfrac{K \cdot W \cdot r^3}{E \cdot I + .061 \cdot E' \cdot r^3} \right)}{D} \cdot 100$$

modified Iowa formula for horizontal deflection, percent, per AWWA (2004), Equation 6-5.

Input

$D := 15 \cdot ft$ $D = 180.00 \cdot in$ penstock inside diameter.

$t := .750 \cdot in$ penstock wall thickness at station 3+00. $r := \dfrac{D + t}{2} = 90.38 \cdot in$ penstock mean radius.

$I := \dfrac{t^3}{12} = \dfrac{(.750 \cdot in)^3}{12}$ $I = 0.0352 \cdot \dfrac{in^4}{in}$ moment of inertia of penstock wall.

$B_c := D + 2 \cdot t$ $B_c = 181.50 \cdot in$ penstock outside diameter.

$D_l := 1.1$ deflection lag factor.

$K := .10$ bedding constant.

$E' := 700 \cdot psi$ soil modulus; coarse-grained soils with little or no fines (SP, SM, GP, GW) AASHTO 85% relative compaction   E' FROM M-11

$H := 5 \cdot ft$ soil cover over top of penstock

$w := 120 \cdot \dfrac{lbf}{ft^3}$ unit weight of soil

BURIED PENSTOCK

Source: The following tables are reprinted from AWWA M11—*Steel Water Pipe—A Guide for Design and Installation* by permission. ©2004 by the American Water Works Association.

Table 6-1   Values* of modulus of soil reaction, E' (psi) based on depth of cover, type of soil, and relative compaction

Type of Soil[†]	Depth of Cover		Standard AASHTO relative compaction[‡]							
			85%		90%		95%		100%	
	ft	(m)	psi	(kPa)	psi	(kPa)	psi	(kPa)	psi	(kPa)
Fine-grained soils with less than 25% sand content (CL, ML, CL-ML)	2–5	(0.06–1.5)	500	(3,450)	700	(4,830)	1,000	(6,895)	1,500	(10,340)
	5–10	(1.5–3.1)	600	(4,140)	1,000	(6,895)	1,400	(9,655)	2,000	(13,790)
	10–15	(3.1–4.6)	700	(4,830)	1,200	(8,275)	1,600	(11,030)	2,300	(15,860)
	15–20	(4.6–6.1)	800	(5,520)	1,300	(8,965)	1,800	(12,410)	2,600	(17,930)
Coarse-grained soils with fines (SM, SC)	2–5	(0.06–1.5)	600	(4,140)	1,000	(6,895)	1,200	(8,275)	1,900	(13,100)
	5–10	(1.5–3.1)	900	(6,205)	1,400	(9,655)	1,800	(12,410)	2,700	(18,615)
	10–15	(3.1–4.6)	1,000	(6,895)	1,500	(10,340)	2,100	(14,480)	3,200	(22,065)
	15–20	(4.6–6.1)	1,100	(7,585)	1,600	(11,030)	2,400	(16,545)	3,700	(25,510)
Coarse-grained soils with little or no fines (SP, SM, GP, GW)	2–5	(0.06–1.5)	700	(4,830)	1,000	(6,895)	1,600	(11,030)	2,500	(17,235)
	5–10	(1.5–3.1)	1,000	(6,895)	1,500	(10,340)	2,200	(15,170)	3,300	(22,750)
	10–15	(3.1–4.6)	1,050	(7,240)	1,600	(11,030)	2,400	(16,545)	3,600	(24,820)
	15–20	(4.6–6.1)	1,100	(7,585)	1,700	(11,720)	2,500	(17,235)	3,800	(26,200)

* Hartley, James D. and Duncan, James M., "E' and its Variation with Depth." *Journal of Transportation*, Division of ASCE, Sept. 1987.

[†] Soil type symbols are from the Unified Classification System.

[‡] *Soil compaction*. When specifying the amount of compaction required, it is very important to consider the degree of soil compaction that is economically obtainable in the field for a particular installation. The density and supporting strength of the native soil should be taken into account. The densification of the backfill envelope must include the haunches under the pipe to control both the horizontal and vertical pipe deflections. Specifying an unobtainable soil compaction value can result in inadequate support and injurious deflection. Therefore, a conservative assumption of the supporting capability of a soil is recommended, and good field inspection should be provided to verify that design assumptions are met.

Table 6-3   Live-load effect

Highway HS-20 Loading*				Railroad E-80 Loading*			
Height of Cover		Load		Height of Cover		Load	
ft	(m)	psf	(kg/m^2)	ft	(m)	psf	(kg/m^2)
1	(0.30)	1,800	(8,788)	2	(0.61)	3,800	(18,553)
2	(0.61)	800	(3,906)	5	(1.52)	2,400	(11,718)
3	(0.91)	600	(2,929)	8	(2.44)	1,600	(7,812)
4	(1.22)	400	(1,953)	10	(3.05)	1,100	(5,371)
5	(1.52)	250	(1,221)	12	(3.66)	800	(3,906)
6	(1.83)	200	(976)	15	(4.57)	600	(2,929)
7	(2.13)	176	(859)	20	(6.10)	300	(1,465)
8	(2.44)	100	(488)	30	(9.14)	100	(488)

* Neglect live load when less than 100 psf; use dead load only.

$w = w_c + w_L$ load w per unit of shell length is sum of soil dead load, w_c, plus surface live load, w_L

1) $W_L = 250 \cdot psf \cdot B_c = 250 \cdot psf \cdot (15 \cdot ft + 2 \cdot .750 \cdot in)$ $\boxed{W_L = 315.10 \cdot \dfrac{lbf}{in}}$ surface live load, assume highway HS-20 live loading; see Table 6-3 from AWWA (2004), excerpt above.

2) $W_c := w \cdot H \cdot B_c = 120 \cdot \dfrac{lbf}{ft^3} \cdot 5 \cdot ft \cdot (15 \cdot ft + 2 \cdot .750 \cdot in)$ $\boxed{W_c = 756.25 \cdot \dfrac{lbf}{in}}$ soil dead load.

3) $W := W_c + W_L$ $\boxed{W = 1071.35 \cdot \dfrac{lbf}{in}}$ surface live load + soil dead load.

$E := 30000 \cdot ksi$ modulus of elasticity of steel.

Calculations

modified Iowa formula for horizontal deflection, in. per AWWA (2004), Equation 6-5.

$$x := D_l \cdot \left(\frac{K \cdot W \cdot r^3}{E \cdot I + .061 \cdot E' \cdot r^3} \right) = 1.1 \cdot \frac{.10 \cdot (W_c + W_L) \cdot r^3}{30000 \cdot ksi \cdot \dfrac{(.750 \cdot in)^3}{12} + .061 \cdot 700 \cdot psi \cdot r^3} = 2.67 \cdot in$$

modified Iowa formula for horizontal deflection, percent per AWWA (2004), Equation 6-5.

$$\Delta\% := \frac{D_l \cdot \left(\dfrac{K \cdot W \cdot r^3}{E \cdot I + .061 \cdot E' \cdot r^3} \right)}{D \cdot \%} = \frac{1.1 \cdot \dfrac{.10 \cdot (W_c + W_L) \cdot r^3}{30000 \cdot ksi \cdot \dfrac{(.750 \cdot in)^3}{12} + .061 \cdot 700 \cdot psi \cdot r^3}}{D \cdot \%} = 1.48 \; \%$$

Conclusion

modified Iowa formula horizontal deflection is $\boxed{x = 2.67 \cdot in}$, or based upon % diameter, $\boxed{\Delta\% = 1.48}$, that is less than allowable deflection for dielectric lined and coated penstock of $\boxed{5\% \cdot D = 9.00 \cdot in}$. Thus penstock is adequate for resisting deflection.

Buckling of Buried Penstock

use buckling formulas at station 3+00 per AWWA (2004) Equations 6-7, 6-8 and 6-9.

Soil Buckling Formula: AWWA (2004), Equation. 6-7, $\quad q_a = \left(\dfrac{1}{FS}\right) \cdot \left(32 \cdot R_w \cdot B' \cdot E' \cdot \dfrac{E \cdot I}{D^3}\right)^{\frac{1}{2}} \quad$ where FS = safety_factor ,

D = pipe_OD , E' = soil_modulus , I = moment_of_inertia_pipe_wall, E = steel_modulus, EI = pipe_wall_stiffness and

$B' = \dfrac{1}{1 + 4 \cdot e^{\left(-.065 \cdot \frac{H}{ft}\right)}}$ empirical coefficient of elastic support where, H = soil_cover_ft. .

For normal pipe installations of external load, AWWA (2004), Equation 6-8 $\quad \boxed{\gamma_w \cdot h_w + R_w \cdot \dfrac{W_c}{D} + P_v \le q_a}$

where h_w = height_water_above_conduit(in) , γ_w = unit_wt_water, P_v = internal_vacuum and

W_c = vertical_soil_load_on_pipe.

For some pipe installations of external live load AWWA (2004), Equation 6-9 $\quad \boxed{\gamma_w \cdot h_w + R_w \cdot \dfrac{W_c}{D} + \dfrac{W_L}{D} \le q_a}$

where h_w = height_water_above_conduit(in) , γ_w = unit_wt_water and W_L = vertical_live_load_on_pipe.

Note from AWWA (2004);

---

*NOTE: Where internal vacuum occurs with cover depth less than 4 ft (1.2 m), but not less than 2 ft (0.6 m), care should be exercised. This is particularly important for large-diameter pipe. In no case shall cover depth be less than 2 ft (0.6 m) for pipe diameters less than 24 in. (600 mm), 3 ft (0.9 m) for pipe diameters 24 in. (600 mm) through 96 in. (2,400 mm), and 4 ft (1.2 m) for pipe over 96 in. (2,400 mm) in diameter.

Input

q_a = to_be_determined   allowable buckling pressure.

h := H   h = 60.00·in   height of ground surface above top of penstock.

D = 180.00·in   diameter of penstock.   FS := 2.0   safety factor.

$805.79 \cdot ft - \left(793.10 \cdot ft + \dfrac{D}{2}\right) = 5.19\,ft$ $h_w := h$ $h_w = 60.00 \cdot in$ height of ground-water surface above top of penstock is equal to soil cover (water surface at grade).

$$R_w := 1 - .33 \cdot \frac{h_w}{h} = 1 - .33 \cdot \frac{H}{H} \qquad \boxed{R_w = 0.6700} \qquad \text{water buoyancy factor.}$$

$$B' := \frac{1}{1 + 4 \cdot e^{-.065 \cdot \frac{H}{ft}}} \qquad \boxed{B' = 0.257} \qquad \text{empirical coefficient of elastic support.}$$

$E = 30000.00 \cdot ksi$ steel elastic modulus.

$E' = 700.00 \, psi$ soil modulus.

$t = 0.75 \cdot in$ penstock wall thickness.

$$I := \frac{t^3}{12} = \frac{(.750 \cdot in)^3}{12} \qquad I = 0.0352 \cdot \frac{in^4}{in} \qquad \text{moment of inertia of penstock wall.}$$

$\gamma_w = 0.0360 \cdot \dfrac{lbf}{in^3}$ unit weight of water.

$P_v := 15 \cdot ft \cdot \gamma_w = 6.48 \cdot psi$ external pressure from 15 ft of head.

$W = W_c + W_L$ load W per unit of shell length is sum of soil dead load, W_c plus surface live load, W_L.

$\boxed{W_L = 315.10 \cdot \dfrac{lbf}{in}}$ HS-20 highway surface live load assumed.

$\boxed{W_c = 756.25 \cdot \dfrac{lbf}{in}}$ soil dead load. $\boxed{W = 1071.35 \cdot \dfrac{lbf}{in}}$ surface live load plus soil dead load.

Calculations

$$q_a := \left(\frac{1}{FS}\right) \cdot \left(32 \cdot R_w \cdot B' \cdot E' \cdot \frac{E \cdot I}{D^3}\right)^{\frac{1}{2}} = \frac{1}{2.0} \cdot \sqrt{32 \cdot \left(1 - .33 \cdot \frac{5 \cdot ft}{5 \cdot ft}\right) \cdot \frac{1}{1 + 4 \cdot e^{-.065 \cdot \frac{5 \cdot ft}{ft}}} \cdot 700 \cdot psi \cdot \frac{30000 \cdot ksi \cdot \frac{(.750 \cdot in)^3}{12}}{(15 \cdot ft)^3}}$$

$\boxed{q_a = 13.21 \, psi}$ allowable buckling pressure, per AWWA (2004), Equation. 6-7.

$$\gamma_w \cdot h_w + R_w \cdot \frac{W_c}{D} + P_v = 62.247 \cdot \frac{lbf}{ft^3} \cdot 5 \cdot ft + \left(1 - .33 \cdot \frac{5 \cdot ft}{5 \cdot ft}\right) \cdot \frac{120 \cdot \frac{lbf}{ft^3} \cdot 5 \cdot ft \cdot (15 \cdot ft + 2 \cdot .750 \cdot in)}{15 \cdot ft} + P_v$$

$$\boxed{\gamma_w \cdot h_w + R_w \cdot \frac{W_c}{D} + P_v = \mathit{11.46\,psi}}$$

$$\gamma_w \cdot h_w + R_w \cdot \frac{W_c}{D} + P_v \le q_a = \mathit{1.00}$$

normal penstock installations of external load, including vacuum, per AWWA (2004), Equation 6-8; note: "1" means statement is true; Equation 6-8 is less than Equation 6-7.

$$\gamma_w \cdot h_w + R_w \cdot \frac{W_c}{D} + \frac{W_L}{D} = 62.247 \cdot \frac{lbf}{ft^3} \cdot 5 \cdot ft + \left(1 - .33 \cdot \frac{5 \cdot ft}{5 \cdot ft}\right) \cdot \frac{120 \cdot \dfrac{lbf}{ft^3} \cdot 5 \cdot ft \cdot (15 \cdot ft + 2 \cdot .750 \cdot in)}{15 \cdot ft} + \frac{250 \cdot psf \cdot (15 \cdot ft + 2 \cdot .750 \cdot in)}{15 \cdot ft}$$

$$\boxed{\gamma_w \cdot h_w + R_w \cdot \frac{W_c}{D} + \frac{W_L}{D} = \mathit{6.73\,psi}}$$

$$\gamma_w \cdot h_w + R_w \cdot \frac{W_c}{D} + \frac{W_L}{D} < q_a = \mathit{1.00}$$

penstock installations of external load, including live load, per AWWA (2004) Eqs. 6-9; note: "1" means statement is true; Eqs. 6-9 is less than Eqs. 6-7

Conclusion

Penstock allowable buckling pressure $\boxed{q_a = \mathit{13.21\,psi}}$ exceeds applied external pressure from above loading combinations. Thus penstock is adequate to resist buckling load combinations.

18.4 Ring Girder

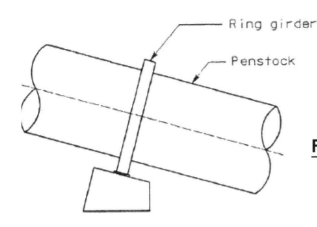

Fig. 18-4. Ring Girder Support of Penstock

▶ Define Units

Input Variables:

$P := 287 \cdot psi$ normal conditions internal pressure. $D := 15ft = 180.00 \cdot in$ penstock inside diameter.

$t_1 := 1.25 \cdot in$ assumed penstock wall thickness between ring girders (internal pressure and bending).

$t_2 := 2.00 \cdot in$ assumed penstock wall thickness at ring girder (thickened to handle local stresses).

$\gamma_s := 490 \cdot \dfrac{lbf}{ft^3}$ unit wt. of steel. $\gamma_w = 62.25 \cdot \dfrac{lbf}{ft^3}$ unit wt. of water.

$F_{mts} := 70 \cdot ksi$ specified minimum tensile stress. $F_y := 38 \cdot ksi$ specified minimum yield stress.

$\begin{pmatrix} \dfrac{F_y}{1.5} \\ \dfrac{F_{mts}}{2.4} \end{pmatrix} = \begin{pmatrix} 25.33 \\ 29.17 \end{pmatrix} \cdot ksi$ $S := min\left(\begin{pmatrix} \dfrac{F_y}{1.5} \\ \dfrac{F_{mts}}{2.4} \end{pmatrix}\right) = 25.33 \cdot ksi$ allowable stress under normal conditions for SA-516, Grade 70 steel per ASME BPV Sec. VIII, Div. 2, and Sec. II, Part D, Table 5A.

$E := 1.0$ joint efficiency for double welded complete joint penetration (CJP) $\gamma_s := 490 \cdot \dfrac{lbf}{ft^3}$ unit weight
butt joints that receive 100% radiographic testing (RT).   of steel.

Penstock Section Properties:

Properties mid span: $t_1 = 1.25 \cdot in$ $OD := D + 2 \cdot t_1 = 182.50 \cdot in$ $ID := D = 180.00 \cdot in$

▶ Moment of Inertia Error

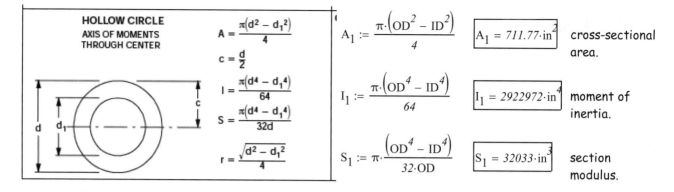

Figure 18-5. Properties of a Hollow Circle

$$\begin{pmatrix} W_{s1} \\ W_w \end{pmatrix} := \begin{bmatrix} \dfrac{\pi \cdot \left(OD^2 - ID^2\right)}{4} \cdot \gamma_s \\ \dfrac{\pi}{4} \cdot ID^2 \cdot \gamma_w \end{bmatrix} = \begin{pmatrix} 2.42 \\ 11.00 \end{pmatrix} \cdot \dfrac{kip}{ft} \qquad w_1 := \sum \begin{pmatrix} W_{s1} \\ W_w \end{pmatrix} \qquad \boxed{w_1 = 13.42 \cdot \dfrac{kip}{ft}} \qquad \text{unit weight of penstock plus water.}$$

Properties at ring girder supports: $t_2 = 2.00 \cdot in$ $OD := D + 2 \cdot t_2 = 184.00 \cdot in$ $ID = 180.00 \cdot in$

$$A_2 := \dfrac{\pi \cdot \left(OD^2 - ID^2\right)}{4} \qquad \boxed{A_2 = 1143.54 \cdot in^2} \qquad \text{cross-sectional area.} \qquad\qquad r := \dfrac{D}{2} = 90.00 \cdot in \qquad \text{radius.}$$

$$I_2 := \dfrac{\pi \cdot \left(OD^4 - ID^4\right)}{64} \qquad \boxed{I_2 = 4735398 \cdot in^4} \qquad \text{moment of inertia.} \quad S_2 := \pi \cdot \dfrac{\left(OD^4 - ID^4\right)}{32 \cdot OD} \qquad \boxed{S_2 = 51472 \cdot in^3} \quad \text{section modulus.}$$

$$\begin{pmatrix} W_{s2} \\ W_w \end{pmatrix} := \begin{bmatrix} \dfrac{\pi \cdot \left(OD^2 - ID^2\right)}{4} \cdot \gamma_s \\ \dfrac{\pi}{4} \cdot ID^2 \cdot \gamma_w \end{bmatrix} = \begin{pmatrix} 3.89 \\ 11.00 \end{pmatrix} \cdot \dfrac{kip}{ft} \qquad w_2 := \sum \begin{pmatrix} W_{s2} \\ W_w \end{pmatrix} \qquad \boxed{w_2 = 14.89 \cdot \dfrac{kip}{ft}} \qquad \text{unit weight of penstock plus water.}$$

Penstock Beam Bending Moment

approximate continuous beam bending moment $M = \dfrac{w \cdot l^2}{9}$.

$w_1 = 13.42 \cdot \dfrac{kip}{ft}$ penstock plus water weight, between ring girders. $l := 100 \cdot ft$ horizontal span between supports.

$\boxed{\alpha := atan(.146) = 8.31 \cdot deg}$ penstock slope.

$\boxed{w := \dfrac{w_1}{\cos(\alpha)} = 13.56 \cdot \dfrac{kip}{ft}}$ unit horizontal weight of penstock plus water weight, adjusted for slope.

$M_1 := \dfrac{w \cdot l^2}{9} = 15071 \cdot kip \cdot ft$ approximate bending moment at mid span and supports.

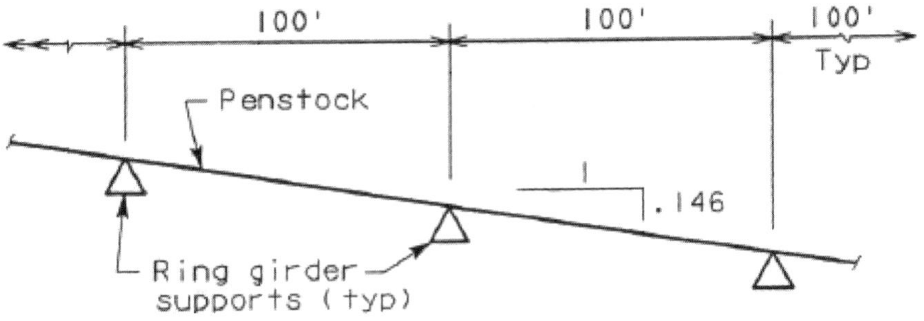

Fig. 18-6. Penstock Support

Longitudinal Bending Stress of Penstock at Mid Span and Supports

Mid Span: $t_1 = 1.25 \cdot in$ thickness. $I_1 = 2922972 \cdot in^4$ moment of inertia.

$$\sigma_y := -\frac{M_1 \cdot r}{I_1} = -5569\,psi$$

longitudinal bending stress; tension (+) at bottom; compression (-) at top.

$$\sigma_x := \frac{P \cdot D}{2 \cdot t_1} = 20664\,psi$$ hoop stress.

$$\sigma_e := \sqrt{\left(\sigma_x\right)^2 - \sigma_x \cdot \sigma_y + \left(\sigma_y\right)^2} = 23939\,psi$$

effective stress is less than allowable; $S = 25333\,psi$
penstock shell thickness is adequate.

Supports: $t_2 = 2.00 \cdot in$ thickness. $I_2 = 4735398 \cdot in^4$ moment of inertia.

$$\sigma_y := -\frac{M_1 \cdot r}{I_2} = -3437\,psi$$

longitudinal bending stress; tension (+)
at top; compression (-) at bottom.

$$f_1 := \frac{M_1 \cdot r}{I_2} = 3437\,psi$$

$$\sigma_x := \frac{P \cdot D}{2 \cdot t_2} = 12915\,psi$$ hoop stress.

$$\sigma_e := \sqrt{\left(\sigma_x\right)^2 - \sigma_x \cdot \sigma_y + \left(\sigma_y\right)^2} = 14933\,psi$$

effective stress is less than allowable; $S = 25333\,psi$
penstock shell thickness is adequate.

Ring Girder Properties (Acting with Shell) $R := \frac{D}{2}$ $R = 90.00 \cdot in$ inside radius.

$t := t_2 = 2.00 \cdot in$ $.78 \cdot \sqrt{R \cdot t} = 10.46 \cdot in$ effective shell distance away from discontinuity.

$x_1 := 14 \cdot in$ $y_1 := t = 2.00 \cdot in$ $x_2 := 1 \cdot in$ $y_2 := 10 \cdot in$ $x_3 := x_2 = 1.00 \cdot in$ $y_3 := y_2 = 10.00 \cdot in$

portion of shell acting with ring as interior flange. $x_4 := 2 \cdot (.78 \cdot \sqrt{R \cdot t}) + x_1 = 34.93 \cdot in$ $y_4 := t = 2.00 \cdot in$

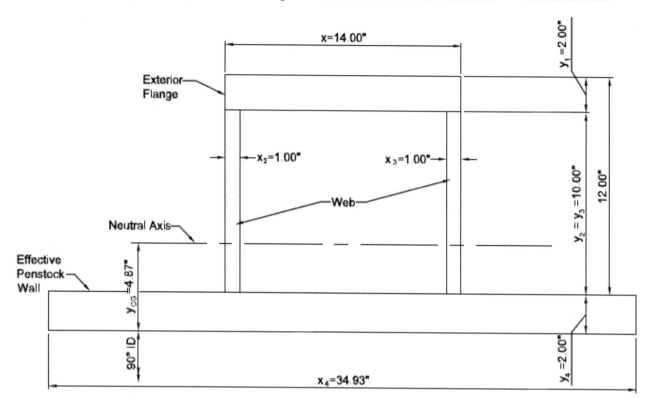

Figure 18-7. Section of Ring Girder

width:

$$b := \begin{pmatrix} x_1 \\ x_2 \\ x_3 \\ x_4 \end{pmatrix} = \begin{pmatrix} 14.00 \\ 1.00 \\ 1.00 \\ 34.93 \end{pmatrix} \cdot in$$

height:

$$h := \begin{pmatrix} y_1 \\ y_2 \\ y_3 \\ y_4 \end{pmatrix} = \begin{pmatrix} 2.00 \\ 10.00 \\ 10.00 \\ 2.00 \end{pmatrix} \cdot in$$

center of gravity of areas:

$$y := \begin{pmatrix} \dfrac{y_1}{2} + y_2 + y_4 \\[2mm] \dfrac{y_2}{2} + y_4 \\[2mm] \dfrac{y_3}{2} + y_4 \\[2mm] \dfrac{y_4}{2} \end{pmatrix} = \begin{pmatrix} 13.00 \\ 7.00 \\ 7.00 \\ 1.00 \end{pmatrix} \cdot in$$

$\boxed{A := b \cdot h = 117.86 \cdot in^2}$ cross sectional area of ring girder and shell.

$n := 4$ number of elements to sum.

$$\boxed{y_{cg} := \dfrac{\displaystyle\sum_{i=1}^{n} \left(b_i \cdot h_i \cdot y_i \right)}{b \cdot h} = 4.87 \cdot in}$$ center of gravity of ring girder and shell.

$$I := \sum_{i=1}^{n} \left[\left(y_i - y_{cg} \right)^2 \cdot \left(b_i \cdot h_i \right) + \frac{b_i \cdot \left(h_i \right)^3}{12} \right] = 3187 \cdot in^4$$

moment of inertia of ring girder and shell.

$$S_{in} := \frac{I}{y_{cg}} = 654.55 \cdot in^3$$

inside section modulus of ring girder and shell.

$$h_w := y_1 + y_2 + y_4 = 14.00 \cdot in$$

total height of ring girder and shell.

$$S_{out} := \frac{I}{h_w - y_{cg}} = 349.03 \cdot in^3$$

outside section modulus of ring girder and shell.

Local Shell Bending Stress

<u>ring area for the local shell bending stress:</u> $x_4 := x_1 = 14.00 \cdot in$ redefine the ring width (shell is no longer effective).

<u>width:</u>

<u>height:</u>

$$b := \begin{pmatrix} x_1 \\ x_2 \\ x_3 \\ x_4 \end{pmatrix} = \begin{pmatrix} 14.00 \\ 1.00 \\ 1.00 \\ 14.00 \end{pmatrix} \cdot in$$

$$h := \begin{pmatrix} y_1 \\ y_2 \\ y_3 \\ y_4 \end{pmatrix} = \begin{pmatrix} 2.00 \\ 10.00 \\ 10.00 \\ 2.00 \end{pmatrix} \cdot in$$

$$A_r := b \cdot h = 76.00 \cdot in^2$$ cross-sectional area of ring.

$$wt := \gamma_s \cdot A_r = 258.61 \cdot \frac{lbf}{ft}$$ unit radial weight of ring girder.

$$f_b := 1.82 \cdot \left(\frac{A_r - x_4 \cdot y_4}{A_r + 1.56 \cdot t_2 \cdot \sqrt{R \cdot t_2}} \right) \cdot \left(\frac{P \cdot R}{t_2} \right) = 9573 \, psi$$

local shell bending stress.

Properties of Ring Girder Support Leg

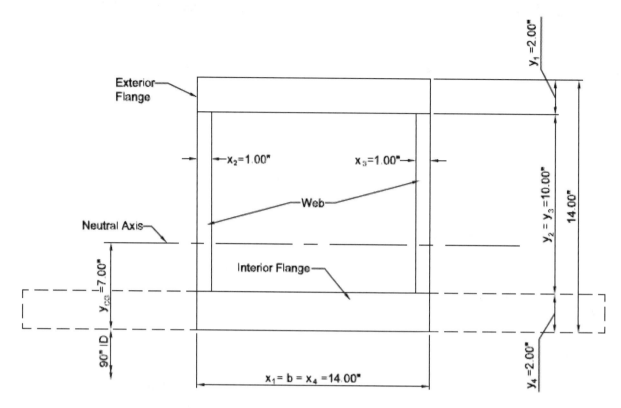

Figure 18-8. Section of Ring Girder Support Leg

$y_{1'} := y_{cg} = 4.87 \cdot in$ center of gravity of ring girder and shell.   Rename variables to match text.

$r = 90.00 \cdot in$ inside radius of penstock.

$R := r + y_{1'} = 94.87 \cdot in$ radius of ring girder centroid. $q := \dfrac{1.285}{\sqrt{r \cdot t}} = 0.0958 \cdot in^{-1}$

$b := x_4 = 14.00 \cdot in$ distance (out-to-out) of ring girder webs.

x distance between centroid of ring girder and centroid of leg supporting ring.

$y_{2'} := \dfrac{h_w}{2}$ $y_{2'} = 7.00 \cdot in$ $\boxed{x := y_{2'} - y_{1'} = 2.13 \cdot in}$ $.04 \cdot R = 3.79 \cdot in$ eccentricity to minimize bending moment.

$\boxed{Q_s := \dfrac{W_{s1} \cdot l}{\cos(\alpha)} = 245 \cdot kip}$ weight of penstock in span. $\boxed{Q := \dfrac{w_1 \cdot l}{\cos(\alpha)} = 1356 \cdot kip}$ weight of penstock plus water in one span. $\dfrac{Q_s}{2 \cdot Q} = 0.0902$

▶ Finite Element Details ─────────────────────────────

Tabular Ring Coefficients (Normal and Intermittent Conditions)

Refer to USBR (1986), Figures 18 and 19.

▶ USBR Tabular Ring Coefficients ─────────────────────

Ring Girder Coefficients (Gravity)

$\mu := .3$ Poisson's ratio for steel. $l = 100.00$ ft   span.

$$K := \frac{r}{l} \cdot \left[\frac{\mu \cdot l^2}{12 \cdot r^2} + \left[\left(1 - \mu^2\right) \cdot \left(1 - \frac{Q_s}{2 \cdot Q}\right) + \left[\frac{(2 + \mu) \cdot l}{4 \cdot q \cdot r^2} \right] \right] \right] = 0.4621$$

$$B := \frac{r}{R} \cdot \left(1 - \frac{2 \cdot K}{q \cdot r}\right) - \frac{x}{R} = 0.8245$$

$$C := \left(\frac{r}{R} - 1\right) \cdot \left(1 - \frac{2 \cdot K}{q \cdot r}\right) - \frac{x}{R} = -0.0683$$

710

ASCE 705 SEISMIC

Seismic Criteria

Refer to ASCE (2010), Chapter 11, Seismic Design Criteria

$S_S := 1.05$ mapped MCE (maximum considered earthquake), 5% damped, spectral response acceleration parameter (at short periods, sec), per Section 11.4.1.

$S_1 := 0.60$ mapped MCE (maximum considered earthquake), 5% damped, spectral response acceleration parameter (at a period of 1 sec,) per Section 11.4.1.

$T_L := 6$ long period transition period per Section 11.4.5.

Site class per list box below. For Site Class F, there is "no application".

SiteClass = "C"   Site Class C Soils per 11.4.2 & Table 20.3-1

▶ Acceleration Based Interpolation Prgm ──────────────────────────────

$S_S = 1.05$ $F_a = 1.00$ interpolated acceleration-based site coefficient:

Table 11.4-1 for $S_S = 1.05$ and SiteClass = "C".

▶ Velocity Based Interpolation Prgm ──────────────────────────────

$S_1 = 0.60$ $F_v = 1.300$ interpolated velocity-based site coefficient:

Table 11.4-2 for $S_1 = 0.60$ and SiteClass = "C".

Maximum considered earthquake (MCE) response spectra, adjusted for site class per Section 11.4.3 and Equations 11.4-1 and 11.4-2:

$S_{MS} := F_a \cdot S_S$ $S_{MS} = 1.05$

$S_{M1} := F_v \cdot S_1$ $S_{M1} = 0.78$

design spectral acceleration parameters per Section 11.4.4 and Equations 11.4-3 and 11.4-4.

$$S_{DS} := \frac{2}{3} \cdot S_{MS}$$ $\boxed{S_{DS} = 0.700}$ Spectral acceleration acceleration at short period

$$S_{D1} := \frac{2}{3} \cdot S_{M1}$$ $\boxed{S_{D1} = 0.520}$ Spectral acceleration acceleration acceleration at 1 second period

<u>Seismic Design Category:</u>   Table 1-1 , Occupancy Category of Buildings and Other Structures for Flood, Wind, Snow Earthquake and Ice Loads.... "Power Generating Stations". Thus Occupancy Category III

$a = 3.00$ $I := \begin{pmatrix} 1.0 \\ 1.0 \\ 1.25 \\ 1.50 \end{pmatrix}_a$ $I = 1.25$ Importance Factor per Table 11.5-1

Table 11.6-1 Seismic Design Category Based on Short Period Response Acceleration Parameter

$$T\_11\_6\_1 := \begin{pmatrix} S_{DS} < .167 & "A" & "A" & "A" \\ .167 \le S_{DS} < .33 & "B" & "B" & "C" \\ .33 \le S_{DS} < .50 & "C" & "C" & "D" \\ .50 \le S_{DS} & "D" & "D" & "E" \end{pmatrix}$$ $$T\_11\_6\_1 = \begin{pmatrix} 0.00 & "A" & "A" & "A" \\ 0.00 & "B" & "B" & "C" \\ 0.00 & "C" & "C" & "D" \\ 1.00 & "D" & "D" & "E" \end{pmatrix}$$

$$T1 := lookup\left(1, T\_11\_6\_1^{\langle 1 \rangle}, T\_11\_6\_1^{\langle a \rangle}\right)_1 = "D"$$ seismic design category per Table 11.6-1 for Occupancy Category III.

Table 11.6-2 Seismic Design Category Based on 1-S Period Response Acceleration Parameter

$$T\_11\_6\_2 := \begin{pmatrix} S_{D1} < .067 & "A" & "A" & "A" \\ .067 \le S_{D1} < .133 & "B" & "B" & "C" \\ .133 \le S_{D1} < .20 & "C" & "C" & "D" \\ .20 \le S_{D1} & "D" & "D" & "D" \end{pmatrix}$$ $$T\_11\_6\_2 = \begin{pmatrix} 0.00 & "A" & "A" & "A" \\ 0.00 & "B" & "B" & "C" \\ 0.00 & "C" & "C" & "D" \\ 1.00 & "D" & "D" & "D" \end{pmatrix}$$

$$T2 := lookup\left(1, T\_11\_6\_2^{\langle 1 \rangle}, T\_11\_6\_2^{\langle a \rangle}\right)_1 = "D"$$ seismic design category per Table 11.6-2 for Occupancy Category III.

$\boxed{Seismic\_Design\_Category := max(T1, T2) = "D"}$ greater seismic design categories of Tables 11.6-1 or 11.6-2

For Chapter 15. Seismic Design Requirements for Nonbuilding Structures; analyze per Section 15.4, Structural Design Requirements, 15.4.1 Design Basis...Where referenced documents are not cited herein, non-building structures shall be designed for compliance with sections 15.5 and 15.6 to resist min. seismic lateral forces that are not less than the requirements of Section 12.8.

1. The seismic force-resisting system shall be selected as follows: for non-building structures similar to buildings, a system shall be selected from among the types indicated in Table 12.2-1 or Table 15.4-1 subject to the system limitations and height limits, based on the seismic design category indicated in the table. The appropriate values of R, Ω_0 and C_d indicated in Table 15.4-1 shall be used in determining base shear, element design forces, and design story drift as indicated in this standard. Design and detailing requirements shall comply with the sections referenced in Table 15.5-1.

$R' := 4.5 \qquad \Omega := 3.0 \qquad C_d := 4.0$ intermediate steel moment frame per Table 15.4-1. Seismic Coefficients for Nonbuilding Structures Similar To Buildings.

$$C_s := \dfrac{S_{DS}}{\dfrac{R'}{I}} = 0.194$$

Equation 12.8-2 seismic response coefficients.

$.044 \cdot S_{DS} \cdot I = 0.038500$ supplement 2 to Equation 12.8-5.

$h_n := 20$ height of structure above base to highest level, ft.

$C_t := .028 \qquad\qquad x" := .80$ values taken from Table 12.8-2 for Structure Type: Steel Moment Resisting Frames.

$T_a := C_t \cdot h_n^{\,x"} = 0.31$ Equation 12.8-7   approximate fundamental period.

coefficient for upper limit on period, per Table 12.8-1, $S_{D1} = 0.52$.

$$T\_12\_8\_1 := \begin{pmatrix} S_{D1} \ge .40 & 1.4 \\ .30 \le S_{D1} < .40 & 1.4 \\ .20 \le S_{D1} < .30 & 1.5 \\ .15 \le S_{D1} < .20 & 1.6 \\ S_{D1} \le .10 & 1.7 \end{pmatrix}$$

$$T\_12\_8\_1 = \begin{pmatrix} 1.00 & 1.40 \\ 0.00 & 1.40 \\ 0.00 & 1.50 \\ 0.00 & 1.60 \\ 0.00 & 1.70 \end{pmatrix}$$

Matrix and logic to select appropriate period

$C_u := \text{lookup}\left(1, T\_12\_8\_1^{\langle 1 \rangle}, T\_12\_8\_1^{\langle 2 \rangle}\right)_1 = 1.40$ $\boxed{T := C_u \cdot T_a = 0.43}$ paragraph 12.8.2 period determination .

$$c_s := \begin{bmatrix} T \le T_L & \dfrac{S_{D1}}{T \cdot \left(\dfrac{R'}{I}\right)} & \text{"Eq 12.8-3"} \\ T > T_L & \dfrac{S_{D1} \cdot T_L}{T^2 \cdot \left(\dfrac{R'}{I}\right)} & \text{"Eq 12.8-4"} \end{bmatrix}$$

Matrix and logic to select appropriate formulas

$$c_s = \begin{pmatrix} 1.00 & 0.34 & \text{"Eq 12.8-3"} \\ 0.00 & 4.67 & \text{"Eq 12.8-4"} \end{pmatrix}$$

$\boxed{C_{smax} := \text{lookup}\left(1, c_s^{\langle 1 \rangle}, c_s^{\langle 2 \rangle}\right)_1 = 0.34}$

$\boxed{\text{lookup}\left(1, c_s^{\langle 1 \rangle}, c_s^{\langle 3 \rangle}\right)_1 = \text{"Eq 12.8-3"}}$

C_s not less than the greater of the following values per Supplement 2, Equation 12.8-5. $\begin{pmatrix} .044 \cdot S_{DS} \cdot I \\ .01 \end{pmatrix} = \begin{pmatrix} 0.04 \\ 0.01 \end{pmatrix}$

$C_{smin} := \max\left(\begin{pmatrix} .044 \cdot S_{DS} \cdot I \\ .01 \end{pmatrix}\right) = 0.04$ C_s minimum $C_{smin} < C_s < C_{smax} = 1.00$ 1" means statement is true; use $\boxed{C_s = 0.194}$

Matrix and logic to determine governing minimum horizontal acceleration, proper equation reference and load combination result.

$$cs := \begin{bmatrix} C_{smin} < C_s < C_{smax} & .7 \cdot \left(\dfrac{S_{DS}}{\frac{R'}{I}}\right) & \text{"Eq 12.8-2"} \\ C_s \geq C_{smax} & .7 \cdot (C_{smax}) & \text{lookup}\left(1, c_s^{\langle 1 \rangle}, c_s^{\langle 3 \rangle}\right)_1 \\ C_s \leq C_{smin} & .7 \cdot (C_{smin}) & \text{"Eq 12.8-5 and Supplement 2"} \end{bmatrix}$$

$$cs = \begin{pmatrix} 1.000 & 0.136 & \text{"Eq 12.8-2"} \\ 0.000 & 0.235 & \text{"Eq 12.8-3"} \\ 0.000 & 0.027 & \text{"Eq 12.8-5 and Supplement 2"} \end{pmatrix}$$

$\boxed{F_h := \text{lookup}\left(1, cs^{\langle 1 \rangle}, cs^{\langle 2 \rangle}\right)_1 = 0.136}$ $\boxed{\text{lookup}\left(1, cs^{\langle 1 \rangle}, cs^{\langle 3 \rangle}\right)_1 = \text{"Eq 12.8-2"}}$ $\boxed{\dfrac{F_h}{\%} = 13.61}$ seismic lateral force, % gravity in Load combination 5

Vertical seismic load per Section 12.4 $\boxed{E_v := .2 \cdot S_{DS} = 0.14}$ $F_v := .7 \cdot E_v = 0.10$

Longitudinal Stress from Gravity and Seismic Forces

assumed coefficient of friction $\beta := .05$

$$\sum F = 0 = \begin{pmatrix} 1 \\ 1 - F_v \\ 1 + F_v \end{pmatrix}_i \cdot W' \cdot \sin(\alpha) - P_1 - \beta \cdot \left[\begin{pmatrix} 1 \\ 1 - F_v \\ 1 + F_v \end{pmatrix}_i \cdot W'\right]$$

Make substitutions, rearrange, and solve for P1

Sum forces parallel with the slope:

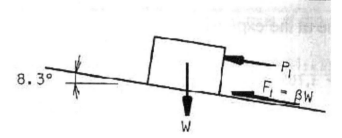

$$fg(W') := \begin{vmatrix} \text{for } i \in 1..3 \\ \quad \begin{vmatrix} P_1 \leftarrow \begin{pmatrix} 1 \\ 1 - F_v \\ 1 + F_v \end{pmatrix}_i \cdot W' \cdot (\sin(\alpha) - \beta) \\ v_{i,1} \leftarrow \begin{pmatrix} \text{"no accel"} \\ \text{"upward accel"} \\ \text{"downward accel"} \end{pmatrix}_i \\ v_{i,2} \leftarrow P_1 \end{vmatrix} \\ v \end{vmatrix}$$

Free-body Diagram
Penstock Support

Fig. 18-9. Free-body Diagram, Penstock Support

axial force on penstock

$$fg(W') \text{ float}, 4 \rightarrow \begin{pmatrix} \text{"no accel"} & 0.09447 \cdot W' \\ \text{"upward accel"} & 0.08521 \cdot W' \\ \text{"downward accel"} & 0.1037 \cdot W' \end{pmatrix}$$

$W' := 500 \cdot ft \cdot w_1 = 6711 \cdot kip$ assume 500 linear ft of penstock plus water weight is acting.

$$fg(W') = \begin{pmatrix} "no \ accel" & 634 \\ "upward \ accel" & 572 \\ "downward \ accel" & 696 \end{pmatrix} \cdot kip$$

direction of seismic force and magnitude acting along axis of penstock.

$$f_1 := \frac{-fg(W')^{\langle 2 \rangle}}{A_2} = \begin{pmatrix} -554 \\ -500 \\ -609 \end{pmatrix} psi$$

seismic stress acting along axis of penstock.

Longitudinal Stress Caused by Gravity acting on Penstock

$P_1 := 0.09447 \cdot W' = 633.99 \cdot kip$ $f_1 := \frac{-P_1}{A_2} = -554 \, psi$

Longitudinal Stress Caused by Frictional Force at Expansion Joint

$P_2 := \frac{500 \cdot lbf \cdot \pi \cdot D}{ft} = 23.56 \cdot kip$ $f_2 := \frac{-P_2}{A_2} = -21 \, psi$ axial stress caused by frictional force at the expansion joint based upon "500 lbs per linear foot of circumference." USBR (1986), page 17.

Longitudinal Stress Caused by Pressure at End of Expansion Joint

$P' := 223 \cdot psi$ normal conditions, pressure at expansion joint #1.

$P_3 := P' \cdot A_1 = 158.72 \cdot kip$ $f_3 := \frac{-P_3}{A_2} = -139 \, psi$ axial force caused by pressure at end of expansion joint.

$f_a := f_1 + f_2 + f_3 = -714 \, psi$ direction of seismic force and total axial stress acting along axis of penstock.

Normal Vertical Forces (Gravity)

compute gravity forces at 270 degrees (-) from crown. $j := 20$ $x'_j = 269.90 \cdot deg$

$N := P \cdot r \cdot \left[b + \frac{2 \cdot \left(1 - \mu^2\right)}{q} \right] = 852448 \, lbf$ tension in ring caused by internal pressure.

$T := Q \cdot \left(K'_{1_j} + B \cdot K'_{2_j} \right) = 339106 \, lbf$ thrust in ring girder, exclusive of N, gravity.

$M := Q \cdot \left(R \cdot K'_{3_j} + x \cdot K'_{4_j} \right) = -722627 \cdot in \cdot lbf$ bending moment in ring girder, gravity.

$S' := Q \cdot \left(K'_{5_j} + C \cdot K'_{6_j} \right) = -137422 \, lbf$ radial shear in ring girder, gravity.

Normal Vertical Stresses (Gravity)

$\dfrac{T}{A} = 2877\,psi$ thrust stress, gravity. $\dfrac{M}{S_{in}} = -1104\,psi$ bending stress inside surface, gravity.

$\sigma_{bg} := \dfrac{M}{S_{out}} = -2070\,psi$ bending stress outside surface, gravity. $\dfrac{N}{A} = 7233\,psi$ tension stress caused by internal pressure.

▶ Effects of "x" on Bending Stress ──────────────────────────────

$\boxed{\sigma_{in} := \dfrac{T}{A} + \dfrac{M}{S_{in}} + \dfrac{N}{A} = 9006\,psi}$ gravity stress, inside surface.

$\boxed{\sigma_{out} := \dfrac{T}{A} - \dfrac{M}{S_{out}} + \dfrac{N}{A} = 12180\,psi}$ gravity stress, outside surface.

$f_{bl} := \dfrac{M_1 \cdot \left(r \cdot \cos\left(x'_j\right)\right)}{I_2} = -6\,psi$ longitudinal bending stress.

$\boxed{f_1 := \dfrac{-P_1}{A_2} = -554\,psi}$ axial vertical seismic stress on penstock.

$\boxed{f_a := f_1 + f_2 + f_3 = -714\,psi}$ total axial vertical seismic stress.

$\sigma_y := f_a + f_b + f_{bl} = 8853\,psi$ total longitudinal stress from axial plus local ring bending plus beam bending.

$\sigma_x := \sigma_{in} = 9006\,psi$ $\sigma_y = 8853\,psi$ $\boxed{\sigma_e := \sqrt{\left(\sigma_x\right)^2 - \sigma_x \cdot \sigma_y + \left(\sigma_y\right)^2} = 8931\,psi}$ effective stress.

Vertical Seismic Forces (Downward Acceleration)

$n := F_h = 0.14$ $F_v = 0.10$ compute vertical seismic forces at 270 degrees (-) from crown. $j := 20$ $x'_j = 269.90 \cdot deg$

$N = 852448\,lbf$ tension in ring girder caused by internal pressure.

$T_v := \left(1 + F_v \right) \cdot Q \cdot \left(K'_{1_j} + B \cdot K'_{2_j} \right) = 372338\,lbf$ thrust in ring girder, exclusive of N, vertical seismic.

$M_v := \left(1 + F_v \right) \cdot Q \cdot \left(R \cdot K'_{3_j} + x \cdot K'_{4_j} \right) = -793444 \cdot in \cdot lbf$ bending moment in ring girder, vertical seismic

$S_v := \left(1 + F_v \right) \cdot Q \cdot \left(K'_{5_j} + C \cdot K'_{6_j} \right) = -150890\,lbf$ radial shear in ring girder, vertical seismic

$P_1 := 0.09447 \cdot \left[\left(1 + F_v \right) \cdot \left(500 \cdot ft \cdot w_1 \right) \right] = 696 \cdot kip$ axial vertical seismic force acting on 500 ft. of penstock.

$M_{lv} := \dfrac{\left(1 + F_v \right) \cdot w \cdot l^2}{9} = 16548.37 \cdot kip \cdot ft$ longitudinal bending vertical seismic moment at ring girder.

Vertical Seismic Stresses (Downward Acceleration)

$\dfrac{T_v}{A} = 3159\,psi$ thrust stress, vertical seismic.

$\dfrac{M_v}{S_{in}} = -1212\,psi$ bending stress inside surface, vertical seismic.

$\dfrac{M_v}{S_{out}} = -2273\,psi$ bending stress outside surface, vertical seismic.

$\dfrac{N}{A} = 7233\,psi$ tension stress caused by internal pressure.

$\boxed{\sigma_{vin} := \dfrac{T_v}{A} + \dfrac{M_v}{S_{in}} + \dfrac{N}{A} = 9180\,psi}$ vertical seismic stress, inside surface.

$\boxed{\sigma_{vout} := \dfrac{T_v}{A} - \dfrac{M_v}{S_{out}} + \dfrac{N}{A} = 12665\,psi}$ vertical seismic stress, outside surface.

$\boxed{f_1 := \dfrac{-P_1}{A_2} = -609\,psi}$ axial vertical seismic stress on penstock.

$\boxed{f_a := f_1 + f_2 + f_3 = -768\,psi}$ total axial vertical seismic stress.

$\boxed{f_{blv} := \dfrac{M_{lv} \cdot \left(r \cdot \cos\left(x'_j \right) \right)}{I_2} = -7\,psi}$ longitudinal bending, vertical seismic stress.

Horizontal Seismic Forces

$x'_j = 269.90 \cdot deg$ $n = 0.14$ seismic coefficient

$b = 14.00 \cdot in$ $q = 0.10 \cdot in^{-1}$ $Q_s = 244766.26\,lbf$ $Q = 1356423.48\,lbf$ $K = 0.46$ $r = 90.00 \cdot in$ $R = 94.87 \cdot in$

$W := 104 \cdot in$ length of leg. $K_{1_j} = -0.0796$ $K_{2_j} = -0.2500$

$\boxed{M_2 := n \cdot Q \cdot \left(R \cdot K_{1_j} + W \cdot K_{2_j} \right) = -6194033 \cdot in \cdot lbf}$ seismic moment in ring girder.

Matrix and logic to determine ring coefficient

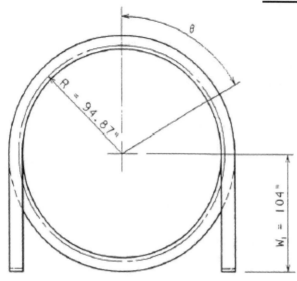

$$b' := \begin{bmatrix} 0\cdot\deg \le x'_j < 90\cdot\deg & \left[\dfrac{r}{R}\cdot\left(1 - \dfrac{2\cdot K}{q\cdot r} \right) + \dfrac{\pi\cdot W}{4\cdot R} \right] \\[2ex] 90\cdot\deg < x'_j < 270\cdot\deg & \left[\dfrac{r}{R}\cdot\left(1 - \dfrac{2\cdot K}{q\cdot r} \right) - \dfrac{\pi\cdot W}{4\cdot R} \right] \\[2ex] 270\cdot\deg < x'_j \le 360\cdot\deg & \left[\dfrac{r}{R}\cdot\left(1 - \dfrac{2\cdot K}{q\cdot r} \right) + \dfrac{\pi\cdot W}{4\cdot R} \right] \end{bmatrix}$$

$$b' = \begin{pmatrix} 0.00 & 1.71 \\ 1.00 & -0.01 \\ 0.00 & 1.71 \end{pmatrix}$$

$$\boxed{B' := \text{lookup}\left(1, b'^{\langle 1 \rangle}, b'^{\langle 2 \rangle} \right)_1 = -0.0140}$$

Fig. 18-10. Ring Girder Elev.

$$K_{3_j} = -0.0796 \qquad K_{4_j} = 0.3183 \qquad T_1 := n\cdot Q\cdot\left(K_{3_j} + B'\cdot K_{4_j} \right) = -15517\,\text{lbf}$$

$$T_2 := T_1 + \frac{n\cdot(Q - Q_s)}{\pi\cdot l}\cdot\left[b + \frac{2}{q}\cdot\left(1 - \mu^2 \right) \right] = -14193\,\text{lbf}$$ thrust in ring girder, horizontal seismic.

$$C_2 := \left(\frac{r}{R} - 1 \right)\cdot\left(1 - \frac{2\cdot K}{q\cdot r} \right) + \frac{\pi\cdot W}{4\cdot R} = 0.8152 \qquad S_h := n\cdot Q\cdot\left(K_{5_j} + C_2\cdot K_{6_j} \right) = 46156\,\text{lbf}$$ radial shear in ring girder, horizontal seismic.

$$M_{lh} := \frac{n\cdot w\cdot l^2}{9} = 2051.38\cdot\text{kip}\cdot\text{ft}$$ longitudinal bending, horizontal seismic moment at ring girder.

Horizontal Seismic Stresses

$$A = 117.86\cdot\text{in}^2 \qquad S_{in} = 654.55\cdot\text{in}^3 \qquad S_{out} = 349.03\cdot\text{in}^3$$

$$\frac{M_2}{S_{in}} = -9463\,\text{psi} \quad \text{inside seismic bending stress.}$$ $$\sigma_{bh} := \frac{M_2}{S_{out}} = -17746\,\text{psi} \quad \text{outside seismic bending stress.}$$

$$\boxed{\sigma_{hin} := \frac{T_2}{A} + \frac{M_2}{S_{in}} = -9583\,\text{psi}}$$ inside stress seismic stress.

$$\boxed{\sigma_{hout} := \frac{T_2}{A} - \frac{M_2}{S_{out}} = 17626\,\text{psi}}$$ outside stress seismic stress.

$$\boxed{f_{blh} := \frac{M_{lh}\cdot\left(r\cdot\sin(x'_j) \right)}{I_2} = -468\,\text{psi}}$$ longitudinal bending horizontal seismic stress.

Vertical (Downward) & Horizontal Seismic Stresses (intermittent Conditions)

$$\boxed{\sigma_y := f_a + f_b + f_{blh} + f_{blv} = 8330\,\text{psi}}$$ total longitudinal horizontal seismic stress from axial plus local ring bending plus longitudinal beam bending in horizontal and vertical directions.

$$\boxed{\sigma'_x := \sigma_{vin} + \sigma_{hin} = -404 \, psi}$$

inside stress vertical and horizontal seismic.

$$\boxed{\sigma_{vout} + \sigma_{hout} = 30291 \, psi}$$

outside stress vertical and horizontal seismic stress is less than allowable stress.

$$\boxed{\sigma'_v := \frac{S_v + S_h}{A} = -889 \, psi}$$

shear stress vertical and horizontal seismic.

$$\boxed{1.33 \cdot S = 33693 \, psi}$$

$$\sigma'_x = -404 \, psi \qquad \sigma_y = 8330 \, psi$$

$$\boxed{\sigma'_e := \sqrt{\left(\sigma'_x\right)^2 - \sigma'_x \cdot \sigma_y + \left(\sigma_y\right)^2} = 8539 \, psi}$$

effective shell stress vertical and horizontal seismic.

Observations

For gravity loads, the outside ring girder bending stress is a maximum value of -2070 psi whereas for horizontal seismic loads, the outside ring girder bending stress is -17746 psi; the seismic bending stress is greater than gravity bending stress by $\dfrac{\sigma_{bh}}{\sigma_{bg}} = 8.57$, nearly a factor of 9.

▶ Prgm To Include Vertical & Horiz. Seismic Accel ——————————————————————————————

▶ Detailed Results Tables-Normally Hidden ——————————————————————————————

Table 18-8. Output Ring Girder Peak Stress Summary

Number	Vert. Seismic Accel.	Type of Stress	Maximum Stress		Minimum Stress		Allow. Stress, psi	Status
			Stress Value, psi	Angular Location, deg	Stress Value, psi	Angular Location, deg		
2	Gravity No Vertical Accel.	Gravity Inside Stress	11298	105	3167	75	25333	OK
3		Gravity Outside Stress	12180	90.1	2285	89.9	25333	OK
4		Effective Stress	11140	0	5244	180	25333	OK
5		Seismic Inside Stress	9606	90.1	-9583	269.9	33693	OK
6		Seismic Outside Stress	17626	269.9	-17603	90.1	33693	OK
7		Gravity + Seismic Inside Stress	19050	105	-4078	75	33693	OK
8		Gravity + Seismic Outside Stress	29806	269.9	-8317	105	33693	OK
9		Gravity + Seismic Shear Stress	248	30	-1166	89.9	33693	OK
10		Gravity + Seismic Effective Stress	16571	105	5041	195	33693	OK
11	Vertical Upward Acceleration	Gravity Inside Stress	10900	105	3566	75	33693	OK
12		Gravity Outside Stress	11695	90.1	2770	89.9	33693	OK
13		Effective Stress	10885	0	5557	180	33693	OK
14		Seismic Inside Stress	9606	90.1	-9583	269.9	33693	OK
15		Seismic Outside Stress	17626	269.9	-17603	90.1	33693	OK
16		Gravity + Seismic Inside Stress	18652	105	-3697	60	33693	OK
17		Gravity + Seismic Outside Stress	29321	269.9	-8203	105	33693	OK
18		Gravity + Seismic Shear Stress	641	15	-1443	89.9	33693	OK
19		Gravity + Seismic Effective Stress	16171	105	5263	195	33693	OK
20	Vertical Downward Acceleration	Gravity Inside Stress	11697	105	2769	75	33693	OK
21		Gravity Outside Stress	12665	90.1	1800	89.9	33693	OK
22		Effective Stress	11395	0	4934	180	33693	OK
23		Seismic Inside Stress	9606	90.1	-9583	269.9	33693	OK
24		Seismic Outside Stress	17626	269.9	-17603	90.1	33693	OK
25		Gravity + Seismic Inside Stress	19448	105	-4476	75	33693	OK
26		Gravity + Seismic Outside Stress	30291	269.9	-8431	105	33693	OK
27		Gravity + Seismic Shear Stress	672	15	-1672	89.9	33693	OK
28		Gravity + Seismic Effective Stress	16905	105	4664	195	33693	OK

Conclusion

All ring girder stresses are less than allowable, thus design is adequate. The highest stress occurs at 270 degrees clockwise rotation from the top centerline, where combined stress from verticle downward seismic, horizontal seismic and gravity loading combine resulting in a peak stress outside flange stress of 30,291 psi.

18.5 Saddle

▶ Define Units

Assumptions: ASME Sec. VIII, Div. 2, Section 4.15.3 *Saddle Supports for Horizontal Vessels* equations are formulated for a horizontal vessel on two saddle supports with heads (as indicated ASME (2010)) and in sketch below; for Penstock example that is continuous on saddle supports without heads, L and a as defined in ASME (2010) have no significance, and it is assumed that $L > 8 \cdot R_m$ and $\dfrac{a}{R_m} > 1.0$. Also, ASME (2010) permits the addition of ring stiffeners in plane of saddle or adjacent to saddle to improve stress distribution however no attempt is made here to design saddle with additional ring stiffeners because it is more efficient to avoid saddles all together in favor of ring girders where stress analysis indicates saddle stiffeners are needed to reduce stress.

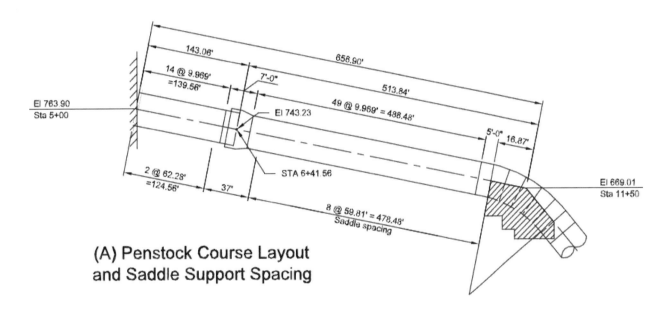

(A) Penstock Course Layout and Saddle Support Spacing

Fig. 18-11. Saddle Supported Penstock Profile

Figure 18-12. Cylindrical Shell Without Stiffening Rings

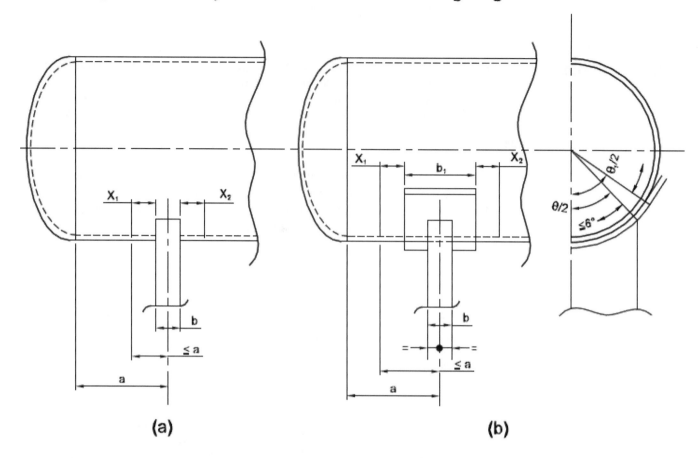

(a)   (b)

Saddle Supported Penstock Design

Stress Coefficients for Horizontal Vessels on Saddle Supports per ASME (2010), Table 4.15.1

$$K_1 = \frac{\Delta + \sin(\Delta)\cdot\cos(\Delta) - \dfrac{2\cdot\sin(\Delta)^2}{\Delta}}{\pi\cdot\left(\dfrac{\sin(\Delta)}{\Delta} - \cos(\Delta)\right)}$$

$$K'_1 = \frac{\Delta + \sin(\Delta)\cdot\cos(\Delta) - \dfrac{2\cdot\sin(\Delta)^2}{\Delta}}{\pi\cdot\left(1 - \dfrac{\sin(\Delta)}{\Delta}\right)}$$

$$K_2 = \frac{\sin(\alpha)}{\pi - \alpha + \sin(\alpha)\cdot\cos(\alpha)}$$

$$K_5 = \frac{1 + \cos(\alpha)}{\pi - \alpha + \sin(\alpha)\cdot\cos(\alpha)}$$

$$K_6 = \frac{\dfrac{3\cdot\cos(\beta)}{4}\cdot\left(\dfrac{\sin(\beta)}{\beta}\right)^2 - \dfrac{5\cdot\sin(\beta)\cdot\cos(\beta)^2}{4\cdot\beta} + \dfrac{\cos(\beta)^3}{2} - \dfrac{\sin(\beta)}{4\cdot\beta} + \dfrac{\cos(\beta)}{4} - \beta\cdot\sin(\beta)\cdot\left[\left(\dfrac{\sin(\beta)}{\beta}\right)^2 - \dfrac{1}{2} - \dfrac{\sin(2\cdot\beta)}{4\cdot\beta}\right]}{2\cdot\pi\cdot\left[\left(\dfrac{\sin(\beta)}{\beta}\right)^2 - \dfrac{1}{2} - \dfrac{\sin(2\cdot\beta)}{4\cdot\beta}\right]}$$

$K_7 = K_6$ is valid since heads or stiffening elements are not located near penstock saddle supports and it is assumed that $\dfrac{a}{R_m} \geq 1$ as defined in ASME (2010).

where θ = saddle contact angle and angles Δ, α, and β are as defined below and as indicated in sketch below.

$$\Delta = \frac{\pi}{6} + \frac{5\cdot\theta}{12} \qquad\qquad \alpha = .95\cdot\left(\pi - \frac{\theta}{2}\right) \qquad\qquad \beta = \pi - \frac{\theta}{2}$$

Nomenclature per ASME (2010), Section 4.15.6

4.15.6   Nomenclature

A	cross-sectional area of the stiffening ring(s) and the associated shell width used in the stress calculation.
a	distance from the axis of the saddle support to the tangent line on the curve for a dished head or to the inner face of a flat cover or tubesheet.
b	width of contact surface of the cylindrical shell and saddle support.
b_1	width of the reinforcing plate welded to the cylindrical shell at the saddle location
c_1, c_2	distance to the extreme axes of the cylinder-stiffener cross section to the neutral axis of the cylinder-stiffener cross-section
E_y	modulus of elasticity.
η	shell to reinforcing plate strength reduction factor.
F_h	saddle horizontal force.
h	spacing between two mounted stiffening rings placed o each side of the saddle support.
h_2	depth of the elliptical head.
I	moment of inertia of cross-sectional area A in relation to its neutral axis that is parallel to the axis of the cylindrical shell.
k	factor to account for the vessel support condition; $k = 1$ is the vessel is resting on the support and $k = 0.1$ is the vessel is welded to the support.
K	factor to set the allowable compressive stress for the cylindrical shell material.
L	length of the cylindrical shell measured from tangent line to tangent line for a vessel with dished heads or from the inner face to inner face for vessels with flat covers or tubesheets.
M_1	net-section maximum longitudinal bending moment at the saddle support; this moment is negative when it results in a tensile stress on the top of the shell.
M_2	net-section maximum longitudinal bending moment between the saddle supports; this moment is positive when it results in a compressive stress on the top of the shell.
P	design pressure, positive for internal pressure and negative for external pressure.
Q	maximum value of the reaction at the saddle support from weight and other loads as applicable.
R_i	inside radius of the spherical dome or a torispherical head.
R_m	mean radius of the cylindrical shell.
S	allowable stress from Annex 3.A for the cylindrical shell material at the design temperature.
S_c	allowable compressive stress for the cylindrical shell material at the design temperature.
S_h	allowable stress from Annex 3.A for the head material at the design temperature.
S_r	allowable stress from Annex 3.A for the reinforcing plate material at the design temperature.
S_s	allowable stress from Annex 3.A for the stiffener material at the design temperature.
t	cylindrical shell or shell thickness, as applicable.
t_h	head thickness.
t_r	reinforcing plate thickness.
T	maximum shear force at the saddle.
θ	opening of the supported cylindrical shell arc.
θ_1	opening of the cylindrical shell arc engaged by a welded reinforcing plate.
x_1, x_2	width of cylindrical shell used in the circumferential normal stress strength calculation.

Input   Steel is SA-516, Grade 70.

$F_{mts} := 70 \cdot ksi$ specified minimum tensile stress. $F_y := 38 \cdot ksi$ specified minimum yield stress.

$$\begin{pmatrix} \dfrac{F_y}{1.5} \\[2mm] \dfrac{F_{mts}}{2.4} \end{pmatrix} = \begin{pmatrix} 25.33 \\ 29.17 \end{pmatrix} \cdot ksi \qquad S := min\left(\begin{pmatrix} \dfrac{F_y}{1.5} \\[2mm] \dfrac{F_{mts}}{2.4} \end{pmatrix} \right) = 25.33 \cdot ksi$$

allowable stress under normal conditions for SA-516, Grade 70 steel per ASME BPV Sec. VIII, Div. 2, and Sec. II, Part D, Table 5A.

$D := 180 \cdot in$ inside diameter of penstock. $t := \dfrac{7}{8} \cdot in$ assumed wall thickness.

$R_m := \dfrac{D+t}{2} = 90.44 \cdot in$ mean radius of penstock. $\theta := 120 \cdot deg$ saddle support angle.

$b := 24 \cdot in$ contact width of saddle. $E_y := 29000 \cdot ksi$ modulus of elasticity.

$$\boxed{\Delta := \dfrac{\pi}{6} + \dfrac{5 \cdot \theta}{12} = 80.00 \cdot deg} \qquad \boxed{\alpha := .95 \cdot \left(\pi - \dfrac{\theta}{2} \right) = 114.00 \cdot deg} \qquad \boxed{\beta := \pi - \dfrac{\theta}{2} = 120.00 \cdot deg}$$

$\gamma_s := 490 \dfrac{lbf}{ft^3}$ steel unit wt. $\gamma_w = 62.25 \cdot \dfrac{lbf}{ft^3}$ water unit wt.

$\alpha_1 := atan(.146) = 8.31 \cdot deg$ penstock slope $s_0 := \dfrac{763.90 - 743.23}{cos(\alpha_1) \cdot 143.06}$ $s_0 = 0.1460$ $atan(s_0) = 8.31 \cdot deg$

slope checks

b)   If the vessel supports are not symmetric, or more than two supports are provided, then the highest moment in the vessel, and the moment and shear force at each saddle location shall be evaluated. The moments and shear force may be determined using strength of materials (i.e. beam analysis with a shear and moment diagram). If the vessel is supported by more than two supports, then differential settlement should be considered in the design.

Calculations

Penstock properties:

$t = 0.88 \cdot \text{in}$ $OD := D + 2 \cdot t = 181.75 \cdot \text{in}$ $ID := D = 180.00 \cdot \text{in}$

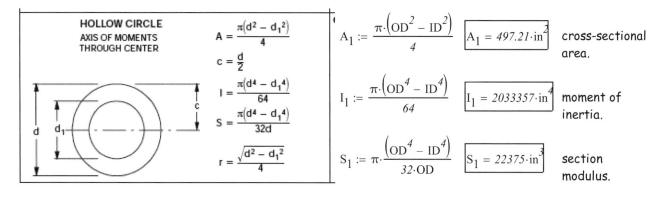

$$A_1 := \frac{\pi \cdot \left(OD^2 - ID^2\right)}{4} \qquad \boxed{A_1 = 497.21 \cdot \text{in}^2} \quad \text{cross-sectional area.}$$

$$I_1 := \frac{\pi \cdot \left(OD^4 - ID^4\right)}{64} \qquad \boxed{I_1 = 2033357 \cdot \text{in}^4} \quad \text{moment of inertia.}$$

$$S_1 := \pi \cdot \frac{\left(OD^4 - ID^4\right)}{32 \cdot OD} \qquad \boxed{S_1 = 22375 \cdot \text{in}^3} \quad \text{section modulus.}$$

Fig. 18-13. Properties of a Hollow Circle

$$\begin{pmatrix} \pi \cdot D \cdot t \cdot \gamma_s \\ \frac{\pi}{4} \cdot D^2 \cdot \gamma_w \end{pmatrix} = \begin{pmatrix} 1.68 \\ 11.00 \end{pmatrix} \cdot \frac{\text{kip}}{\text{ft}} \qquad w_1 := \sum \begin{pmatrix} \pi \cdot D \cdot t \cdot \gamma_s \\ \frac{\pi}{4} \cdot D^2 \cdot \gamma_w \end{pmatrix} \qquad w_1 = 12.68 \cdot \frac{\text{kip}}{\text{ft}} \quad \text{linear unit weight on slope.}$$

$$\boxed{w := \frac{w_1}{\cos(\alpha_1)}} \qquad w = 12.82 \cdot \frac{\text{kip}}{\text{ft}} \qquad \text{unit horizontal weight of penstock plus water weight, adjusted for slope.}$$

$$\left[500 + 1 \cdot \left(\cos(\alpha_1) \cdot 62.28\right)\right] \cdot \text{ft} = 561.63 \, \text{ft} \quad \text{Station at second saddle (middle support below)}$$

determine saddle coordinates:

$$\text{augment}\left[\begin{bmatrix} 0 \\ \cos(\alpha_1) \cdot 62.28 \\ 2 \cdot \left(\cos(\alpha_1) \cdot 62.28\right) \\ \cos(\alpha_1) \cdot 143.06 \end{bmatrix} \cdot \text{ft} + 500 \cdot \text{ft}, \, 763.90 \cdot \text{ft} - \begin{bmatrix} 0 \\ \cos(\alpha_1) \cdot 62.28 \\ 2 \cdot \left(\cos(\alpha_1) \cdot 62.28\right) \\ \cos(\alpha_1) \cdot 143.06 \end{bmatrix} \cdot \text{ft} \cdot s_0 \right] = \begin{pmatrix} 500.00 & 763.90 \\ 561.63 & 754.90 \\ 623.25 & 745.90 \\ 641.56 & 743.23 \end{pmatrix} \text{ft} \quad \begin{array}{l} \text{coordinates} \\ \text{of saddle} \\ \text{supports} \end{array}$$

Figure 18-14

Shear Forces at Supports

$V_1 := 431 \cdot \text{kip}$ maximum shear near middle saddle.

$Q := 844 \cdot \text{kip}$ middle saddle reaction.

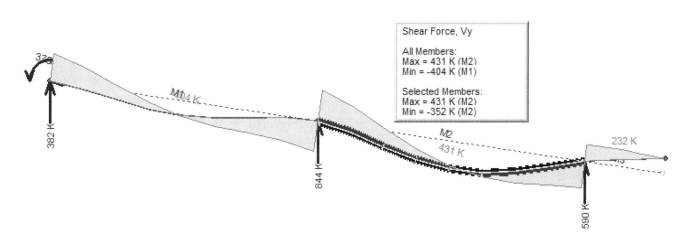

Figure 18-15

Bending Moments at Supports and Mid-span

$M_1 := -4604 \cdot \text{kip} \cdot \text{ft}$ $M_2 := 2771 \cdot \text{kip} \cdot \text{ft}$

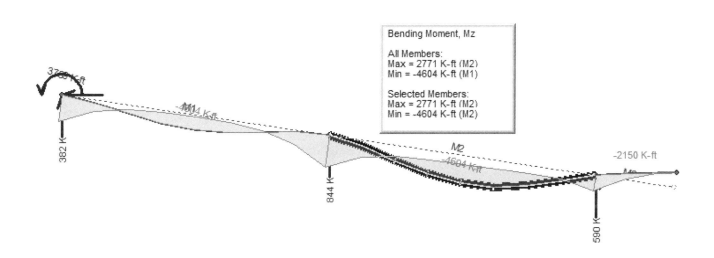

Normal Conditions Hoop Stress

P = *212*·psi   normal condition pressure at middle saddle (see internal design pressure calculations)

k := *1.0*   allowable stress design increase factor, normal condition

D = *15.00* ft   penstock diameter   S = *25.33*·ksi   allowable stress   E := *1.0*   joint efficiency for double CJP Butt Joint Welds & 100% RT.

$\frac{P \cdot D}{2 \cdot k \cdot S}$ = *0.75*·in   required normal condition pipe wall is less than provided, $t = \frac{7}{8}$·in .

Longitudinal Stress

The longitudinal membrane plus bending stresses in the cylindrical shell between supports; where, M_2 = *2771.00*·kip·ft , is the net section maximum longitudinal bending moment between middle and right saddle support. This moment is positive when it results in a compressive stress on the top of the shell.

4.15.3.3   Longitudinal Stress

a)   The longitudinal membrane plus bending stresses in the cylindrical shell between the supports are given by the following equations.

$$\sigma_1 = \frac{PR_m}{2t} - \frac{M_2}{\pi R_m^2 t} \qquad \left(top\ of\ shell\right) \qquad\qquad (4.15.6)$$

$$\sigma_2 = \frac{PR_m}{2t} + \frac{M_2}{\pi R_m^2 t} \qquad \left(bottom\ of\ shell\right) \qquad\qquad (4.15.7)$$

$$\sigma_1 = \frac{P \cdot R_m}{2 \cdot t} - \frac{M_2}{\pi \cdot R_m^2 \cdot t} \qquad \text{top of shell between supports , Equation 4.15.6}$$

$$\sigma_2 = \frac{P \cdot R_m}{2 \cdot t} + \frac{M_2}{\pi \cdot R_m^2 \cdot t} \qquad \text{bottom of shell between supports, Equation 4.15.7}$$

however, because heads or valves are not located near the saddle, the first term $\frac{P \cdot R_m}{2 \cdot t}$ becomes zero and Equations 4.15.6 and 4.15.7 simplify to the following:

$$\boxed{\sigma_1 := -\frac{M_2}{\pi \cdot R_m^2 \cdot t} = -1.48 \cdot ksi}$$ compression in top of shell between supports, Equation 4.15.6 modified.

$$\boxed{\sigma_2 := \frac{M_2}{\pi \cdot R_m^2 \cdot t} = 1.48 \cdot ksi}$$ tension in bottom of shell between supports, Equation 4.15.7 modified.

<u>At the support (un stiffened shell):</u> where $M_1 = -4604.00 \cdot kip \cdot ft$ is the net section maximum longitudinal bending moment at the middle saddle support, this moment is negative when it results in a tensile stress on the top of the shell.

$$\Delta = 80.00 \cdot deg$$

$$K_1 := \frac{\Delta + \sin(\Delta) \cdot \cos(\Delta) - \dfrac{2 \cdot \sin(\Delta)^2}{\Delta}}{\pi \cdot \left(\dfrac{\sin(\Delta)}{\Delta} - \cos(\Delta) \right)} \qquad \boxed{K_1 = 0.11}$$

$$K'_1 := \frac{\Delta + \sin(\Delta) \cdot \cos(\Delta) - \dfrac{2 \cdot \sin(\Delta)^2}{\Delta}}{\pi \cdot \left(1 - \dfrac{\sin(\Delta)}{\Delta} \right)} \qquad \boxed{K'_1 = 0.19}$$

...

2) Unstiffened Shell – The maximum values of longitudinal membrane plus bending stresses at the saddle support are given by the following equations. The coefficients K_1 and K_1^* are given in Table 4.15.1.

$$\sigma_3^* = \frac{PR_m}{2t} - \frac{M_1}{K_1 \pi R_m^2 t} \qquad (\textit{points A and B in Figure 4.15.5}) \qquad (4.15.10)$$

$$\sigma_4^* = \frac{PR_m}{2t} + \frac{M_1}{K_1^* \pi R_m^2 t} \qquad (\textit{bottom of shell}) \qquad (4.15.11)$$

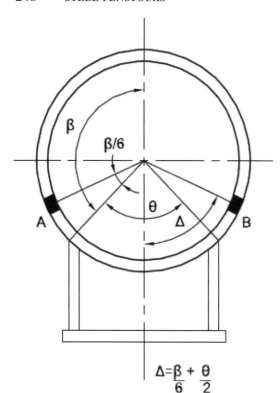

(a)

Figure 18-16. Locations at Support of Maximum Tension

similar to above equations for longitudinal stress between the supports , because heads or valves are not located near the saddle, the first term $\dfrac{P \cdot R_m}{2 \cdot t}$ becomes zero and Equations 4.15.10 and 4.15.11 simplify to the following.

$$\sigma'_3 := -\frac{M_1}{K_1 \cdot \pi \cdot R_m^2 \cdot t} = 23.05 \cdot ksi$$

tension stress at Points A and B in Fig 18-16, per Equation 4.15.10, modified

$$\sigma'_4 := \frac{M_1}{K'_1 \cdot \pi \cdot R_m^2 \cdot t} = -12.78 \cdot ksi$$

compression stress at bottom of shell, per Equation 4.15.11 modified

Allowable Longitudinal Stress

$K := 1.0$ normal conditions $E_y = 29000.00 \cdot ksi$

$$S_c := \frac{K \cdot t \cdot E_y}{16 \cdot R_m} = 17.54 \cdot ksi$$

normal allowable compression stress as limited by buckling, per Section 4.15.12 in ASME (2010)

Compare stresses

at mid span: at top of shell
$\sigma_1 = -1.48 \cdot ksi$

top of shell, Equation 4.15.6 modified

at bottom of shell
$\sigma_2 = 1.48 \cdot ksi$

bottom of shell, Equation 4.15.7 modified

at saddle: at Points A and B $\sigma'_3 = 23.05 \cdot ksi$

Points A and B in Figure 18-16, Equation 4.15.10 modified

at bottom of shell
$\sigma'_4 = -12.78 \cdot ksi$

bottom of shell, Equation 4.15.11 modified

c) Acceptance Criteria

1) The absolute value of σ_1, σ_2, and σ_3, σ_4 or σ_3^*, σ_4^*, as applicable shall not exceed S.

2) If the any of the stresses in paragraph 4.15.3.3.c.1 above are negative, the absolute value of the stress shall not exceed S_c that is given by Equation (4.15.12) where $K = 1.0$ for normal operating conditions and $K = 1.35$ for exceptional operating or hydrotest condition.

$$S_c = \frac{KtE_y}{16R_m}$$

(4.15.12)

the following program checks acceptance criteria for longitudinal stresses

$$
f := \left|
\begin{array}{l}
\text{for } i \in 1..4 \\
\quad \left|
\begin{array}{l}
\sigma \leftarrow \begin{pmatrix} \sigma_1 \\ \sigma_2 \\ \sigma'_3 \\ \sigma'_4 \end{pmatrix}_i \\[20pt]
ss \leftarrow \left| \begin{array}{l} S_c \ \text{if } \sigma < 0 \\ S \ \text{otherwise} \end{array} \right. \\[14pt]
v_{i,1} \leftarrow \dfrac{\sigma}{\text{ksi}} \\[10pt]
v_{i,2} \leftarrow \dfrac{ss}{\text{ksi}} \\[8pt]
\text{"test 1: absolute value of sigma} < S\text{"} \\
v_{i,3} \leftarrow \left| \begin{array}{l} \text{"OK" if } |\sigma| \le S \\ \text{"NG" otherwise} \end{array} \right. \\[10pt]
\text{"test2: for any negative sigma, the absolute value is} < Sc\text{"} \\
v_{i,4} \leftarrow \left| \begin{array}{l} \text{"OK" if } \sigma < 0 \wedge |\sigma| < S_c \\ \text{"N/A" if } \sigma \ge 0 \\ \text{"NG" otherwise} \end{array} \right.
\end{array} \right. \\[30pt]
\text{augment}\left[\begin{pmatrix} \text{"}\sigma 1\text{"} \\ \text{"}\sigma 2\text{"} \\ \text{"}\sigma`3\text{"} \\ \text{"}\sigma`4\text{"} \end{pmatrix}, v \right]
\end{array} \right.
$$

Table 18-9. Longitudinal Stress Summary

Longitudinal Stress	Actual Stress, ksi	Allowable Stress, ksi	Test 1: is absolute value < S?	Test 2: for negative sigma, is absolute value < Sc?
σ1	-1.48	17.54	OK	OK
σ2	1.48	25.33	OK	N/A
σ`3	23.05	25.33	OK	N/A
σ`4	-12.78	17.54	OK	OK

Longitudinal Stress Summary With and Without PA Force

$$ff := \begin{vmatrix} \text{for } i \in 1..4 \\ \quad \text{for } j \in 1..2 \\ \qquad P \leftarrow \begin{pmatrix} 0 \\ 212 \cdot psi \end{pmatrix}_j \\ \qquad \sigma \leftarrow \begin{pmatrix} \dfrac{P \cdot R_m}{2 \cdot t} - \dfrac{M_2}{\pi \cdot R_m^2 \cdot t} \\[2ex] \dfrac{P \cdot R_m}{2 \cdot t} + \dfrac{M_2}{\pi \cdot R_m^2 \cdot t} \\[2ex] \dfrac{P \cdot R_m}{2 \cdot t} - \dfrac{M_1}{K_1 \cdot \pi \cdot R_m^2 \cdot t} \\[2ex] \dfrac{P \cdot R_m}{2 \cdot t} + \dfrac{M_1}{K'_1 \cdot \pi \cdot R_m^2 \cdot t} \end{pmatrix}_i \\ \qquad ss \leftarrow \begin{vmatrix} S_c & \text{if } \sigma < 0 \\ S & \text{otherwise} \end{vmatrix} \\ \qquad v_{i,\,4 \cdot j - 3} \leftarrow \left[\dfrac{\begin{pmatrix} 0 \\ \frac{P \cdot R_m}{2 \cdot t} \end{pmatrix}}{ksi} \right]_j \\ \qquad v_{i,\,4 \cdot j - 2} \leftarrow \dfrac{\sigma}{ksi} \\ \qquad v_{i,\,4 \cdot j - 1} \leftarrow \dfrac{ss}{ksi} \\ \qquad v_{i,\,4 \cdot j} \leftarrow \begin{vmatrix} \text{"OK"} & \text{if } \sigma < 0 \wedge |\sigma| < \min(S, S_c) \\ \text{"OK"} & \text{if } \sigma \geq 0 \wedge \sigma < S \\ \text{"NG"} & \text{otherwise} \end{vmatrix} \\ \text{augment} \left[\begin{pmatrix} \text{"σ1"} \\ \text{"σ2"} \\ \text{"σ`3"} \\ \text{"$\sigma$`4"} \end{pmatrix}, v \right] \end{vmatrix}$$

As and aid to the reader, this program evaluates the effect of longitudinal tension force ("PA" force) that is often present in penstocks because of a bulkhead or closed valve. The example problem does not include this effect because bulkheads or valves are not located near saddle supports.

Iteration program to find shell thickness to satisfy tension allowable stress at $\sigma 3$

$$t' := \begin{vmatrix} \text{while } \dfrac{P \cdot R_m}{2 \cdot t} - \dfrac{M_1}{K_1 \cdot \pi \cdot R_m^2 \cdot t} > S \\ \quad t \leftarrow t + .01 \cdot in \\ t \end{vmatrix}$$

Required shell thickness to satisfy allowable stress for full PA force at σ_3 is $\boxed{t' = 1.18 \cdot in}$

Table 18-10. Longitudinal Stress Summary With and Without PA Force

Longitudinal Stress	No PA Force				Full PA Force			
	Membrane Stress	Combined Stress	Allow. Stress	Status	Membrane Stress	Combined Stress	Allow. Stress	Status
$\sigma 1$	0	-1.48	17.54	OK	10.96	9.48	25.33	OK
$\sigma 2$	0	1.48	25.33	OK	10.96	12.43	25.33	OK
$\sigma`3$	0	23.05	25.33	OK	10.96	34.01	25.33	NG
$\sigma`4$	0	-12.78	17.54	OK	10.96	-1.82	17.54	OK

Shear Stress

$T := V_1 = 431.00 \cdot kip$

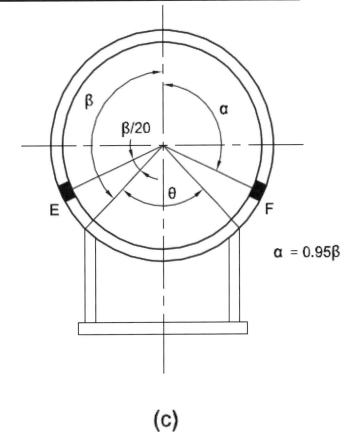

(c)

Figure 18-17   Locations at Support of Maximum Shear Stress

$\theta = 120.00 \cdot deg$ $\beta = 120.00 \cdot deg$ $\alpha = 114.00 \cdot deg$ $t = 0.88 \cdot in$ $K_2 := \dfrac{\sin(\alpha)}{\pi - \alpha + \sin(\alpha) \cdot \cos(\alpha)}$ $\boxed{K_2 = 1.17}$

$\boxed{\tau_2 := \dfrac{K_2 \cdot T}{R_m \cdot t} = 6.38 \cdot ksi}$

maximum shear stress in cylindrical shell without stiffening rings that is not stiffened by a formed head, flat cover, or tube sheet is at points E & F of Figure 18-17. $\boxed{\tau_2 = 6.38 \cdot ksi}$ is less than allowable stress $.80 \cdot S = 20.27 \cdot ksi$.

Circumferential Stress

$Q = 844.00 \cdot kip$ maximum reaction

$$K_6 := \frac{\frac{3 \cdot \cos(\beta)}{4} \cdot \left(\frac{\sin(\beta)}{\beta}\right)^2 - \frac{5 \cdot \sin(\beta) \cdot \cos(\beta)^2}{4 \cdot \beta} + \frac{\cos(\beta)^3}{2} - \frac{\sin(\beta)}{4 \cdot \beta} + \frac{\cos(\beta)}{4} - \beta \cdot \sin(\beta) \cdot \left[\left(\frac{\sin(\beta)}{\beta}\right)^2 - \frac{1}{2} - \frac{\sin(2 \cdot \beta)}{4 \cdot \beta}\right]}{2 \cdot \pi \cdot \left[\left(\frac{\sin(\beta)}{\beta}\right)^2 - \frac{1}{2} - \frac{\sin(2 \cdot \beta)}{4 \cdot \beta}\right]}$$

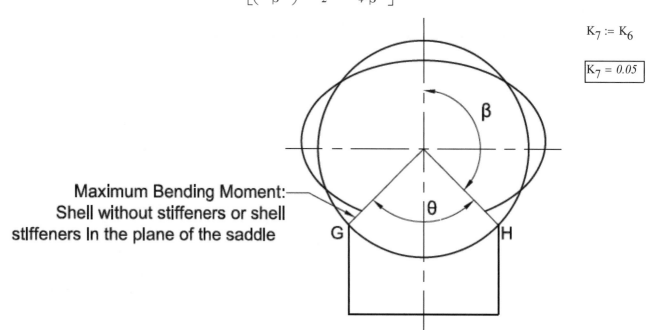

$K_7 := K_6$

$\boxed{K_7 = 0.05}$

Maximum Bending Moment:
Shell without stiffeners or shell
stiffeners In the plane of the saddle

Figure 18-18. Location at Support of Maximum Bending Moment

$\boxed{M_\beta := K_7 \cdot Q \cdot R_m = 336.18 \cdot kip \cdot ft}$

The maximum circumferential bending moment of the cylindrical shell at the horn of saddle without a stiffening ring or stiffening in the plane of the saddle as shown in Figure 18-18

Circumferential Compressive Membrane Stress at Base of Saddle

tributary width of shell $t = 0.88 \cdot in$ $x_1 := .78 \cdot \sqrt{R_m \cdot t} = 6.94 \cdot in$ $x_2 := x_1 = 6.94 \cdot in$

$\alpha = 114.00 \cdot deg$ $K_5 := \frac{1 + \cos(\alpha)}{\pi - \alpha + \sin(\alpha) \cdot \cos(\alpha)}$ $\boxed{K_5 = 0.76}$

c) Circumferential stresses in the cylindrical shell without stiffening ring(s)

1) The maximum compressive circumferential membrane stress in the cylindrical shell at the base of the saddle support shall be computed using Equation (4.15.23). The coefficient K_5 is given in Table 4.15.1.

$$\sigma_6 = \frac{-K_5 Q k}{t(b + x_1 + x_2)} \qquad (4.15.23)$$

$k := 1.0$ This definition accounts for the saddle not welded to the vessel; ($k = .10$ if it is welded to the vessel).

$$\boxed{\sigma_6 := \frac{-K_5 \cdot Q \cdot k}{t \cdot (b + x_1 + x_2)} = -19.36 \cdot ksi}$$ The maximum compressive circumferential membrane stress in the cylindrical shell at the base of the saddle support, assuming that the the saddle is not welded to vessel, $k = 1.00$, Equation 4.15.23; $\boxed{|\sigma_6| = 19.36 \cdot ksi}$ stress is less than allowable stress $S = 25.33 \cdot ksi$.

Circumferential Compressive Membrane Plus Bending Stress @ Points G and H (horn) in Figure 18-18

2) The circumferential compressive membrane plus bending stress at Points G and H of Figure 4.15.6 Sketch (a) is determined as follows. The coefficient K_7 is given in Table 4.15.1.

i) If $L \geq 8R_m$, then the circumferential compressive membrane plus bending stress shall be computed using Equation (4.15.24).

$$\sigma_7 = \frac{-Q}{4t(b + x_1 + x_2)} - \frac{3K_7 Q}{2t^2} \qquad (4.15.24)$$

$b = 24.00 \cdot in$ $t = 0.88 \cdot in$ $Q = 844.00 \cdot kip$ $\boxed{K_7 = 0.053}$

$$\sigma_7 := \frac{-Q}{4 \cdot t \cdot \left(b + 1.56 \cdot \sqrt{R_m \cdot t}\right)} - \frac{3 \cdot K_7 \cdot Q}{2 \cdot t^2} = -93.76 \cdot \text{ksi}$$

For the circumferential compressive membrane plus bending stress at points G and H in Figure 18-18, modified Equation 4.15.24 stress at the horn $\left| \sigma_7 \right| = 93.76 \cdot \text{ksi}$ is greater than the allowable stress $1.25 \cdot S = 31.67 \cdot \text{ksi}$

Allowable Stress

f) Acceptance Criteria

1) The absolute value of σ_6 or $\sigma_{6,r}$, as applicable, shall not exceed S.

2) The absolute value of σ_6^*, as applicable, shall not exceed $\min\left[S, S_r\right]$.

3) The absolute value of σ_7, σ_7^*, $\sigma_{7,r}$, $\sigma_{7,r}^*$, $\sigma_{7,1}$, $\sigma_{7,1}^*$, σ_8, σ_8^*, σ_{10}, and σ_{10}^*, as applicable, shall not exceed $1.25S$.

4) The absolute value of σ_9, σ_9^*, σ_{11}, and σ_{11}^*, as applicable, shall not exceed $1.25S_s$.

Compare Allowable Stresses

Compressive circumferential membrane stress $\sigma_6 = -19.36 \cdot \text{ksi}$.

$$\text{if}\left(\left| \sigma_6 \right| < S, \text{"stress is OK"}, \text{"stress is too high"}\right) = \text{"stress is OK"}$$

maximum compressive circumferential membrane stress in cylindrical shell at base of saddle support, Equation 4.15.23.

Compressive circumferential membrane plus bending stress $\sigma_7 = -93.76 \cdot \text{ksi}$.

$$\text{if}\left(\left| \sigma_7 \right| < 1.25 \cdot S, \text{"stress is OK"}, \text{"stress is too high"}\right) = \text{"stress is too high"}$$

circumferential compressive membrane plus bending stress at points G and H in Fig. 18-18 and Equation 4.15.24.

Reduce Horn Stress   Try thickening shell locally

$$t := \left| \begin{array}{l} \text{while } \left| \frac{-Q}{4 \cdot t \cdot \left(b + 1.56 \cdot \sqrt{R_m \cdot t}\right)} - \frac{3 \cdot K_7 \cdot Q}{2 \cdot t^2} \right| > 1.25 \cdot S = 1.53 \cdot \text{in} \\ \quad t \leftarrow t + .01 \cdot \text{in} \\ t \end{array} \right.$$

For the required shell thickness at saddle, $t = 1.53 \cdot \text{in}$ round upward to the nearest 1/8 in.

$$t := \text{Ceil}\left(\frac{t}{\text{in}}, \frac{1}{8}\right) \cdot \text{in} = 1.63 \cdot \text{in}$$ Locally thicken shell to $t = 1.63 \cdot \text{in}$

Conclusion

The saddle design appears adequate without stiffeners; however the shell must be thickened locally to $t = 1.63 \cdot in$.

Wear Plate option

$t := \dfrac{7}{8} \cdot in$ shell thickness at saddle. $t_r := \dfrac{3}{4} \cdot in$ trial wear plate thickness .

as an optional design, a wear plate can be installed to reduce horn stress.

$S_r := S = 25.33 \cdot ksi$ assume the wear plate allowable stress is the same as the penstock. $\eta := min\left(\dfrac{S_r}{S}, 1.0\right) = 1.00$ Equation 4.15.29.

$b := 24 \cdot in$ the width of the contact surface of the shell and the saddle.

$b_1 := \left(b + 1.56 \cdot \sqrt{R_m \cdot t}\right)$ $b_1 = 37.88 \cdot in$ the width of saddle reinforcing plate per modified Equation 4.15.1.

$\boxed{\theta_1 := \theta + \dfrac{\theta}{12} = 130.00 \cdot deg}$ the arc length of the saddle reinforcing plate per Equation 4.15.2

3) The stresses σ_6 , σ_7 , and σ_7^* may be reduced by adding a reinforcement or wear plate at the saddle location that is welded to the cylindrical shell that satisfies the requirements of paragraph 4.15.3.1.c. The stress can be computed using the equations shown below.

$$\sigma_{6,r} = \frac{-K_5 Q k}{b_1 \left(t + \eta t_r\right)} \tag{4.15.26}$$

$$\sigma_{7,r} = \frac{-Q}{4\left(t + \eta t_r\right) b_1} - \frac{3 K_7 Q}{2\left(t + \eta t_r\right)^2} \tag{4.15.27}$$

$$\sigma_{7,r}^* = \frac{-Q}{4\left(t + \eta t_r\right) b_1} - \frac{12 K_7 Q R_m}{L\left(t + \eta t_r\right)^2} \tag{4.15.28}$$

where

$$\eta = min\left[\frac{S_r}{S}, 1.0\right] \tag{4.15.29}$$

$K_5 = 0.76$

$$\sigma_{6r} := \frac{-K_5 \cdot Q \cdot k}{b_1 \cdot (t + \eta \cdot t_r)} = -10.42 \cdot ksi$$

Equation 4.15.26.

$K_7 = 0.053$

$$\sigma_{7r} := \frac{-Q}{4 \cdot (t + \eta \cdot t_r) \cdot b_1} - \frac{3 \cdot K_7 \cdot Q}{2 \cdot (t + \eta \cdot t_r)^2} = -28.77 \cdot ksi$$

Equation 4.15.27.

The circumferential compressive membrane plus bending stress at points G and H in Figure 18-18, $|\sigma_{7r}| = 28.77 \cdot ksi$ is less than the allowable stress $1.25 \cdot S = 31.67 \cdot ksi$. The stress at the horn is less than the allowable stress; thus wear plate is adequate.

Conclusion

Wear plate Option; Saddle design appears adequate using a wear plate of thickness $t_r = 0.75 \cdot in$, arc length $\theta_1 = 130.00 \cdot deg$, and width $b_1 = 37.88 \cdot in$.

Saddle Base Stability

$P := Q = 844.00 \cdot kip$ vertical reaction. $N := 12 \cdot ft$ $B := 30 \cdot in$

assumed base plate length and width.

$15 \cdot \% \cdot Q = 126.60 \cdot kip$ The approximate lateral load from seismic forces.

$M := 15 \cdot \% \cdot P \cdot (R_m + 2 \cdot ft) = 1207.32 \cdot kip \cdot ft$ The overturning moment of the base plate.

$D = 15.00 \cdot ft$ $R_m = 7.54 \cdot ft$ $R_m + 2 \cdot ft = 9.54 \cdot ft$ the vertical distance from the center of penstock to the base plate.

$15 \cdot \% \cdot (R_m + 2 \cdot ft) = 17.17 \cdot in$ $\frac{N}{6} = 24.00 \cdot in$ The location of the resultant falls in middle 1/3 of the base. Thus, there is no uplift on the anchor bolts.

$\frac{P}{N \cdot B} = 28.13 \cdot ksf$ $\frac{M}{\left(\frac{B \cdot N^2}{6}\right)} = 20.12 \cdot ksf$

$$f_1 := \frac{P}{N \cdot B} + \frac{M}{\left(\frac{B \cdot N^2}{6}\right)} = 48.26 \cdot ksf$$ toe pressure is positive.

$$f_2 := \frac{P}{N \cdot B} - \frac{M}{\left(\frac{B \cdot N^2}{6}\right)} = 8.01 \cdot ksf$$ heel pressure is positive.

Conclusion

The preliminary base width and length dimensions, $B = 2.50 \cdot ft$ and $N = 12.00 \cdot ft$, indicate positive toe and heel pressures and are adequate for saddle base stability.

18.6 Reducing Bend

▶ Define Units

Reducing bend dimensions: Inside diameter of large end = D_1. $D_1 := 15\,ft$ $D_1 = 180.00 \cdot in$

Inside diameter of small end = D_n $D_n := 12\,ft$ $D_n = 144.00 \cdot in$

Radius of bend = R   In this case, the bend radius is defined by the design requirements as $4*D_1$ therefore,

$R := 4 \cdot D_1 = 60.00\,ft$ Central angle = bend angle = Δ $\Delta := 39.6927°$

Calculations

Tangent length = TL $TL := R \cdot \tan\left(\dfrac{\Delta}{2}\right) = 21.66\,ft$ $\boxed{TL = 259.87 \cdot in}$

Number of miter cuts in the bend * 2 = n $\boxed{n := 2 \cdot 3 = 6.00}$

Division angle = ρ $\boxed{\rho := \dfrac{\Delta}{n} = 6.6155 \cdot °}$

Angle of inclination of right circular cone = θ $\boxed{\theta := \mathrm{asin}\left[\dfrac{D_1 - D_n}{2(n-2)R \cdot \tan(\rho)}\right] = 3.0892 \cdot °}$

Inside radius of large end = r_1 $r_1 := \dfrac{D_1}{2} = 7.50\,ft$ $r_1 = 90.00 \cdot in$

Inside radius of small end = r_n $r_n := \dfrac{D_n}{2} = 6.00\,ft$ $r_n = 72.00 \cdot in$

Number of divisions from the point of curvature (PC) to the point of evaluation = x

Inside radius at any division = r_x $r_x = r_1 - (x - 1)R \cdot \tan(\rho) \cdot \sin(\theta)$

Inside diameter at any division = D_x $D_x = \dfrac{D_1 - 2(x - 1)R \cdot \tan(\rho) \cdot \sin(\theta)}{\cos(\theta)}$

Therefore, the following equation is used to calculate the inside diameter at division D_2:

$x := 2$ $D_2 := \dfrac{D_1 - 2(x - 1)R \cdot \tan(\rho) \cdot \sin(\theta)}{\cos(\theta)} = 14.27\,\text{ft}$ $\boxed{D_2 = 171.25 \cdot \text{in}}$

And to calculate the inside diameter at division D_4:

$x := 4$ $D_4 := \dfrac{D_1 - 2(x - 1)R \cdot \tan(\rho) \cdot \sin(\theta)}{\cos(\theta)} = 12.77\,\text{ft}$ $\boxed{D_4 = 153.22 \cdot \text{in}}$

Miter cut angle $= \phi$ $\phi := \operatorname{atan}\left(\dfrac{\sin(2\rho)}{\cos(2\rho) + \cos(\theta)}\right) = 6.62 \cdot^{\circ}$

Offset dimension at large end of elbow $= z_1$ $z_1 := \dfrac{r_1 \cdot \sin(\theta)}{\cos(2\rho) + \cos(\theta)} = 0.20\,\text{ft}$ $\boxed{z_1 = 2.46 \cdot \text{in}}$

Offset dimension at small end of elbow $= z_n$ $z_n := \dfrac{r_n \cdot \sin(\theta)}{\cos(2\rho) + \cos(\theta)} = 0.16\,\text{ft}$ $\boxed{z_n = 1.97 \cdot \text{in}}$

Intermediate offset dimension at any division $= y_x$ $y_x = \dfrac{r_x \cdot \sin(\theta)}{\cos(\rho)}$

Therefore, the following equation is used to determine the intermediate offset at $x = 1$:

$y_1 := \dfrac{r_1 \cdot \sin(\theta)}{\cos(\rho)} = 0.41\,\text{ft}$ $\boxed{y_1 = 4.88 \cdot \text{in}}$

And to calculate the intermediate offset at $x=3$: $x := 3$ $r_3 := r_1 - (x - 1) \cdot R \cdot \tan(\rho) \cdot \sin(\theta) = 6.75\,\text{ft}$

$y_3 := \dfrac{r_3 \cdot \sin(\theta)}{\cos(\rho)} = 0.37\,\text{ft}$ $\boxed{y_3 = 4.39 \cdot \text{in}}$

And to calculate the intermediate offset at $x=5$: $x := 5$ $r_5 := r_1 - (x - 1) \cdot R \cdot \tan(\rho) \cdot \sin(\theta) = 6.00\,\text{ft}$

$y_5 := \dfrac{r_5 \cdot \sin(\theta)}{\cos(\rho)} = 0.33\,\text{ft}$ $\boxed{y_5 = 3.91 \cdot \text{in}}$

Shell Wall Thickness of Reducing Bend

Refer to Section 4.5.3 for a discussion of stress analysis.

Input $D := 15.25\,\text{ft}$ Outside diameter of the reducing bend (approximate) = D

$P := 331\,\text{psi}$ Design pressure at PI #3 (in this case this is at the normal condition) = P

$R = 60.00\,\text{ft}$ Radius of reducing bend=R

Allowable tensile stress (in this case this is at the normal condition) = f   Steel is SA-516, Grade 70;

Calculations

$F_{mts} := 70\cdot\text{ksi}$ specified minimum tensile stress. $F_y := 38\cdot\text{ksi}$ specified minimum yield stress.

$$\begin{pmatrix} \dfrac{F_y}{1.5} \\ \dfrac{F_{mts}}{2.4} \end{pmatrix} = \begin{pmatrix} 25.33 \\ 29.17 \end{pmatrix}\cdot\text{ksi} \qquad f := \min\left(\begin{pmatrix} \dfrac{F_y}{1.5} \\ \dfrac{F_{mts}}{2.4} \end{pmatrix}\right) = 25.33\cdot\text{ksi}$$

allowable stress under normal conditions for SA-516, Grade 70 steel per ASME BPV Sec. VIII, Div. 2, and Sec. II, Part D, Table 5A.

$$t := \left(\frac{P\cdot D}{2\cdot f}\right)\cdot\left(1 + \frac{D}{3R - 1.5D}\right) = 1.31\cdot\text{in}$$

The required wall thickness of the reducing bend = t.

18.7 Transition

Derivation   Combine two equations into a single formula for shell thickness and support spacing.

$M = \dfrac{p \cdot L^2}{9}$ approximate maximum bending moment, M, in a continuous, uniformly loaded beam with equal spans where p = pressure and L = span lengths.

$S = \dfrac{6 \cdot M}{t^2}$ bending stress where S = stress, M = bending moment and t = shell thickness.

$S = \dfrac{6 \cdot \left(\dfrac{p \cdot L^2}{9}\right)}{t^2}$ simplify $\rightarrow$ $S = \dfrac{2 \cdot L^2 \cdot p}{3 \cdot t^2}$ or $t^2 = \dfrac{2 \cdot L^2 \cdot p}{3 \cdot S}$ or $t = \sqrt{\dfrac{2 \cdot L^2 \cdot p}{3 \cdot S}}$ or $t = \sqrt{\dfrac{2 \cdot p}{3 \cdot S}} \cdot L$

or $\dfrac{t}{L} = \sqrt{\dfrac{2 \cdot p}{3 \cdot S}}$ or $\boxed{L = \dfrac{t}{\sqrt{\dfrac{2 \cdot p}{3 \cdot S}}}}$

▶ Define Units

Input

The transition is square (15 x 15 ft) to circular (15 ft diameter) located at stations 0+00 to 0+10 (see Pressure Design Example).

The calculated thickness based on minimum handling is $t := .625 \cdot in$. The external pressure is an exceptionally high 1,165 - 815 = 350 feet (151 psi); therefore a stiffened design is required.

$h := (1165 \cdot ft - 815 \cdot ft) = 350.00\, ft$ $\gamma_w = 62.25 \cdot \dfrac{lbf}{ft^3}$ unit weight of water. $p := \gamma_w \cdot h$ $\boxed{p = 151.29\, psi}$

The derived formula for determining the stiffener spacing for the transition is: $\boxed{L = \dfrac{t}{\sqrt{\dfrac{2 \cdot p}{3 \cdot S}}}}$

where L = spacing between stiffeners on surface of transition, t = shell thickness, p = external pressure and S = allowable stress for a flat, rectangular shape (see Section 3.4.7) where S = f_n*1.5.

Allowable stress; assume that the penstock steel is SA-516, Grade 70;

$F_{mts} := 70 \cdot ksi$ specified minimum tensile stress. $F_y := 38 \cdot ksi$ specified minimum yield stress.

$$\begin{pmatrix} \dfrac{F_y}{1.5} \\ \dfrac{F_{mts}}{2.4} \end{pmatrix} = \begin{pmatrix} 25.33 \\ 29.17 \end{pmatrix} \cdot ksi$$ $$S := \min\left(\begin{pmatrix} \dfrac{F_y}{1.5} \\ \dfrac{F_{mts}}{2.4} \end{pmatrix}\right) = 25.33 \cdot ksi$$ allowable stress under normal conditions for SA-516, Grade 70 steel per ASME BPV Sec. VIII, Div. 2, and Sec. II, Part D, Table 5A.

$\boxed{S := 1.5 \cdot S = 38.00 \cdot ksi}$ allowable steel stress.

Calculations

$t = 0.625 \cdot in$ $p = 151.29 \, psi$ $S = 38.00 \cdot ksi$ $L := \dfrac{t}{\sqrt{\dfrac{2 \cdot p}{3 \cdot S}}}$ $\boxed{L = 12.13 \cdot in}$ span length.

Try several values of commercially available shell thickness and find the span length.

$t := \dfrac{5}{8} \cdot in, \dfrac{11}{16} \cdot in \, .. \, \dfrac{3}{4} \cdot in$ $L(t) := \dfrac{t}{\sqrt{\dfrac{2 \cdot p}{3 \cdot S}}}$

Thickness;		Span;	
$t =$		$L(t) =$	
0.625	$\cdot in$	12.13	$\cdot in$
0.688		13.34	
0.750		14.56	

Conclusion
Use a 5/8 in. thickness shell and 12 in. stiffener spacing.

Consider studs on 12 in. centers each way.

Required area per stud = A_s Allowable stud stress $s := 10 \cdot ksi$. $L := 12in$ stud spacing.

$A_s = \dfrac{\pi \cdot d^2}{4}$ solve, $d \rightarrow \begin{pmatrix} \dfrac{2 \cdot \sqrt{A_s}}{\sqrt{\pi}} \\ -\dfrac{2 \cdot \sqrt{A_s}}{\sqrt{\pi}} \end{pmatrix}$ or $d = \dfrac{2 \cdot \sqrt{A_s}}{\sqrt{\pi}}$ or $\dfrac{2 \cdot \sqrt{A_s}}{\sqrt{\pi}}$ float, $4 \rightarrow 1.128 \cdot \sqrt{A_s}$ stud diameter

$A_s := \dfrac{L^2 \cdot p}{s}$ $A_s = 2.18 \cdot in^2$ required area of stud. $d := 1.128 \cdot \sqrt{A_s}$ $\boxed{d = 1.66 \cdot in}$ required stud diameter.

<u>Conclusion</u>

The required stud diameter is $\boxed{d = 1.66 \cdot in}$. This measurement is prohibitively large, so use continuous tee-ring frames. The design of the tee-ring frame is tied to the design of the surrounding concrete. Concrete design is outlined in IBC (2009) Chapter 19 and ACI (2008).

18.8 Expansion Joints and Sleeve Couplings

▶ Define Units

Mechanical Expansion Joint Design

(1) Wall Thickness

Determine sleeve thickness (t) using the Barlow hoop stress formula: $t = \dfrac{P \cdot r}{S_H}$

Where: Penstock diameter $D := 180 \cdot in$ $r := \dfrac{D}{2} = 90.00 \cdot in$

Allowable stress: assume the middle ring of the coupling is SA-516, Grade 70 steel.

$\left(F_{mts} := 70 \cdot ksi\right)$ specified minimum tensile stress $\left(F_y := 38 \cdot ksi\right)$ specified minimum yield stress

$$\left(\left(\begin{array}{c} \dfrac{F_y}{1.5} \\ \dfrac{F_{mts}}{2.4} \end{array}\right)\right) = \left(\begin{array}{c} 25.33 \\ 29.17 \end{array}\right) \cdot ksi \qquad S_H := \min\left(\left(\begin{array}{c} \dfrac{F_y}{1.5} \\ \dfrac{F_{mts}}{2.4} \end{array}\right)\right) = 25.33 \cdot ksi$$

allowable stress under normal conditions for SA-516, Grade 70 steel per ASME BPV Sec. VIII, Div. 2, and Sec. II, Part D, Table 5A.

Wall thickness of the middle ring at expansion joint number 2:

$P := 296 \cdot psi$ see internal design calculations. $r = 90.00 \cdot in$ $S_H = 25.33 \cdot ksi$ $\boxed{t := \dfrac{P \cdot r}{S_H} = 1.05 \cdot in}$ round upward to nearest 1/8 in.

$\boxed{t := Ceil\left(\dfrac{t}{in}, \dfrac{1}{8}\right) \cdot in = 1.13 \cdot in}$ **Conclusion**   The required middle ring thickness is $t = 1.13 \cdot in$.

(2) Number of Bolts Required   Determine minimum number of bolts required:

minimum number of bolts = N_{bmin} $N_{bmin} := \dfrac{D}{in} \cdot .4384 = 78.91$ (try using 80 bolts) $\left(N_b := 80\right)$

The number of bolts required may vary substantially depending on the manufacturer or the gasket material used. Bolts are ASTM A307 Gr B, 7/8 in. diameter, with an allowable bolt force of 7700 pounds per bolt assuming an allowable stress of 16.67 ksi on tensile stress area of $A_k = .462 \cdot in^2$ Bolts with different strengths are available. Consult the bolt manufacturer for available strengths.

(3) Gasket Pressure

$D = 180.00 \cdot in$ $gw := 1 \cdot in$ trial gasket width $P = 296.00\,psi$ normal pressure

Determine gasket pressure:

Try a $gw = 1.00 \cdot in$ wide gasket. Consult the manufacturer for available gasket widths.

Gasket area = G_a = π * (D + gasket width) * (gasket width)

$$G_a := gw \cdot (D + gw) \cdot \pi = 568.63 \cdot in^2$$

Gasket pressure = G_p = (N_b * bolt force)/G_a $N_b = 80.00$ $F = 7.70 \cdot kip$

$$G_p := \frac{N_b \cdot F}{G_a} = 1083 \cdot psi$$

$$\frac{G_p}{P} = 3.66$$

$$if\left(\frac{G_p}{P} > 3.5, "Acceptable", "Not\ Acceptable"\right) = "Acceptable"$$

Conclusion

A gasket pressure, 3.5 times the working pressure or higher is acceptable. Therefore, the trial gasket width is acceptable. The gasket pressure may vary substantially, depending on the manufacturer or gasket material used.

Bolted Sleeve Type Coupling (BSTC) Design

(1) Wall Thickness of Middle Ring at BSTC

Determine sleeve thickness (t) using the Barlow hoop stress formula: $t = \dfrac{P \cdot r}{S_H}$

where   Penstock diameter $D := 144 \cdot in$ $r := \dfrac{D}{2} = 72.00 \cdot in$ $S_H = 25.33 \cdot ksi$

$P := 334 \cdot psi$ pressure at BSTC (end of penstock); see internal design calculations.

$$t := \frac{P \cdot r}{S_H} = 0.95 \cdot in$$ round upward to nearest 1/8 in. $$t := Ceil\left(\frac{t}{in}, \frac{1}{8}\right) \cdot in = 1.00 \cdot in$$

Conclusion   The required middle ring thickness of BSTC is $t = 1.00 \cdot in$

(2) Number of Bolts Required

Determine the minimum number of bolts required.

minimum number of bolts = N_{bmin} $N_{bmin} := \dfrac{D}{in} \cdot .4384 = 63.13$ (try using 64 bolts) $N_b := 64$

The number of bolts required may vary substantially depending on the manufacturer or the gasket material used. Bolts are ASTM A307 Gr B, 7/8 in. diameter, with an allowable bolt force of 7700 pounds per bolt assuming an allowable stress of 16.67 ksi on tensile stress area of $A_k = .462 \cdot in^2$ Bolts with different strengths are available. Consult the bolt manufacturer for available strengths.

(3) Gasket Pressure

$D = 144.00 \cdot in$ $gw := \dfrac{5}{8} \cdot in$ trial gasket width. $P = 334.00\, psi$ normal pressure.

Determine gasket pressure.

Try a $gw = 0.63 \cdot in$ wide gasket. Consult the manufacturer for available gasket widths.

Gasket area = G_a = π * (D + gasket width) * (gasket width) $\boxed{G_a := gw \cdot (D + gw) \cdot \pi = 283.97 \cdot in^2}$

Gasket pressure = G_p = (N_b * bolt force) / G_a $N_b = 64.00$ $F = 7.70 \cdot kip$ $\boxed{G_p := \dfrac{N_b \cdot F}{G_a} = 1735 \cdot psi}$

$\boxed{\dfrac{G_p}{P} = 5.19}$ $\boxed{if\left(\dfrac{G_p}{P} > 3.5, "Acceptable", "Not\ Acceptable"\right) = "Acceptable"}$

Conclusion

A gasket pressure, 3.5 times the working pressure or higher is acceptable. Therefore, the trial gasket width is acceptable. The gasket pressure may vary substantially, depending on the manufacturer or gasket material used.

18.9 Anchor Block

The anchor block is at bend no. 2 of the exposed/supported penstock shown in Figure 18-1. See figures 18-19 through 18- 21.

Given Data

Undefined Units: Cubic feet per second: $CFS \equiv \dfrac{ft^3}{s}$

Gravity Acceleration: $Accel \equiv 32.2 \dfrac{ft}{s^2}$

Global Definitions:

Weight of water (Kips/ft³): $WTw \equiv .0624 \dfrac{kip}{ft^3}$

Weight of concrete (Kips/ft³): $WTconc \equiv .150 \dfrac{kip}{ft^3}$

Weight of steel (Kips/ft³): $WTs \equiv .490 \dfrac{kip}{ft^3}$

Geometry:

Upstream pipe slope (S_1) = -0.146

$\theta := 8.3065°$

Downstream pipe slope (S_2) = -0.830

$\alpha := 39.6927°$

Penstock diameter: $D := 15ft$

$Area := \pi \dfrac{D^2}{4} = 176.715 \, ft^2$

Flow and Velocity:

Flow rate (cfs): $Flow := 2600 \cdot CFS$

Velocity (fps): $Vel := \dfrac{Flow}{Area} = 14.713 \dfrac{ft}{s}$

Friction Loads:

Saddle bearing friction coefficient: $fr := 0.12$

Expansion joint friction (per lineal foot of circumference): $EJF := 500plf$

Miscellaneous Loads:

Dead load of water: $Dwater := WTw \cdot Area = 11.027 \cdot \dfrac{kip}{ft}$

Penstock OD Areas (for miscellaneous calculations):

Wall thickness 1-1/8 inch: $\text{Shell1} := \left(\pi\, ft^2 \cdot \dfrac{15.1875^2}{4} \right) = 181.16\, ft^2$

Wall thickness 1-3/16 inch: $\text{Shell2} := \left(\pi\, ft^2 \cdot \dfrac{15.1979^2}{4} \right) = 181.41\, ft^2$

Wall thickness 1-1/4 inch: $\text{Shell3} := \left(\pi\, ft^2 \cdot \dfrac{15.2083^2}{4} \right) = 181.66\, ft^2$

The generalized design method for the anchor at Bend No. 2 follows Chapter 8. Loads acting are grouped by the system given in Chapter 8 and, in particular, Figs. 8-3 and 8-4.

(1) Assume a design for the upper part of the anchor block (encasement of penstock). Apply penstock bend loads entirely to this part of the anchor (see Figure 18-20).

(2) Apply loading from groups "A" and "B" to the assumed upper part; consolidate all loads into two prime resultants and one moment about the point of intersection (PI).

(3) Add additional mass of concrete (or other anchor systems) under the assumed upper part until an acceptable design is achieved.

(4) It is assumed that the foundation system under the anchor block consists of reinforced concrete forming a rigid connection between the block and the first downstream saddle support. This mass concrete is not considered as contributing to the anchor block (at designer's discretion).

(5) Check the stability of the total anchor under seismic loadings, for other upsurge conditions, and for hydrotesting during construction.

Reference Figures:

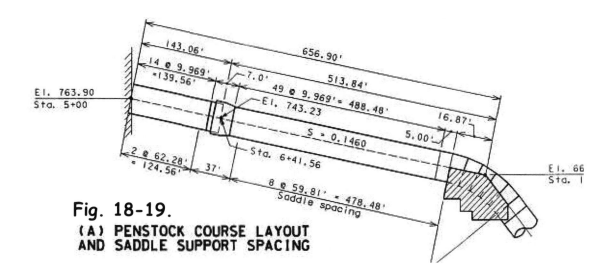

Fig. 18-19.
(A) PENSTOCK COURSE LAYOUT
AND SADDLE SUPPORT SPACING

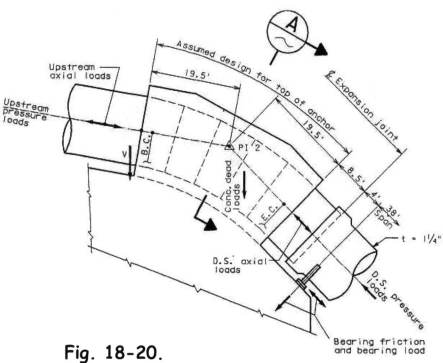

Fig. 18-20.
Anchor Block Section Detail

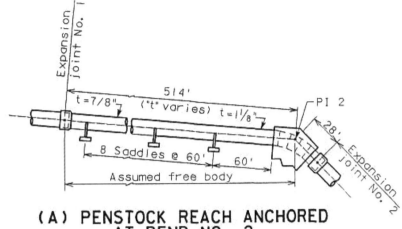

**(A) PENSTOCK REACH ANCHORED
AT BEND NO. 2**

NOTE:

Dimensions simplified for
this example.

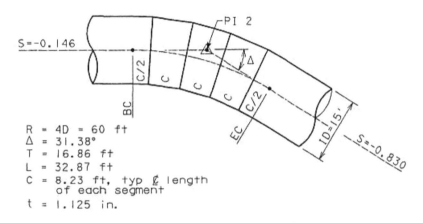

R = 4D = 60 ft
Δ = 31.38°
T = 16.86 ft
L = 32.87 ft
C = 8.23 ft, typ ₵ length
    of each segment
t = 1.125 in.

(B) DETAIL OF PENSTOCK BEND

Fig. 18-21.
Penstock Bend Detail

Group A:

(1) Axial Loads Upstream of Anchor

- Static water column at the upstream expansion joint is 421.77 ft.

- Length of penstock between upstream expansion joint and upper block face is
  $508.44/\cos\theta$ - 19.5 = 494.33 ft.

- Thickness of penstock is 1-1/8 in.

$$UPsteel := WTs \bullet 494.33ft \cdot (Shell1 - Area) = 1076.8 \bullet kip$$

$$UPwater := 494.33ft \bullet Dwater = 5451.0 \bullet kip$$

$$UPvertical := UPsteel + UPwater = 6527.8 \bullet kip$$

(a) Axial load

$$UPaxial := (UPsteel + UPwater) \bullet \sin(\theta) = 943.1 \bullet kip$$

(b) Friction load on upstream saddles

$$UPsaddle := fr \bullet (UPsteel + UPwater) \bullet \cos(\theta) = 775.1 \bullet kip$$

(c) Friction caused by upstream expansion joint

$$UPexp := EJF \bullet (\pi \bullet ft15.1875) = 23.9 \bullet kip$$

(d) Pressure thrust acting at upstream expansion joint

$$P1 := 421.77ft \bullet Area \bullet WTw = 4650.9 \bullet kip$$

(e) Pressure head acting on the pipe shell ends inside upstream expansion joint

$$Pshellup := 421.77ft \bullet WTw \bullet (Shell1 - Area) = 117.0 \bullet kip$$

(f) Total upstream axial loads

$$UPtotal := UPaxial + UPsaddle + UPexp + P1 + Pshellup$$

$$\boxed{UPtotal = 6509.9 \bullet kip}$$

(2) Fixed End Shear and Moment Loads at Upstream Face of Block

- Shear dead load for the first span between upstream block face and first support is 1/2 the span load from span to support; moment arm is 19.5 ft. from block PI.

- First span length is 60 ft.

- Beam fixed end moment is $(wL^2)/12$.

(a) Shear at face of block

$$Vup := \left(\frac{1}{2}\right) \bullet 60ft \bullet \cos(\theta) \bullet [Dwater + WTs \bullet (Shell1 - Area)] = 392.0 \bullet kip$$

$$Vupmom := Vup \bullet 19.5ft = 7644.0 \bullet ft\cdot kip$$

(b) Fixed end moment caused by beam action in first upstream span:

$$FEMup := \left[\frac{(60ft)^2}{12}\right] \bullet \cos(\theta) \bullet [Dwater + WTs \bullet (Shell1 - Area)] = 3920.0 \bullet ft \bullet kip$$

Saddle friction moment for supports upstream of the block are ignored in this analysis. This situation would be expected to contribute (to a limited extent) to the beam FEM, but it would be small in magnitude.

(3) Gravity Loads from Anchor Block (Embedded Penstock)

- Bend segments within anchor are approximated as single bevel located at the PI.

- Upstream leg from PI to edge of block is 19.5 ft.

- Downstream leg from PI to expansion joint adjacent to block face is 19.5 ft.

- Thickness of penstock below PI is 1-3/16 in. (above PI is 1-1/8 in.).

(a) Upstream force and fixed end moment components of block (pipe and water):

$$UPvert := 19.5ft \bullet [Dwater + WTs \bullet (Shell1 - Area)] = 257.5 \bullet kip$$

$$FEMun := UPvert \bullet -\cos(\theta) \bullet \frac{19.5}{2}ft = -2484.3 \bullet ft \bullet kip$$

(b) Downstream force and fixed end moment component of block (pipe and water):

$$DNvert := 19.5ft \bullet [Dwater + WTs \bullet (Shell2 - Area)] = 259.9 \bullet kip$$

$$FEMdn := DNvert \bullet \cos(\alpha) \bullet \frac{19.5}{2}ft = 1949.7 \bullet ft \bullet kip$$

(4) Gravity Loads from Upper Anchor Block (Concrete Encasement)

- Weight of concrete block is determined by dividing it into components
  (A through F) (Table 18-11).

- Each component is approximated as area times average chord length less half of
  pipe volume (upper or lower) times weight of concrete.

- Calculations are summarized in the following spreadsheet generated table

Table 18-11. Gravity Load Component Summary

Seg	Short Side (ft)	Long Side (ft)	Top Bevel (ft)	Height (ft)	Block Width (ft)	Pipe OD (ft)	Block Weight (kips)	dcg-PI (ft)	Mseg (kip-ft)
A	10.75	12.20	4.50	10.50	21.00	15.1875	196.4	-12.36	-2,427.7
B	9.30	10.75	0.00	10.50	21.00	15.1875	185.5	-15.40	-2,857.0
C	16.85	19.74	4.50	10.50	21.00	15.1927	316.0	2.15	679.5
D	13.95	16.85	0.00	10.50	21.00	15.1927	280.3	-3.67	-1,028.5
E	10.75	12.20	4.50	10.50	21.00	15.1979	196.2	15.34	3,010.0
F	9.30	10.75	0.00	10.50	21.00	15.1979	185.3	6.96	1,289.8

Total Weight: **1359.8**   MNET: **-1,333.9**

$$WTa := 196.4\,kip \qquad WTb := 185.5\,kip \qquad WTc := 316.0\,kip$$

$$WTd := 280.3\,kip \qquad WTe := 196.2\,kip \qquad WTf := 185.3\,kip$$

Total weight of block: $WTblock := WTa + WTb + WTc + WTd + WTe + WTf$

$$\boxed{WTblock = 1359.7 \cdot kip}$$

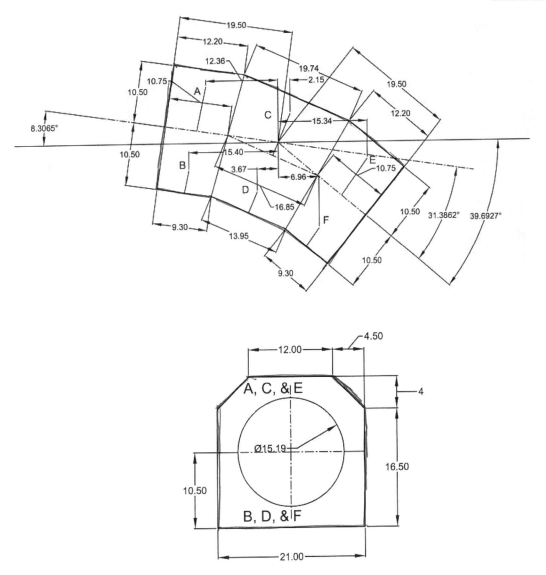

Fig. 18-22. Anchor Block Geometry

Moment arms determined from coordinates of block components:

- Coordinates for each block component determined from graphic layout or geometry (Table 18-11).
- Centroid of each block component approximated at center of each component.
- Ordinate of coordinates set to PI of anchor (PI 2).
- X value of each coordinate pair represents moment arm for each block component, perpendicular to vertical weight vector.
- Sign conventions set temporarily to conventional X-Y convention.

$$CC := 1ft \bullet \begin{pmatrix} -12.36 & 7.04 \\ -15.40 & -3.00 \\ 2.15 & 2.63 \\ -3.67 & -6.96 \\ 15.34 & -4.83 \\ 6.96 & -9.05 \end{pmatrix} \qquad WT := \begin{pmatrix} WTa \\ WTb \\ WTc \\ WTd \\ WTe \\ WTf \end{pmatrix} \qquad WT = \begin{pmatrix} 196.4 \\ 185.5 \\ 316 \\ 280.3 \\ 196.2 \\ 185.3 \end{pmatrix} \bullet kip$$

$$MOM := \overrightarrow{\left(WT \bullet CC^{\langle 0 \rangle}\right)} \qquad MOM = \begin{pmatrix} -2427.5 \\ -2856.7 \\ 679.4 \\ -1028.7 \\ 3009.7 \\ 1289.7 \end{pmatrix} \bullet ft \cdot kip$$

$$\sum MOM = -1334.1 \bullet ft \cdot kip \qquad NETARM := \sum MOM \div WTblock$$

$$\boxed{NETARM = -0.98\,ft} \qquad \text{This is 0.98 ft. left of block PI.}$$

$$BLOCKMOM := NETARM \bullet WTblock \qquad \boxed{BLOCKMOM = -1334.1 \bullet ft \cdot kip} \quad \text{Equation Checks}$$

(5) Axial Loads Downstream of Anchor

- Static water column at the downstream expansion joint is 513.91 ft.
- Length of penstock between block PI to downstream expansion joint is 28 ft. (19.5+8.5)
- Thickness of penstock is 1-3/16 in.
- Length of exposed penstock from downstream block face to expansion joint is 8.5 ft.
- Thickness of penstock downstream of expansion joint is 1-1/4 in.

(a) Vertical load of exposed penstock to downstream expansion joint

$$DNsteel := WTs \bullet (8.5)ft \cdot (Shell2 - Area) = 19.5 \bullet kip$$

$$DNwater := (8.5)ft \bullet Dwater = 93.7 \bullet kip$$

$$DNvertical := DNsteel + DNwater = 113.3 \bullet kip$$

$$FMupsaddle := DNvertical \bullet \left(19.5 + \frac{8.5}{2}\right) \bullet 1ft \bullet \cos(\alpha) = 2070.2 \bullet ft \cdot kip$$

(b) Axial load

$$\text{DNaxial} := (-\text{DNvertical}) \bullet \sin(\alpha) = -72.3 \bullet \text{kip}$$

(c) Friction load on downstream saddle adjacent to expansion joint [*]:

- Vertical load assumed to be expansion joint 2 to saddle and 1/2 length of next support span (expansion joint isolates system from downstream beam shear and moment).

$$\text{DNsaddlesteel} := \text{WTs} \bullet \left(4 + \frac{38}{2}\right) \text{ft} \cdot (\text{Shell3} - \text{Area}) = 55.7 \bullet \text{kip}$$

$$\text{DNsaddlewater} := \left(4 + \frac{38}{2}\right) \text{ft} \bullet \text{Dwater} = 253.6 \bullet \text{kip}$$

$$\text{DNsaddlevert} := \text{DNsaddlesteel} + \text{DNsaddlewater} = 309.3 \bullet \text{kip}$$

$$\text{DNsaddle} := \text{fr} \bullet (\text{DNsaddlevert}) \bullet \cos(\alpha) = 28.6 \bullet \text{kip}$$

(d) Friction caused by downstream expansion joint:

$$\text{DNexp} := \text{EJF} \bullet (\pi \bullet \text{ft}15.198) = 23.9 \bullet \text{kip}$$

(e) Pressure thrust acting at downstream expansion joint:

$$\text{P2} := 513.88\text{ft} \bullet \text{Area} \bullet \text{WTw} = 5666.5 \bullet \text{kip}$$

(f) Pressure head acting on the pipe shell ends inside downstream expansion joint:

$$\text{Pshelldn} := 513.91\text{ft} \bullet \text{WTw} \bullet (\text{Shell2} - \text{Area}) = 150.5 \bullet \text{kip}$$

(g) Total downstream axial loads (using established horizontal sign convention):

$$\text{DNtotal} := \text{DNaxial} + \text{DNsaddle} + \text{DNexp} + \text{P2} + \text{Pshelldn}$$

$$\boxed{\text{DNtotal} = 5797.2 \bullet \text{kip}}$$

[*] *Load calculations apply with the assumption that upper block and first downstream*

saddle support are rigidly connected by the supporting foundation system.

(6) Moment Loads at First Downstream Saddle Support Below Block

- Moment acts as a vertical down force acting at the saddle multiplied by the span length from the PI to the saddle: 32 ft.
- Moment arm for saddle friction assumed to act from centerline of penstock to friction point at saddle bearing ~ 10.5 ft.

(a) Fixed end moment caused by dead load in first downstream span from expansion joint:

$$\text{FEMdnsaddle} := (-\text{DNsaddlevert}) \cdot \cos(\alpha) \cdot 32 \cdot \text{ft}$$

$$\boxed{\text{FEMdnsaddle} = -7616.4 \cdot \text{ft·kip}}$$

(b) Moment caused by saddle friction:

$$\text{FEMfriction} := -\text{DNsaddle} \cdot 10.5\text{ft}$$

Group B:

$$\boxed{\text{FEMfriction} = -299.9 \cdot \text{ft·kip}}$$

(1) Upsurge Pressures

- Upsurge reaction head at the upstream expansion joint is 95.05 ft. above the static reservoir hydraulic grade line (HGL) (Fig. 18-1).
- Upsurge reaction head at the downstream expansion joint is 172.59 ft. above the static reservoir HGL (Fig. 18-1).
- Upsurge reaction head at PI 2 is 170.37 ft. above the static reservoir HGL (Fig. 18-1).

(a) Upsurge pressure force caused by pipe shell end area at upstream expansion joint:

$$\text{UPupsurge} := 95.05\text{ft} \cdot (\text{Shell1} - \text{Area}) \cdot \text{WTw} = 26.4 \cdot \text{kip}$$

(b) Upsurge pressure force caused by pipe shell end area at downstream expansion joint:

$$\text{DNupsurge} := 172.59\text{ft} \cdot (\text{Shell3} - \text{Area}) \cdot \text{WTw} = 53.2 \cdot \text{kip}$$

(c) Upsurge pressure force at PI 2:

$$\text{PI2upsurge} := 170.37 \cdot 1\text{ft} \cdot \text{Area} \cdot \text{WTw} = 1878.7 \cdot \text{kip}$$

(d) Total upsurge pressure force at PI 2 up and down:

$$\text{P1upsurge} := \text{PI2upsurge} + \text{UPupsurge} = 1905.0 \cdot \text{kip}$$

$$\text{P2upsurge} := \text{PI2upsurge} + \text{DNupsurge} = 1931.9 \cdot \text{kip}$$

(2) Momentum Force

Momentum through the bend (water mass * flow * velocity change) with positive vector in direction of flow leaving minus flow incoming

$$\text{MomForce} := \left(\frac{\text{WTw}}{\text{Accel}}\right) \bullet \text{Flow} \bullet \text{Vel} = 74.1 \bullet \text{kip}$$

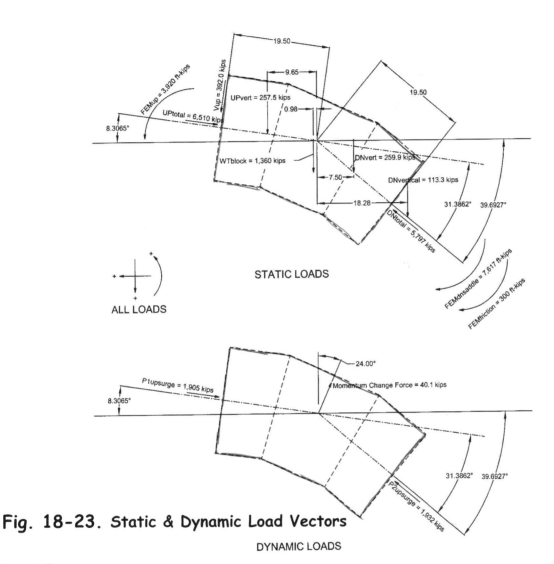

Fig. 18-23. Static & Dynamic Load Vectors

DYNAMIC LOADS

Load Summary

- The summary of loads and moments is included in Table 18-12.
- Components associated with friction effects (expansion joints and saddles) are assigned sign direction based on the assumption that the penstock is elongating away from the anchor block (both sides) because of thermal expansion. This association is more conservative than the contraction assumption, and the assignments were made on this basis.
- Summation of forces shown in the table, along with combined moment, is the basis for development of pressure distributions.

Pressure Distributions for Final Design

- See Fig. 18-24 and Table 18-12 for summation of forces and anticipated pressure distrib
- The subfoundation (under the anchor block) is assumed to be stepped along the contact surface with vertical and horizontal contact areas for easier construction and anchoring.
- Final design would need to consider appropriate safety factors as suggested in Chapter 8. A geotechnical analysis would be appropriate to determine these factors.
- Additional loads (such as seismic) are not included in this example and would have to be considered as required for finalizing the design.
- A typical next step for design of the anchor block would be to consider rock bolts or an expanded lower block to overcome the uplift, shear, and overturning moments.

Projected horizontal area of the block is based on the cosine of the angles of each of the three major block components multiplied by its width (21 ft.):

$$\text{LengthProjection} := 9.3 \bullet ft \bullet \cos(\theta) + 9.3 \bullet ft \bullet \cos(\alpha) + 13.95 \bullet ft \bullet \cos\left[\frac{(\theta + \alpha)}{2}\right] = 29.1\,ft$$

(a) Horizontal Shear Distribution (P/A):

$$\text{Shear} := (-2395 \bullet kip) \div (\text{LengthProjection} \bullet 21 \bullet ft) = -3919 \bullet psf$$

(b) Vertical Uplift Pressure Distribution (P/A):

$$\text{Uplift} := (-1375 \bullet kip) \div (\text{LengthProjection} \bullet 21 \bullet ft) = -2250 \bullet psf$$

(c) Counterclockwise Overturning Moment Pressure Distribution
   (using fixed end arch: M/S or $6M/lt^2$ for S term):

$$\text{Overturn} := 3446.2 \bullet ft\cdot kip \div \frac{\left[21 \bullet ft \bullet (\text{LengthProjection})^2\right]}{6} = 1163 \bullet psf$$

(d) Combined Pressure Distribution (Shear acting upward on left side of block and downward on right):

$$\text{PressureL} := \text{Uplift} + \text{Overturn} = -1087 \bullet psf$$

$$\text{PressureR} := \text{Uplift} - \text{Overturn} = -3412 \bullet psf$$

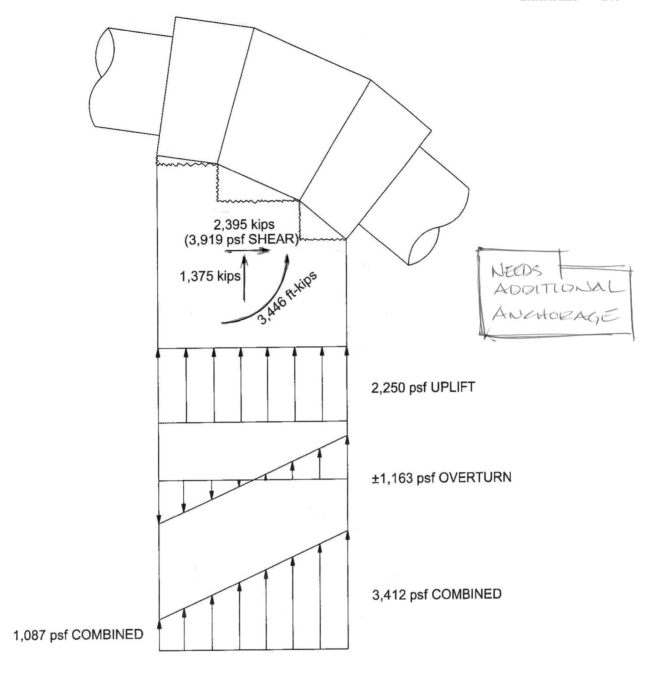

Fig. 18-24. Pressure Distribution Summary

Load Desig.	Description of Load	Vertical (k)	Horiz (k)	Normal (k)	Moment Arm (ft)	Axials Loads Vert Dir Calc	Axial Loads Horiz Dir Calc	Mom @ PI (ft-k)
	Exp. Jt. 1 to AB face							
UPvertical, UPaxial	Pipe and Water	6527.8				943.1	-943.1	
UPsaddle	Saddle Friction	0.0	0.0	6459.3		775.1	-775.1	
UPexp	Expansion Joint Friction	0.0	0.0			23.9	-23.9	
Pshellup	Head Pres on end of shell	0.0	0.0			117.0	-117.0	
P₁	Pressure Thrust at PI	0.0	0.0			4650.9	-4650.9	
FEMup	FEM		0.0			0.0	0.0	3920.0
Vup	End Reaction u/s anchor	392.0	0.0		19.50	0.0	0.0	7644.0
Uptotal	Subtotal (1)	392.0				6509.9	6509.9	11564.0
	At Anchor Block							
UPvert	Pipe and Water	257.5						
FEMup	FEM							2484.3
DNvert	Pipe and Water	259.9						
FEMdn	FEM							-1949.7
WTblock, NETARM, BLOCKMOM	Anchor Block Weight	1359.7			0.98			1334.1
	Subtotal (2)	1877.1			0.98			1868.7
	AB face to Exp. Jt. 2							
DNvertical, DNaxial, FEMupsaddle	Pipe and Water	113.3			18.28	72.3	-72.3	-2070.2
FEMdnsaddle	Saddle pipe extenstion on AB				32.00			-7616.4
DNsaddle	Saddle Friction			238.0	10.5	-28.6	28.6	-299.9
DNexp	Expansion Joint Friction	0.0	0.0			-23.9	23.9	
Pshelldn	Head Pres on end of shell	0.0	0.0			-150.5	150.5	
P₂	Pressure Thrust at PI	0.0	0.0			-5666.6	5666.6	
DNtotal	Subtotal (3)	113.0				-5797.3	5797.3	-9986.5
	Total (1)	392.0	0.0			6509.9	-6509.9	11564.0
	Total (2)	1877.1	0.0			0.0	0.0	1868.7
	Total (3)	113.0	0.0			-5797.3	5797.3	-9986.5
	Group A Dgn Loads	-379.4	-1980.5		0.98			3446.2

Load Desig.	Description of Load	Vertical (k)	Horiz (k)	Normal (k)	Moment Arm (ft)	Axial Loads Vert Dir Calc	Axial Loads Horiz Dir Calc	Mom @ PI (ft-k)
MomForce	Momentum Load					74.1	74.1	
P1upsurge	Normal Operating (u/s) Hd					1905.0	-1905.0	
P2upsurge	Normal Operating (d/s) Hd					-1932.2	1932.2	
	Change Axial to Vert/Horiz Ld	-995.3	-414.5					

	Total Design Loads							
	Group A & B	-1374.7	-2395.0		0.98			3446.2

Group A Loads=   Static Levels   Resultant Load =   2761.5
Group B Loads=   Normal Operating Loads (waterhammer)   Resultant Angle =   29.9

Table 18-12. Overall Load Summary and Resultant Forces

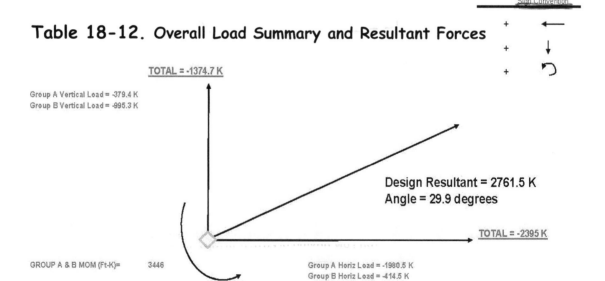

Group A Vertical Load = -379.4 K
Group B Vertical Load = -995.3 K

TOTAL = -1374.7 K

Design Resultant = 2761.5 K
Angle = 29.9 degrees

TOTAL = -2395 K

GROUP A & B MOM (Ft-K)=   3446

Group A Horiz Load = -1980.5 K
Group B Horiz Load = -414.5 K

Sign Convention
+ ←
+ ↓
+ ↻

18.10 Riveted Penstock Joints

▶ Define Units

General

The following calculations do not apply to the design example listed in Section 18. Calculations are presented only to provide examples of how riveted joints have been designed in the past and to provide guidance for evaluating existing joints for verification.

The efficiency of riveted joints is defined as the ratio of the strength of a unit length of a riveted joint to the strength of the same unit length of the parent plate material. The efficiency of a joint is calculated by the genera method illustrated in the following examples.

Single-Riveted Lap Joint (Longitudinal or Circumferential)

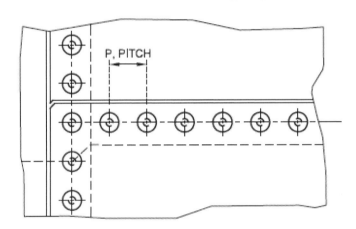

Figure 18-25. Single-Riveted Lap Joint (Longitudinal or Circumferential)

The following values are given:

$T_s := 55000$psi   tensile strength stamped on plate, psi = T_s

$t := 0.25$in   thickness of plate, in. = t

$P := 1.625$in   pitch of rivets (on row having greatest pitch), in. = P d := 0.6875in   diameter of rivet after driving (equal to diameter of rivet hole), in. = d

a := $\frac{\pi}{4} \cdot d^2 = 0.3712 \cdot in^2$ cross-sectional area of rivet after driving, in.^{2} = a

s := 44000psi   shearing strength of rivet in single shear, psi = s

c := 95000psi   crushing strength of mild steel, psi = c

n := 1 number of rivets in single shear in a unit length of joint = n

Substituting in the values gives the following:

A := $P \cdot t \cdot T_s = 22344$ lbf   strength of solid plate = A = P * t * T$_s$

B := $(P - d) \cdot t \cdot T_s = 12891$ lbf   strength of plate between rivet holes = B = (P-d) * t * T$_s$

C := $n \cdot s \cdot a = 16334 \cdot$lbf   shearing strength of one rivet in single shear = C = n * s * a

D := $n \cdot d \cdot t \cdot c = 16328$ lbf   crushing strength of plate in front of one rivet = D = n * d * t * c

▶ Sorting Prgm ───

Table 18-13. Sorted Values of B, C, and D

Load Path Check	
Load Case	Strength, lbs
B	12891
D	16328
C	16334

Find least value of B, C, and D, then divide by A. $\min(B, C, D) = 12891 \cdot$lbf

Joint efficiency then equals $e := \frac{\min(B, C, D)}{A} = 0.577$

Double-Riveted Lap Joint (Longitudinal or Circumferential)

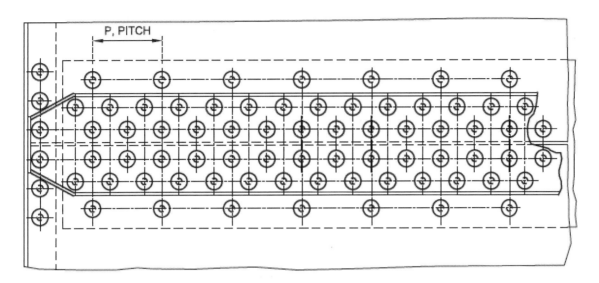

Figure 18-26. Double-Riveted Lap Joint (Longitudinal or Circumferential)

The following values are given:

$T_s := 55000$ psi

tensile strength stamped on plate, psi = T_s

$t := 0.3125$ in

thickness of plate, in. = t

$P := 2.875$ in

pitch of rivets (on row having greatest pitch), in. = P

$d := 0.750$ in

diameter of rivet after driving (equal to diameter of rivet hole), in. = d

$a := \dfrac{\pi}{4} \cdot d^2 = 0.4418 \cdot in^2$

cross-sectional area of rivet after driving, in.^{2} = a

$s := 44000$ psi

shearing strength of rivet in single shear, psi = s

$c := 95000$ psi

crushing strength of mild steel, psi = c

$n := 2$

number of rivets in single shear in a unit length of joint = n

Substituting in the values gives the following:

$A := P \cdot t \cdot T_s = 49414$ lbf

strength of solid plate = A = P * t * T_s

$B := (P - d) \cdot t \cdot T_s = 36523 \, \text{lbf}$ strength of plate between rivet holes = B = (P-d) * t * T_s

$C := n \cdot s \cdot a = 38877 \cdot \text{lbf}$ shearing strength of one rivet in single shear = C = n * s * a

$D := n \cdot d \cdot t \cdot c = 44531 \, \text{lbf}$ crushing strength of plate in front of one rivet = D = n * d * t * c

▶ Sorting Prgm ────────────────────────────────────

Table 18-14. Sorted Values of B, C, and D

Load Path Check	
Load Case	Strength, lbs
B	36523
C	38877
D	44531

Find least value of B, C, and D, then divide by A.

$\min(B, C, D) = 36523 \cdot \text{lbf}$

Joint efficiency then equals

$$e := \frac{\min(B, C, D)}{A} = 0.739$$

Double-Riveted Butt and Double Strap Joint

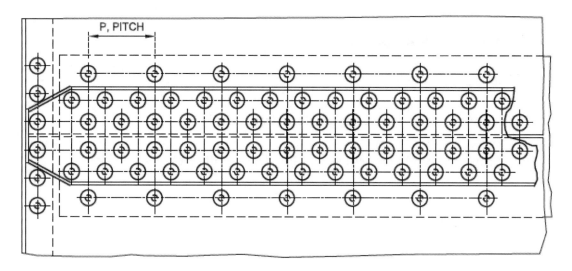

Figure 18-27. Double-Riveted Butt and Double Strap Joint

The following values are given:

$T_s := 55000 \text{psi}$ tensile strength stamped on plate, psi = T_s

$t := 0.375 \text{in}$ thickness of plate, in. = t

$b := .3125 \cdot \text{in}$ thickness of butt strap, in. = b

$P := 4.875 \text{in}$ pitch of rivets (on row having greatest pitch), in. = P

$d := 0.875 \text{in}$ diameter of rivet after driving (equal to diameter of rivet hole), in. = d

$a := \dfrac{\pi}{4} \cdot d^2 = 0.6013 \cdot \text{in}^2$ cross-sectional area of rivet after driving, in.^{2} = a

$s := 44000 \text{psi}$ shearing strength of rivet in single shear, psi = s

$S := 88000 \cdot \text{psi}$ shearing strength of rivet in double shear, psi = S

$c := 95000 \text{psi}$ crushing strength of mild steel, psi = c

$n := 1$ number of rivets in single shear in a unit length of joint = n

$N := 2$ number of rivets in double shear in a unit length of joint = N

Substituting in the values gives the following:

$A := P \cdot t \cdot T_s = 100547 \, \text{lbf}$ strength of solid plate = A = P * t * T_s

$B := (P - d) \cdot t \cdot T_s = 82500 \, \text{lbf}$ strength of plate between rivet holes in the outer plate = B = (P-d) * t * T_s

$C := N \cdot S \cdot a + n \cdot s \cdot a = 132291 \, \text{lbf}$ shearing strength of two rivets in double shear, plus the shearing strength of one rivet in single shear = C = N * S * a + n * s * a

$D := (P - 2d) \cdot t \cdot T_s + n \cdot s \cdot a = 90911 \, \text{lbf}$ strength of plate between rivet holes in the second row, plus the shearing strength of one rivet in single shear in the outer row = D = (P-2d) * t * T_s + n * s * a

$E := (P - 2d) \cdot t \cdot T_s + d \cdot b \cdot c = 90430 \, \text{lbf}$ strength of plate between rivet holes in the second row, plus the crushing strength of butt strap in front of one rivet in the outer row = E = (P-2d) * t * T_s + d * b * c

$F := N \cdot d \cdot t \cdot c + n \cdot d \cdot b \cdot c = 88320 \, \text{lbf}$

crushing strength of plate in front of two rivets, plus the crushing strength of butt strap in front of one rivet = F = N * d * t * c + n * d * b * c

$G := N \cdot d \cdot t \cdot c + n \cdot s \cdot a = 88802 \, \text{lbf}$

crushing strength of plate in front of two rivets, plus the shearing strength of one rivet in single shear = G = N * d * t * c + n * s * a

$H := (P - 2d) \cdot 2 \cdot b \cdot T_s = 107422 \, \text{lbf}$

strength of butt straps between rivet holes in the inner row = H = (P-2d) * 2 * b * T_s

▶ Sorting Prgm ──

Table 18-15. Sorted Values of B, C, D, E, F, G, and H

Load Path Check	
Load Case	Strength, lbs
B	82500
F	88320
G	88802
E	90430
D	90911
H	107422
C	132291

Find least value of B, C, D, E, F, G, and H, then divide by A.

$$\min(B, C, D, E, F, G, H) = 82500 \cdot \text{lbf}$$

Joint efficiency then equals:

$$e := \frac{\min(B, C, D, E, F, G, H)}{A} = 0.821$$

Triple-Riveted Butt and Double Strap Joint

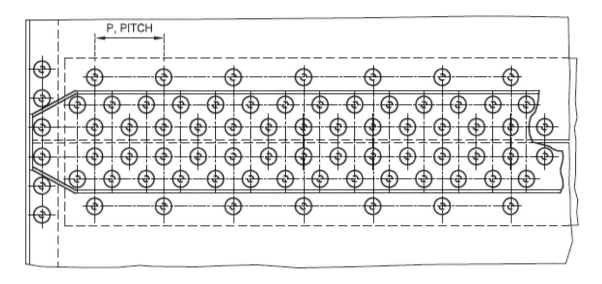

Figure 18-28. Triple-Riveted Butt and Double Strap Joint

The following values are given:

$T_s := 55000$psi   tensile strength stamped on plate, psi = T_s

$t := 0.375$in   thickness of plate, in. = t

$b := .3125 \cdot$in   thickness of butt strap, in. = b

$P := 6.50$in   pitch of rivets (on row having greatest pitch), in. = P

$d := 0.8125$in   diameter of rivet after driving (equal to diameter of rivet hole), in. = d

$a := \dfrac{\pi}{4} \cdot d^2 = 0.5185 \cdot in^2$ cross-sectional area of rivet after driving, in.^{2} = a

$s := 44000\,\text{psi}$ shearing strength of rivet in single shear, psi = s

$S := 88000 \cdot \text{psi}$ shearing strength of rivet in double shear, psi = S

$c := 95000\,\text{psi}$ crushing strength of mild steel, psi = c

$n := 1$ number of rivets in single shear in a unit length of joint = n

$N := 4$ number of rivets in double shear in a unit length of joint = N

Substituting in the values gives the following:

$A := P \cdot t \cdot T_s = 134062\,\text{lbf}$ strength of solid plate = A = P * t * T_s

$B := (P - d) \cdot t \cdot T_s = 117305\,\text{lbf}$ strength of plate between rivet holes in the outer plate = B = (P-d) * t * T_s

$C := N \cdot S \cdot a + n \cdot s \cdot a = 205320 \cdot \text{lbf}$ shearing strength of two rivets in double shear, plus the shearing strength of one rivet in single shear = C = N * S * a + n * s * a

$D := (P - 2d) \cdot t \cdot T_s + n \cdot s \cdot a = 123360\,\text{lbf}$ strength of plate between rivet holes in the second row, plus the shearing strength of one rivet in single shear in the outer row = D = (P-2d) * t * T_s + n * s * a

$E := (P - 2d) \cdot t \cdot T_s + d \cdot b \cdot c = 124668\,\text{lbf}$ strength of plate between rivet holes in the second row, plus the crushing strength of butt strap in front of one rivet in the outer row = E = (P-2d) * t * T_s + d * b * c

$F := N \cdot d \cdot t \cdot c + n \cdot d \cdot b \cdot c = 139902\,\text{lbf}$ crushing strength of plate in front of two rivets, plus the crushing strength of butt strap in front of one rivet = F = N * d * t * c + n * d * b * c

$G := N \cdot d \cdot t \cdot c + n \cdot s \cdot a = 138595\,\text{lbf}$ crushing strength of plate in front of two rivets, plus the shearing strength of one rivet in single shear = G = N * d * t * c + n * s * a

$H := (P - 2d) \cdot 2 \cdot b \cdot T_s = 167578\,\text{lbf}$ strength of butt straps between rivet holes in the inner row = H = (P-2d) * 2 * b * T_s

▶ Sorting Prgm

Table 18-16. Sorted Values of B, C, D, E, F, G, and H

Load Path Check	
Load Case	Strength, lbs
B	117305
D	123360
E	124668
G	138595
F	139902
H	167578
C	205320

Find least value of B, C, D, E, F, G, and H, then divide by A.

$$\min(B, C, D, E, F, G, H) = 117305 \cdot \text{lbf}$$

Joint efficiency then equals

$$e := \frac{\min(B, C, D, E, F, G, H)}{A} = 0.875$$

INDEX